NANOMATERIALS: SYNTHESIS, CHARACTERIZATION, HAZARDS, AND SAFETY

NANOMATERIALS: SYNTHESIS, CHARACTERIZATION, HAZARDS, AND SAFETY

Edited by

MUHAMMAD BILAL TAHIR

Department of Physics, Khwaja Fareed University of Engineering and Information Technology, RYK, Pakistan

MUHAMMAD SAGIR

Department of Chemical Engineering, Khwaja Fareed University of Engineering and Information Technology, RYK, Pakistan

ABDULLAH M. ASIRI

Director of the Centre of Excellence for Advanced Materials Research, King Abdulaziz University, Jeddah, Saudi Arabia

ELSEVIER

Elsevier
Radarweg 29, PO Box 211, 1000 AE Amsterdam, Netherlands
The Boulevard, Langford Lane, Kidlington, Oxford OX5 1GB, United Kingdom
50 Hampshire Street, 5th Floor, Cambridge, MA 02139, United States

Library of Congress Cataloging-in-Publication Data
A catalog record for this book is available from the Library of Congress

British Library Cataloguing-in-Publication Data
A catalogue record for this book is available from the British Library

ISBN: 978-0-12-823823-3

For information on all Elsevier publications
visit our website at https://www.elsevier.com/books-and-journals

Publisher: Matthew Deans
Acquisitions Editor: Simon Holt
Editorial Project Manager: Mariana L. Kuhl
Production Project Manager: Prasanna Kalyanaraman
Cover Designer: Matthew Limbert

Typeset by SPi Global, India
Transferred to Digital Printing 2021

Contents

Contributors

Waleed Mumtaz Abbasi
Department of Soil Science, Faculty of Agriculture and Environmental Sciences, The Islamia University of Bahawalpur, Bahawalpur, Pakistan

Farman Ali
Department of Chemistry, Hazara University, Mansehra, Pakistan

Nisar Ali
Key Laboratory for Palygorskite Science and Applied Technology of Jiangsu Province, National & Local Joint Engineering Research Center for Deep Utilization Technology of Rock-salt Resource, Faculty of Chemical Engineering, Huaiyin Institute of Technology, Huaian, China

Shafaqat Ali
Department of Environmental Sciences and Engineering, Government College University, Faisalabad, Pakistan

Abdullah G. Al-Sehemi
Department of Chemistry, College of Science, King Khalid University, Abha, Saudi Arabia

Rabia Amen
Institute of Soil and Environmental Sciences, University of Agriculture Faisalabad, Faisalabad, Pakistan

Tabinda Athar
Institute of Soil and Environmental Sciences, Faculty of Agriculture, University of Agriculture Faisalabad, Faisalabad, Pakistan

Asma Ayub
Department of Physics, University of Gujrat, Gujrat, Pakistan

Muhammad Ashar Ayub
Institute of Soil and Environmental Sciences, Faculty of Agriculture, University of Agriculture Faisalabad, Faisalabad, Pakistan

Muhammad Azhar
Institute of Soil and Environmental Sciences, Faculty of Agriculture, University of Agriculture Faisalabad, Faisalabad, Pakistan

Muhammad Babar
Institute of Soil and Environmental Sciences, University of Agriculture Faisalabad, Faisalabad, Pakistan

Ayesha Baig
Department of Chemistry, University of Gujrat, Gujrat, Pakistan

Maria Batool
Department of Chemistry, Faculty of Science, University of Gujrat, Gujrat, Pakistan

Saira Batool
Department of Physics, University of Punjab, Lahore, Pakistan

Muhammad Bilal
School of Life Science and Food Engineering, Huaiyin Institute of Technology, Huaian, China

Muhammad Azmi Bustam
Institute of Soil and Environmental Sciences, University of Agriculture Faisalabad, Faisalabad, Pakistan

Mujahid Farid
Department of Environmental Sciences, University of Gujrat, Gujrat, Pakistan

Zia Ur Rahman Farooqi
Institute of Soil and Environmental Sciences, Faculty of Agriculture, University of Agriculture Faisalabad, Faisalabad, Pakistan

Hina Fatima
School of Applied Biosciences, Kyungpook National University, Daegu, South Korea

Muhammad Mahroz Hussain
Institute of Soil and Environmental Sciences, Faculty of Agriculture, University of Agriculture Faisalabad, Faisalabad, Pakistan

Irfan Iftikhar
Institute of Soil and Environmental Sciences, Faculty of Agriculture, University of Agriculture Faisalabad, Faisalabad, Pakistan

Predrag Ilic
PSRI Institute for Protection and Ecology of the Republic of Srpska, Banja Luka, Banja Luka, Republic of Srpska, Bosnia and Herzegovina

Hafiz M.N. Iqbal
Tecnologico de Monterrey, School of Engineering and Sciences, Monterrey, Mexico

Shazia Iqbal
Institute of Soil & Environmental Sciences, University of Agriculture, Faisalabad, Pakistan

Adnan Khan
Institute of Chemical Sciences, University of Peshawar, Peshawar, Khyber Pakhtunkhwa, Pakistan

Hamayun Khan
Department of Chemistry, Islamia College University, Peshawar, Pakistan

Hassnain Abbas Khan
Clean Combustion Research Centre, Division of Physical Sciences and Engineering, King Abdullah University of Science and Technology (KAUST), Thuwal, Saudi Arabia

Syed Ejaz Hussain Mehdi
Institute of Soil and Environmental Sciences, University of Agriculture Faisalabad, Faisalabad, Pakistan

Ahmad Mukhtar
Department of Chemical Engineering, Universiti Teknologi PETRONAS, Perak, Malaysia

Muhammad Nadeem
Institute of Soil and Environmental Sciences, Faculty of Agriculture, University of Agriculture Faisalabad, Faisalabad, Pakistan

Muhammad Faizan Nazar
Department of Chemistry, Faculty of Science, University of Gujrat, Gujrat, Pakistan

Madiha Nisar
Institute of Soil and Environmental Sciences, Faculty of Agriculture, University of Agriculture Faisalabad, Faisalabad, Pakistan

Abdul Qadeer
Institute of Soil and Environmental Sciences, Faculty of Agriculture, University of Agriculture Faisalabad, Faisalabad, Pakistan

Bakhtawar Razzaq
Department of Physics, University of Gujrat, Gujrat, Pakistan

Umair Riaz
Soil and Water Testing Laboratory for Research, Bahawalpur, Pakistan

M. Rizwan
Department of Physics, University of Gujrat, Gujrat, Pakistan

Muhammad Sagir
Department of Chemical Engineering, Khwaja Fareed University of Engineering and Information Technology, RYK, Pakistan

Tayyaba Samreen
Institute of Soil & Environmental Sciences, University of Agriculture, Faisalabad, Pakistan

Sidra Saqib
Department of Chemical Engineering, COMSATS University Islamabad, Lahore, Pakistan

Falak Shafiq
Department of Physics, University of Gujrat, Gujrat, Pakistan

Laila Shahzad
Sustainable Development Study Center, Government College University, Lahore, Pakistan

Aleena Shoukat
Department of Physics, University of Gujrat, Gujrat, Pakistan

Ayesha Siddiqui
Department of Botany, University of Agriculture, Faisalabad, Faisalabad, Pakistan

Muhammad Irfan Sohail
Institute of Soil and Environmental Sciences, Faculty of Agriculture, University of Agriculture Faisalabad, Faisalabad, Pakistan

Muhammad Bilal Tahir
Department of Physics, Khwaja Fareed University of Engineering and Information Technology, RYK, Pakistan

Sami Ullah
Department of Chemistry, College of Science, King Khalid University, Abha, Saudi Arabia

Wajid Umar
Institute of Soil and Environmental Sciences, Faculty of Agriculture, University of Agriculture Faisalabad, Faisalabad, Pakistan; Doctoral School of Environmental Sciences, Szent Istvan University, Gödöllő, Hungary

Muneeb ur Rahman
Department of Physics, Islamia College Peshawar (Public Sector University), Peshawar, Pakistan

Aisha A. Waris
Institute of Soil and Environmental Sciences, Faculty of Agriculture, University of Agriculture Faisalabad, Faisalabad, Pakistan

Nukshab Zeeshan
Institute of Soil and Environmental Sciences, Faculty of Agriculture, University of Agriculture Faisalabad, Faisalabad, Pakistan

Muhammad Zia ur Rehman
Institute of Soil and Environmental Sciences, Faculty of Agriculture, University of Agriculture Faisalabad, Faisalabad, Pakistan

Muhammad Zubair
Department of Chemistry, University of Gujrat, Gujrat, Pakistan

Introduction

Falak Shafiq[a], Muhammad Bilal Tahir[b], Muhammad Tayyub Nazir[c], Anum Shafiq[d], and Muhammad Sagir[e]
[a]Department of Physics, University of Gujrat, Gujrat, Pakistan
[b]Department of Physics, Khwaja Fareed University of Engineering and Information Technology, RYK, Pakistan
[c]Department of Electrical Engineering, Bahria University, Islamabad, Pakistan
[d]Department of Management Sciences, University of Education Lower Mall Campus, Lahore, Pakistan
[e]Department of Chemical Engineering, Khwaja Fareed University of Engineering and Information Technology, RYK, Pakistan

1. Introduction

A vast range of technologies that manipulate, measure, or incorporate materials with dimensions in the 1–100nm range can be referred to as nanotechnology (Boverhof and David, 2010). The materials at this scale are known as nanomaterials, and exhibit versatile properties that are different from bulk materials (Warheit, 2018; Thomas et al., 2006). Nanomaterials can be one-, two-, or three-dimensional. Fig. 1 (Yokel and MacPhail, 2011) shows some materials that are at different scales in the nanometer range with respect to shape and size (Dowling, 2004). Nanomaterials can be metals, metal oxides, metal sulfides, nitrides, etc., and can be synthesized via various physical and chemical methods, depending on the required properties of the nanomaterials. Color, shape, size, increased strength, optical or magnetic properties, conductivity, reactivity, and flexibility are properties that can be unique at nanoscale. The surface area of the synthesized materials is also enhanced at this scale (Rafique et al., 2019, 2020; Iqbal et al., 2019). Due to these inimitable properties, nanomaterials have become a focus of widespread research in different fields such as energy production, electronics, materials engineering, and energy conservation, as well as in the biomedical field. Nanomaterials possess a wide range of everyday applications for the convenience of human beings. However, while offering great benefits, nanomaterials also present risks to human health and the safety of the environment. Nanomaterials can be toxic in various ways (Holsapple et al., 2005; Oberdörster et al., 2005). Therefore, it is also necessary to address the potential hazards to human health, as well as to the environment, of synthesized nanomaterials.

2. Nanomaterial characterization

Although a substantial amount of work is available in the field of advanced nanotechnology and its applications, understanding of the risks to the environment and human health of these nanomaterials is still in its primitive stages. It is challenging for companies and regulators to ensure human health and environmental safety in commercializing nanomaterials of various kinds, as used in different fields. The literature regarding

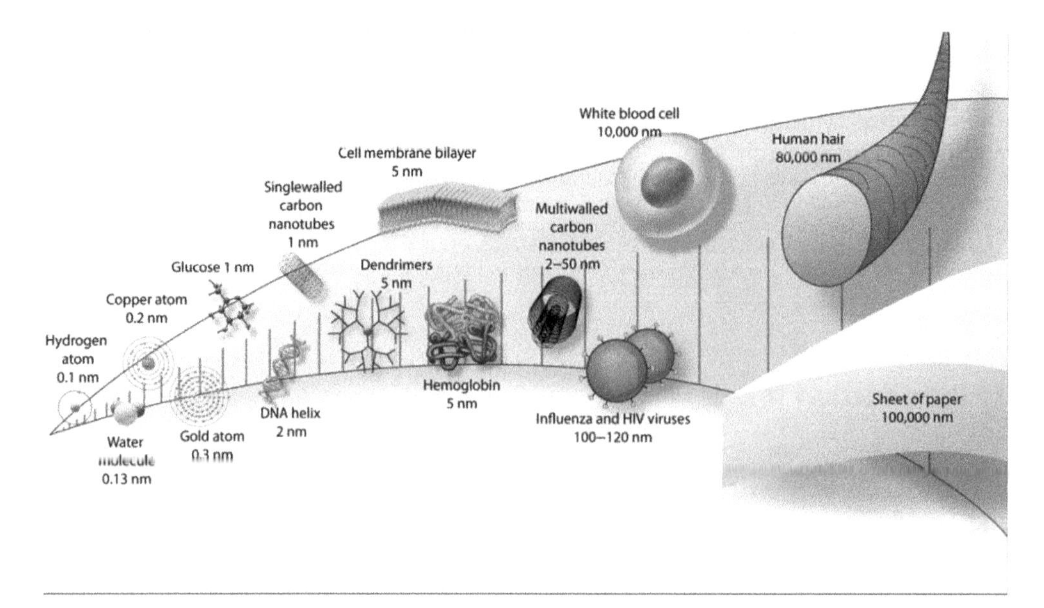

Fig. 1 The sizes and shapes of some synthesized nanomaterials compared to more familiar materials.

characterization of nanomaterials is not fully provided by researchers so that manufacturers can analyze the risk of a nanomaterial's exposure to the environment. Proper characterization of nanomaterials is vital for safety evaluation. For instance, in the past decades various research strategies and challenges have been proposed to ensure the safety and exposure of different nanoparticles in various forms. However, the biggest problem is that the safety and environmental effects evaluation is not possible for different kinds of synthesized nanomaterials, owing to (Warheit, 2018):

(1) the variety of nanomaterials;
(2) the variations of different kinds within the same nanomaterials; i.e., carbon exists in various kind of nanoscale forms (carbon, carbon black, single-walled carbon nanotubes, multiwalled carbon nanotubes, fullerenes, carbon nanofibers, etc.);
(3) the huge costs required to evaluate all of the individual nanomaterials; and
(4) the testing of each individual particle requiring a lot of time.

In this regard, data provided by researchers on the toxicology of nanomaterials are not efficient and sufficient, as they do not consider the effects of nanomaterials on biological systems. Therefore, characterization of nanomaterials is necessary, as at nanometer scale the physiochemical properties alter, which may have adverse effects on the testing systems. Brunner et al. showed that dose metric and dose response can only be understood by the proper characterization of nanomaterials in a simple cytotoxicity assay. The material at bulk may be nontoxic, but at nanometer scale can be toxic (Nel et al., 2009; Adlakha-Hutcheon et al., 2009; Kuhlbusch et al., 2009).

3. Consideration and needs for hazards assessment

Synthesized nanomaterials are known as a group of multiple substances, in many forms and sizes, which may be formed with a number of surface coatings (Borm et al., 2006; Seaton et al., 2010; Maynard et al., 2004; Maynard and Aitken, 2007). Recently, the interaction between synthesized nanomaterials and biological systems has increased, but its impacts on the systems have not been evaluated. For precise assessment of risks of nanomaterials, an understanding of these interactions is necessary. The dynamic interaction interfaces of nanomaterials and biological systems include different parameters that are crucial to comprehend. Nel et al. mentioned important material parameters such as chemical composition, shape, size, surface functionalization, and other surface parameters, which are important to study in interaction with a testing system (Nel et al., 2009). Different nanomaterials behave differently due to the unique properties of individual nanomaterials.

Nanoparticles can cause health hazards such as pulmonary inflammation (Donaldson et al., 2002, 2006). Many studies show that exposure of the biological organ system to nanoparticles can cause strong lung inflammation even at very low doses. For instance, titanium dioxide nanoparticles and carbon nanotubes, when interacting with the lungs of mice via either intratracheal aspiration, intratracheal installation, or through any other means, show strong pulmonary response. Although this method of synthesized nanoparticles exposure was criticized recently, the nanoparticles exhibited a strong response to the organs of the tested systems (Lam et al., 2004, 2006; Bermudez et al., 2004; Shvedova et al., 2005). There are other toxic effects of nanomaterials that need to be measured, such as nanomaterials translocation into the body, in vivo and in vitro toxicity assessment, genotoxicity of synthesized nanomaterials (Li et al., 2007; Muller et al., 2008), carcinogenic effects of nanomaterials, nanomaterials effect on circulation, etc., to avoid potential health hazards (Lam et al., 2006; Takagi et al., 2008).

4. Hazards and risk assessment strategies for nanoparticles exposures

The development of nanomaterials has recently increased significantly. With the development of novel nanomaterials, it has become necessary to evaluate the unique risks associated with them (Paik et al., 2008; Savolainen et al., 2010). However, proper strategies have still not been developed to assess the exposure of nanomaterials; i.e., which methods we should use to assess the toxicology of nanomaterials, and metrics required for their systematic use.

A tiered approach is required for the evaluation of the risks presented by nanomaterials. This approach allows separation of low and high toxicity nanomaterials, thus providing a way of protecting workers from nanoparticles exposure, reducing the risk to health and safety. Fig. 2 (Savolainen et al., 2010) shows a tiered approach for the toxicity

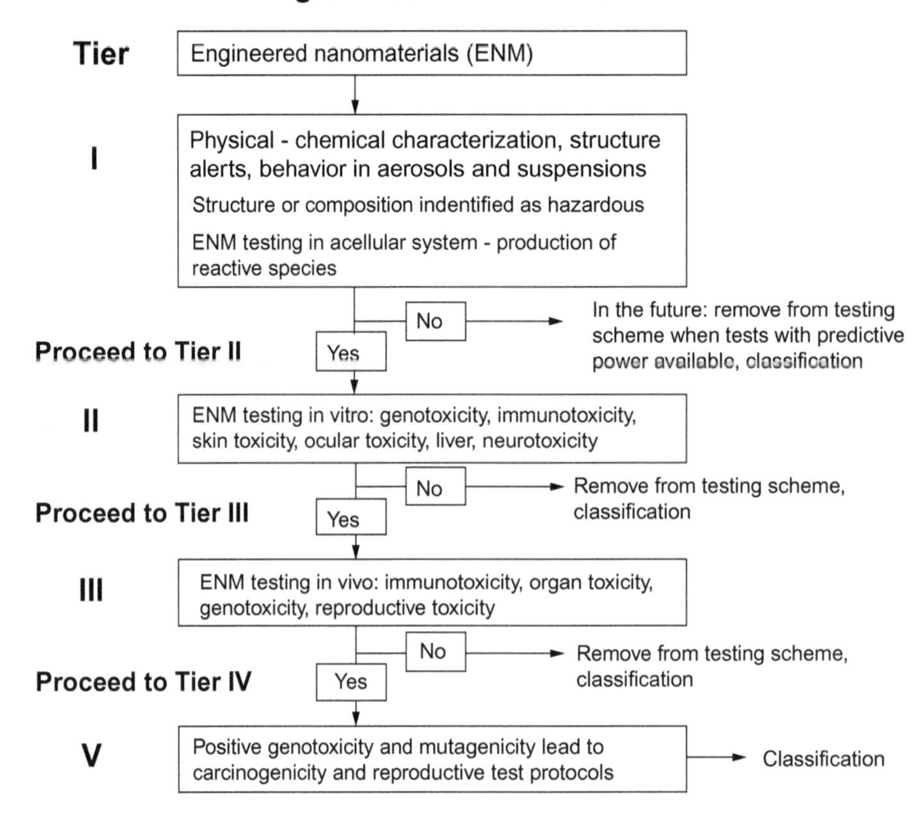

Fig. 2 A schematic representation of the proposed tiered testing approach for synthesized nanomaterials.

assessment of nanomaterials exposure. There are some fundamental elements of risk assessment that should be noted, including:

(1) recognition of hazards present in nanomaterials;

(2) characterization of these hazards;

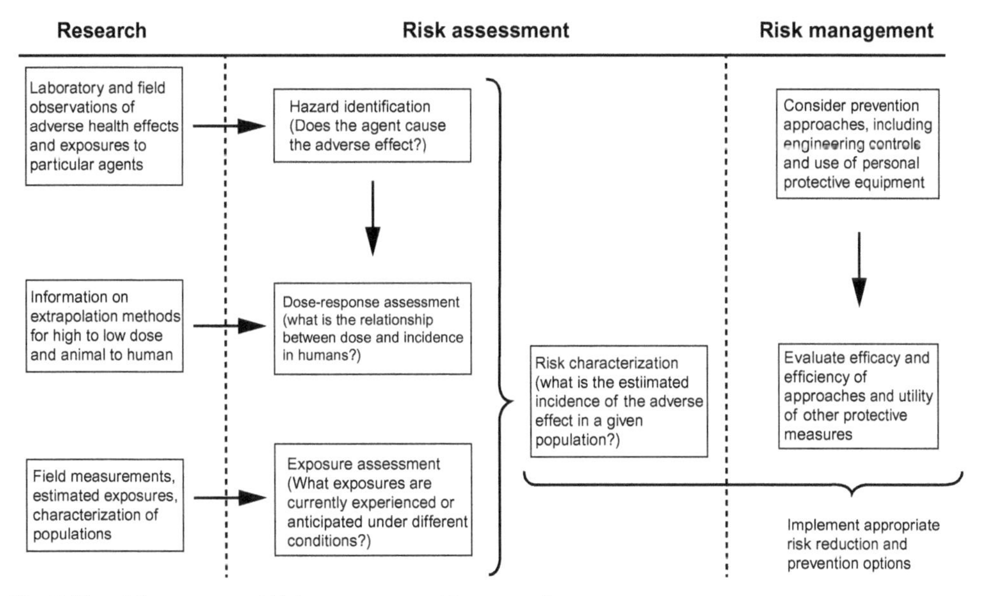

Fig. 3 The risk assessment/risk management framework.

(3) assessment of exposure to these nanomaterials; and
(4) risk characterization of nanomaterials exposure.

Fig. 3 (Yokel and MacPhail, 2011) shows a flow chart of research on nanomaterials using different testing systems and their risk assessment. After assessing the hazards of nanomaterials exposure, it is possible to take precautionary measures to manage these risks for the health and safety of workers.

5. Safety evaluation

For the safety evaluation of nanoparticles, it has become necessary to test commercially the individual nanoparticle types (i.e., nanomaterials to which humans are exposed at any level). This will help in determining the health hazards and risk to the environment and biological systems. Dose level and dose response characterization must be evaluated carefully to provide authentic data for the safety of workers that are directly linked with nanomaterials production and commercialization. In addition, it is necessary to provide protective equipment for the safety of workers (Linkov et al., 2007; Linkov and Satterstrom, 2008). The synthesis process of materials should be contained in gloves boxes, biological safety cabinets, or externally ventilated units. There should also be administrative controls, such as proper planning for nanomaterials engineering. When using nanomaterials, workers should use respirators and gloves to reduce contact with the nanomaterials. Biological monitoring and medical examination must be present in the working units. Using these norms, health hazards can be minimized (Kuhlbusch et al., 2009).

6. Conclusion

Nanomaterials are the building blocks of nanotechnology, of which risk assessment is the biggest current challenge. The risk assessment of nanomaterials is difficult because of their unique properties at nanoscale. The fundamental risk assessment elements are: (i) hazards recognition; (ii) hazards characterization; (iii) assessment of exposure; and (iv) risk characterization. These are the four major steps of risk and safety assessment. However, the characterization of the nanomaterials—to know their physical and chemical properties, which alter at nanoscale—is also a great challenge. Many strategies are used to assess the toxicity of nanomaterials, which may cause a threat to human health and safety. However, there is still a need for the development of a tiered assessment method to analyze the risk and establish safe use of these nanomaterials in future for the development of technology.

References

Adlakha-Hutcheon, G., et al., 2009. Nanomaterials, nanotechnology. In: Nanomaterials: Risks and Benefits. Springer, pp. 195–207.

Bermudez, E., et al., 2004. Pulmonary responses of mice, rats, and hamsters to subchronic inhalation of ultrafine titanium dioxide particles. Toxicol. Sci. 77 (2), 347–357.

Borm, P.J., et al., 2006. The potential risks of nanomaterials: a review carried out for ECETOC. Part. Fibre Toxicol. 3 (1), 11.

Boverhof, D.R., David, R.M., 2010. Nanomaterial characterization: considerations and needs for hazard assessment and safety evaluation. Anal. Bioanal. Chem. 396 (3), 953–961.

Donaldson, K., et al., 2002. The pulmonary toxicology of ultrafine particles. J. Aerosol. Med. 15 (2), 213–220.

Donaldson, K., et al., 2006. Carbon nanotubes: a review of their properties in relation to pulmonary toxicology and workplace safety. Toxicol. Sci. 92 (1), 5–22.

Dowling, A.P., 2004. Development of nanotechnologies. Mater. Today 7 (12), 30–35.

Holsapple, M.P., et al., 2005. Research strategies for safety evaluation of nanomaterials, part II: toxicological and safety evaluation of nanomaterials, current challenges and data needs. Toxicol. Sci. 88 (1), 12–17.

Iqbal, T., et al., 2019. Review on green synthesis and characterization of cadmium oxide nanoparticles. Int. J. Biochem. Biomol. 5 (1), 20–29.

Kuhlbusch, T., Fissan, H., Asbach, C., 2009. Nanotechnologies and environmental risks. In: Nanomaterials: Risks and Benefits. Springer, pp. 233–243.

Lam, C.-W., et al., 2004. Pulmonary toxicity of single-wall carbon nanotubes in mice 7 and 90 days after intratracheal instillation. Toxicol. Sci. 77 (1), 126–134.

Lam, C.-W., et al., 2006. A review of carbon nanotube toxicity and assessment of potential occupational and environmental health risks. Crit. Rev. Toxicol. 36 (3), 189–217.

Li, J.G., et al., 2007. Comparative study of pathological lesions induced by multiwalled carbon nanotubes in lungs of mice by intratracheal instillation and inhalation. Environ. Toxicol. 22 (4), 415–421.

Linkov, I., Satterstrom, F.K., 2008. Nanomaterial risk assessment and risk management. In: Real-Time and Deliberative Decision Making. Springer, pp. 129–157.

Linkov, I., et al., 2007. Multi-criteria decision analysis and environmental risk assessment for nanomaterials. J. Nanopart. Res. 9 (4), 543–554.
Maynard, A.D., Aitken, R.J., 2007. Assessing exposure to airborne nanomaterials: current abilities and future requirements. Nanotoxicology 1 (1), 26–41.
Maynard, A.D., et al., 2004. Exposure to carbon nanotube material: aerosol release during the handling of unrefined single-walled carbon nanotube material. J. Toxic. Environ. Health A 67 (1), 87–107.
Muller, J., et al., 2008. Clastogenic and aneugenic effects of multi-wall carbon nanotubes in epithelial cells. Carcinogenesis 29 (2), 427–433.
Nel, A.E., et al., 2009. Understanding biophysicochemical interactions at the nano–bio interface. Nat. Mater. 8 (7), 543–557.
Oberdörster, G., et al., 2005. Principles for characterizing the potential human health effects from exposure to nanomaterials: elements of a screening strategy. Part. Fibre Toxicol. 2 (1), 8.
Paik, S.Y., Zalk, D.M., Swuste, P., 2008. Application of a pilot control banding tool for risk level assessment and control of nanoparticle exposures. Ann. Occup. Hyg. 52 (6), 419–428.
Rafique, M., et al., 2019. Eco-friendly green and biosynthesis of copper oxide nanoparticles using Citrofortunella microcarpa leaves extract for efficient photocatalytic degradation of Rhodamin B dye form textile wastewater. Optik 208, 164053.
Rafique, M., et al., 2020. Plant-mediated green synthesis of zinc oxide nanoparticles from *Syzygium cumini* for seed germination and wastewater purification. Int. J. Environ. Anal. Chem., 1–16.
Savolainen, K., et al., 2010. Risk assessment of engineered nanomaterials and nanotechnologies—a review. Toxicology 269 (2–3), 92–104.
Seaton, A., et al., 2010. Nanoparticles, human health hazard and regulation. J. R. Soc. Interface 7 (suppl_1), S119–S129.
Shvedova, A.A., et al., 2005. Unusual inflammatory and fibrogenic pulmonary responses to single-walled carbon nanotubes in mice. Am. J. Phys. Lung Cell. Mol. Phys. 289 (5), L698–L708.
Takagi, A., et al., 2008. Induction of mesothelioma in p53+/− mouse by intraperitoneal application of multi-wall carbon nanotube. J. Toxicol. Sci. 33 (1), 105–116.
Thomas, K., et al., 2006. Research strategies for safety evaluation of nanomaterials, part VIII: international efforts to develop risk-based safety evaluations for nanomaterials. Toxicol. Sci. 92 (1), 23–32.
Warheit, D.B., 2018. Hazard and risk assessment strategies for nanoparticle exposures: how far have we come in the past 10 years? F1000Res. 7.
Yokel, R.A., MacPhail, R.C., 2011. Engineered nanomaterials: exposures, hazards, and risk prevention. J. Occup. Med. Toxicol. 6 (1), 7.

CHAPTER 1

History and development of nanomaterials

Rabia Amen[a], Ahmad Mukhtar[b], Sidra Saqib[c], Sami Ullah[d], Abdullah G. Al-Sehemi[d], Syed Ejaz Hussain Mehdi[a], Muhammad Babar[a], and Muhammad Azmi Bustam[a]
[a]Institute of Soil and Environmental Sciences, University of Agriculture Faisalabad, Faisalabad, Pakistan
[b]Department of Chemical Engineering, Universiti Teknologi PETRONAS, Perak, Malaysia
[c]Department of Chemical Engineering, COMSATS University Islamabad, Lahore, Pakistan
[d]Department of Chemistry, College of Science, King Khalid University, Abha, Saudi Arabia

1. Introduction and historical background

The background of nanotechnology can be seen from the dawn of research; artisans in Mesopotamia made their first recorded usage of nanoparticles (NPs) in the 9th century (Mason, 1997). The goal was to produce a shiny effect on a pot's surface. The sparkling surface of the ceramics was attributed to the inclusion of silver and copper NPs in the glassy matrix (Sattler, 2019). Today, many examples of medieval and Renaissance pottery still maintain a distinctive metallic shine of gold or copper glitter. This luster includes Ag and Cu NPs, which are spread evenly in the ceramic glaze's glassy matrix. In those days, of course, the term nanoparticles (NPs) was not known to the craftsmen (Pérez-Arantegui et al., 2001).

There are many cases of ancient objects utilizing nanocomposites. The Lycurgus cup, created around 400 AD by the Romans, was an impressive example and made with a glass that alters its color due to reflection of light. The glass contains Au-Ag alloyed NPs that are arranged in such a way that the glass appears green; however, when light passes through the cup, a bright red is exposed (Barber and Freestone, 1990). Between CE 300 and CE 1700, swordsmiths in Damascus used NPs to construct steel swords with incredible strength, crushing resistance, and extraordinarily sharp edges. Maya Blue, a corrosion-resistant azure dye discovered in CE 800 by the preColumbian Mayan town of Chichen Itza, was also used by our ancestors (Arnold, 2005). The pigment is a versatile medium comprising nanopores in clay that chemically mixes with indigo dye to produce a pigment that is stable in the environment (Heiligtag and Niederberger, 2013).

Nanomaterials: Synthesis, Characterization, Hazards and Safety
https://doi.org/10.1016/B978-0-12-823823-3.00008-2

2. Differences between nanomaterials and bulk materials

Two major factors are responsible for the substantial disparity between nanomaterials and bulk materials: the surface effect, which induces smooth characteristic scaling of atoms on the surface, and the quantum effect, which exhibits discontinuous behavior, owing to quantum containment influences in materials with remote electrons—this influences the chemical reactivity and electronic, physical, electrical, and magnetic properties of materials.

In contrast to microparticles or bulk materials, the proportion of atoms on the surface of the nanoparticle is increased. Nanoparticles have a large surface area and a large number of particles per unit mass compared with microparticles. For instance, one 60 m diameter carbon microparticle has a 0.3 g mass and 0.01 mm^2 surface area. With each particle having a diameter of 60 nm, the equivalent surface area of carbon in nanoparticular form is 11.3 mm^2 in surface area and consists of 1×10^9 nanoparticles. A particle with a diameter of 60 nm has a ratio of surface area to volume or mass 1000 times higher compared to a particle with a diameter of 60 m. As the nanoparticle has a much greater surface area, the reactivity is improved almost 1000-fold. However, the chemical reactivity normally improves as particle size decreases; surface coating and other changes may have complicating effects, and often have lower reactivity as component size decreases in a few examples.

At the surface, the atoms have fewer neighbors than bulk atoms, which contribute to decrease the binding strength with the decreasing size of the particle. Following the Gibbs-Thomson theorem, the effect of decreased atomic binding energy per atom is a decrease in melting point with decrease in the radius of the particle (Roduner, 2006).

Quantum dots—synthesized nanostructures of sizes as low as a few nanometers—are an instance of a type of materials that specifically exploit quantum effects. Quantum dots are analogous to the electrical activity of individual atoms or small molecules and quantum dots are observed as being equivalent to artificial atoms. A quantized energy spectrum is formed corresponding to the confining of electrons to fundamental points in all three spatial directions. The presence of magnetic moments of nonmagnetic material, such as gold, platinum, or palladium in nanoparticles, is often a consequence of the quantum containment effect. Magnetic moments occur from multiple irregular electron spins in nanoparticles made of some hundred atoms (Kouwenhoven et al., 2001).

Quantum confining contributes to quantified improvements often expressed in the catalytic capacity to accept or donate an electric charge or electron affinity. For instance, in the decomposition of N_2O, the reactivity of cationic platinum clusters is determined by several atoms of the cluster: 6–9, 11, 12, 15, and 20 are highly reactive atomic clusters, whereas 10, 13, 14, and 19 are weak in reactivity.

3. Nano, from its beginning to the present

One of the first observations of nanoscale particles was gold-related. The colloidal gold preparation called "divided metals" was demonstrated by Faraday. He called the particles "the divided state of gold" in his account on April 2, 1856, and his solutions can now be found at the Royal Institution in Mayfair, London, UK. In 1890, German microbiologist Robert Koch proved that bacterial growth is impeded by gold-based compounds, which contributed to his winning the 1905 Nobel Prize for Medicine. In reality, gold usage in medicine is not recent, and there are examples of gold use for medicinal purposes throughout history. For example, gold for memory medications, known as Sarawatharishtam, has long been prepared in India. A gold coin was used in China in cooking rice, a custom that was believed to help regenerate the body's gold deficit (Kharlamov et al., 2011).

In history books, the study of nanoscale structures was not explored, not until the lecture to the American Physical Society in 1959 by Richard Feynman, named: "There's plenty of room at the bottom." During his speech, he said: "The rules of physics do not prohibit the idea of maneuvering things atom by atom as far as I can see." In a way, this was the first recommendation of the bottom-up technique for nanomaterial synthesis. Richard Feynman added It is noteworthy that a physicist may synthesize some chemical material published by the chemist. Provide the instructions and synthesizes by a physicist. What is it like? Put atoms down and you create the things, where the chemists say.

It may be good to solve the problems related to biology and chemistry if, in the end, our capacity to look at what we do, and to do it at the atomic level, is improved—a change that cannot be stopped, I think.

Nevertheless, after 1981 the tools were available to check such a hypothesis, with the invention of scanning tunneling microscopy (STM). This method permitted the conception and operation of materials at the nanoscale. These capacities inevitably contributed to the increasing excitement and development in nanotechnological research that soon started to evolve, with the rapid introduction of modern nanotechnological technologies to the market (Feynman, 1960).

4. Defining the nano-dimension

The trend in materials in the current environment is primarily related to the research of Luis Brus in the 1980s, in which he postulated that the bandgap of the basic direct bandgap semiconductor must be based on its size until its dimensions were less than the Bohr radius (Fig. 1). Furthermore, since nanotechnology and nanomaterials work has evolved quickly with a lot of uncertainty in terms, it is appropriate to add fundamental definitions:

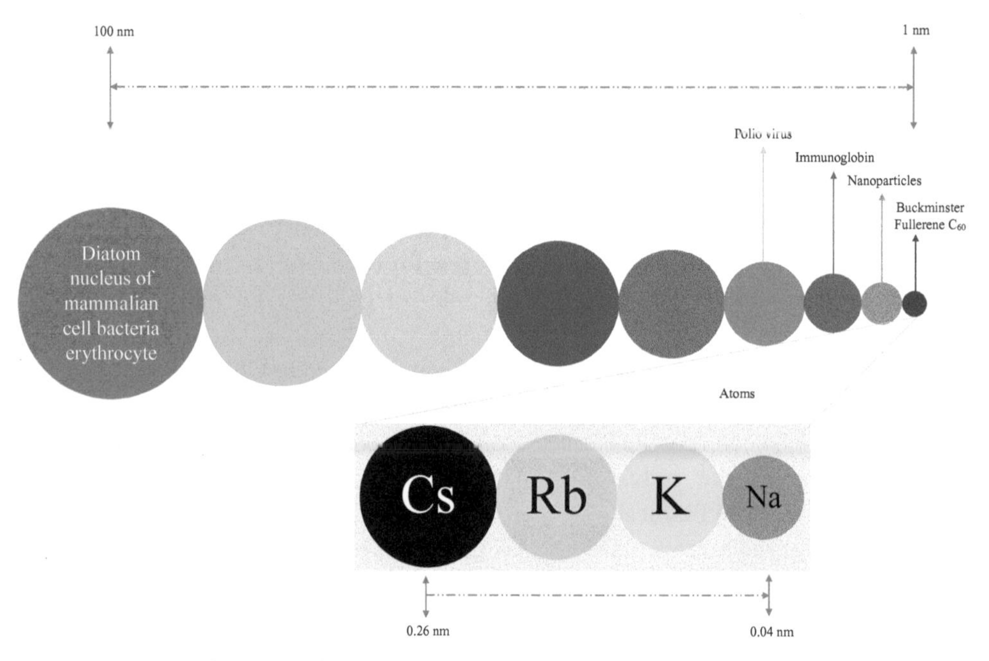

Fig. 1 Comparison of the size of nanoparticles.

- Colloid: stable state of liquid comprising 1–1000 nm particles. The particles in the range of 1–1000 nm may be colloidal particles.
- Nanoparticle: a rigid particle of 1–1000 nm that may relate to a noncrystalline material, to a crystallite aggregation, or even to one crystallite.
- Quantum dot: any material that has at least one dimension with a quantization impact
- Nanomaterial: a solid substance with a nanometer dimension.

5. Physical chemistry of nanoparticles

The overriding influence of the surface area to volume ratio is a prevalent characteristic of nanomaterials, as materials reach nanoscale dimensions. Often, nanomaterials have a large area-to-volume ratio that influences the impacts of such materials. The surface/volume $3/r$ ratio and the surface ratio of the constituent atoms all increase as the radium r of the spherical particle decreases. The strong interatomic bond that occurs in broad crystals is not fulfilled at the atom surface and thus increases mobility and reactivity, such that surface chemistry governs overall characteristics of the nanomaterial. The following nanoparticles are based on a metallic form, although most of the features mentioned also refer to nonmetal materials. This is described in the following context:

1. In aqueous states.
2. In a combination of two phases (e.g., ceramic metal or a polymer-metal combination).
3. As a discontinuous metal film (DMTF) of nanoparticles on a surface of an insulating substrate.

The nanoparticles are generally based on spherical particles in cases 1 and 2, but particle structure may be interrupted for DMTF. The foundation for development and growth of nanoparticles is formed by the classic nucleation principle; however, the calculation of essential nucleus measurements of the subnanometer range is incompatible with the usage of bulk thermodynamic properties in the classic model. The atomic nucleation theory is appropriate for nuclear groups that contain just a few atoms, as is generally used in all three of these systems.

6. Structure of nanoparticles

Nanoparticles could present as pure crystal; meanwhile, in very brief amounts of time, impurities and grids move onto the surface. Within a liquid, the shape (i.e., cones, prisms, rods, circles, etc.) and size of nanoparticles may be regulated by changing precipitation conditions.

Concerning DMTF, the nanoparticles tend to prolate slightly in the deposition, since r^2 is proportional to mass (or time) rather than r^3-proportional to quasi-spherical production. The prevailing particle shape is oblate, with the potential prompts of electrostatic or adatom capture of the substrate being inadequately heat-energized to reach spherical equilibrium. The strength of crystallization and the comparatively poor interaction of the surface atoms are another consideration in deciding the structure of nanoparticles, allowing rapid motion and constant rapid crystallographic reconstruction in the size range 1–10 nm (Kortan et al., 1990).

7. Synthetic methods for nanomaterials synthesis

7.1 Bottom-up methods

7.1.1 Organic ligand-assisted growth

The capping ligands are the ligands that direct the growth of the crystal during the synthesis of colloidal ions/atoms and modulate it. The ligand normally consists of more than one functional group, which could attach to the 2D metal surface, as well as a chemically inert body that enhances the stability of the steric length of the metal complex. Such surface ligands are commonly used in bottom-up wet chemical synthesis to guide the 2D growth of nanocrystals. In the colloidal synthesis of 2D metal nanostructures, ligands usually perform more than one function. The structure of metal-ligand relativity can contribute in three ways to the creation of the 2D morphology (Ortiz and Skrabalak, 2014). This involves: (i) reducing the surface energy of the detected metal base, which considerably lessens the ligand's free energy; (ii) 2D impulsive packaging that contributes toward a soft templating effect, allowing the formation of 2D types of metal-ligand complexes (Ah et al., 2005); and (iii) long chains, which give excellent steric stability with the organic ligand.

Essentially, many specific classes have various linking forces and tastes for the metal/atoms, enabling them to perform many functions, such as moderate reducers, faceting agents, modulators of surface characteristics, and highly viscous solvents. Ligands are usually divided into three classes: organic ion, polymer, and organic nonionic. Typically, ligands are classified into three groups: ionic organic, nonionic organic, and polymer, depending on the form of metallurgical contact (Xu et al., 2017; Li et al., 2011; Jang et al., 2010).

7.1.2 Mediated synthesis of small molecules and ions

In a synthetic wet chemical framework, in addition to metal precursors, reduction, and ligand defense, small numbers of molecules and ions, for example, carbon monoxide (CO) (Hwang et al., 2014; Zhao et al., 2015), hydrogen peroxide (H_2O_2) (Tsuji et al., 2012), and oxygen (O_2) (Guo et al., 2008), halide ions (mainly referring to I^-, Cl^-, and Br^-) (Lohse et al., 2014; DuChene et al., 2013; Millstone et al., 2008; Tang et al., 2009; HoáKim et al., 2014), and metal ions (e.g., Ag^+, Cu^{2+}, and Fe^{3+}), may play a significant role in developing the growth of 2D metal nano-crystalline. Such small molecules/ions deliberately operate to harmonize metal precursors' redox potential and surface energy from various crystal facets.

7.1.3 2D template-confined growth

In general, every 2D or 3D substrate may be used as the base for the production of such 2D nanomaterials. They may either interact with metals (or with their precursors, intermediate reduction) to achieve chemically 2D development, or physically confine 2D development to their interlayers or surfaces. To date, a number of 2D materials or substrates, including graphene oxide (GO) (Huang et al., 2010a, 2011a, b; Yang et al., 2014a; Jiang et al., 2016a; Lou et al., 2015); graphites; 199 – 20, graphene's (Bai et al., 2008; Li et al., 2005) and reduced GO (r-GO) (Shirai et al., 2000; Walter et al., 2000; Walter, 2000), zwitterionic vesicles (Gao et al., 2014); indium tin oxide (ITO) (Li et al., 2009; Lee et al., 2012), n-type GaAs (Sun et al., 2008), vesicles and interface (Gao et al., 2014) has been investigated.

7.1.4 Seeded growth

Seed growth seems to be a synthetic two-step technique that affectively distinguishes between nucleation and ensuing production. Seed engineering can attain fine-tuned structure, with only a minimal scale distribution, by utilizing presynthesized seeds to focus only on the kinetic control of growth (Kan et al., 2010; Li et al., 2014; Millstone et al., 2006; Niu et al., 2012; Liu et al., 2014; Xia et al., 2017). The main factor of semen-mediated processing is the assurance of high-quality seed. For example, the influence of poly(sodium styrene sulfonate) (PSSS) on the suppression of various seed sizes and forms for high-efficiency growth of Ag nano-prisms has been reported. (Aherne et al., 2008). The existence of PSSS in the development of seed will result in a yield of Ag nano-prisms with an abundance of stacking defects of up to 95%. Moreover, there is a strong need to monitor the overcrowding direction in seed production. For example, citrate ligands were found to restrict the overcrowding of Ag on the surfaces efficiently (Park et al., 2006) and allow just lateral development (Goebl et al., 2012; Zhang et al., 2010). Subsequently, thin Ag-nanoplates with a thickness of less than 10 nm and lateral development of four micrograms were effectively obtained by gradual reductions in Ag-citrate complexes. Essentially, the propagation of seeds can have various modes of processing. In the case of different coating agents such as sodium citrate and PVP, Ag can be specifically coated onto various aspects of plants, leading to Ag nanoplates of <200 nm in thickness and Ag nanosheets of $\sim 55\,\mu m$ in lateral size, respectively. For instance, Xia et al. reported vertical and lateral overgrowth of Ag with triangular nanoscale Ag nanoplates as plants.

7.1.5 Photochemical synthesis

In recent years, photochemical synthesis has been used as a specific and efficient method for the production of two-dimensional nanostructures. Light irradiation is used for the development of metal precursors as an efficient process that helps 2D metal nanostructures to shape.

For instance, high-energy UV light may permit radicals in superoxides and ethoxies to emerge as powerful reducing agents for metallic ions from the ethanol molecules, converting $AuCl_4^-$ metal ions. Production by inserting Ag^+ ions as shape orientation agents of AU-triangular nanoplates in mesoporous silicon shells can be achieved (Kajimoto et al., 2013; Redjala et al., 2008; Huang et al., 2010b; Cha et al., 2007; Shaik et al., 2017).

7.2 Hydro/solvothermal methods

Due to their simple setup and fast activity, hydrothermal and solvothermal methods are widely used to manufacture various nanomaterials (Feng and Xu, 2001; Lai et al., 2015; Xu and Wang, 2012). The synthesis is usually done in aqueous (hydrothermal) or solvothermal solutions packed in Teflon lined stainless steel, autoclaved above environmental temperature and pressure.

The reaction will take place above the usual boiling point of solvents at a temperature that is unavailable in other methods of synthesis. Reactivity and solubility may be improved at extremely high temperatures and elevated pressures, resulting in physiochemical properties improvements for solvents. The nucleation and growth processes of nanocrystals in metals can be controlled to create standardized materials with a fine-grain composition, following the optimal mixture of reactants, for instance ligands, precursors, and reductants. In comparison, hydrothermal and solvothermal methods require extremely high energy usage compared to other wet chemical approaches.

The hydrothermal process that uses water as a solvent has been used successfully in synthesizing many 2D metal nanomaterials, such as Cu nanoplatelets, Ag nanoplates (Yang et al., 2006, 2014b), Co nanosheets (Gao et al., 2016), Ni nanoplate and nanoplates, Ru nanoplates, Co nanoplates (Xu et al., 2007), and Rh nanosheets (Jiang et al., 2016b; Bai et al., 2016).

7.2.1 Crystal phase transformation

In comparison to conventional synthetic methods, crystal-phases transition can create novel 2D metal nanostructures from presynthesized 2D metal nanostructures with distinct crystal structures (Fan et al., 2015). Such uncommon 2D nanostructures may also be used as active models, in which processes of surface alteration, such as ligand exchanges and coating, can be used to induce partial or nonparticular coatings.

7.2.2 Biological synthesis

Natural compounds or biomimetic hybrid methods are used in biological techniques to generate metal nanostructures. In comparison, giant biomolecules including proteins and fibrils may aid in creating macroscale 2D metal nanostructures and are often advantageous in terms of ease of access, expense, and low environmental impacts. Mezzenga et al. reported the templating activity of the β-lactoglobulin-like amyloid Fibrille as a popular example in a micro-micrometric synthesis of triangular and hexagonal Au nanoplates (Bolisetty et al., 2011). Twisted lactoglobulin filaments are protein-drift nanofibrils of a diameter of 1 – 6 nm and length of more than 1.0 μm. From another study, unique critical Au micro flocks with a planar surface area of over 104 mm^2 with a mixture of chloroaurate ions with amyloid fiber with an appropriate proportion of 60°C and a pH of 2 have been successfully compiled by one research team (Zhou et al., 2015). Because of their abundant biological ability, they are adapted for large-scale synthesis of 2D metal nanostructures. Qian et al. demonstrated that a simple pathway for hexagonal mass production of Au nanosheets with a hexane thickness of <50 nm is a cellulose-biopolymer that could operate as either a stable or reducing agent, using polysaccharide chitosan. In addition, after extraction, the synthesis of 2D metal nanostructures may also use human body proteins.

7.2.3 Nanoparticle assembly

Different from the previous approaches, the array builds 2D metallic nanostructures while employing metal nanocrystals as building blocks (Li et al., 2016; Lee et al., 2015). The 2D self-assembly of nanoparticles is typically activated by guiding the anisotropic connections between nanoparticles at interfaces, such as the solid-liquid interface

(Santhanam et al., 2003; Ciesielski et al., 2010), liquid-liquid interface (Duan et al., 2004; Böker et al., 2007; Reincke et al., 2004; Wu et al., 2013, 2014, 2015), and liquid-air interface (Korgel et al., 1998; Cheng et al., 2009; Bigioni et al., 2006). As an example of solvent vaporization at the liquid-air crossing stage, the elimination of solvents will interrupt the stabilization of the colloidal suspension and contribute to the formation of monolayer islands, which eventually combine into constant superlattice films via an effective drying cycle. Focused on this technique, the aggregation efficiency of nanoparticles could be regulated while using specific surface ligands to regulate the attraction and repulsion forces between nanoparticles (Martin et al., 2000; Wu et al., 2017).

7.3 Top-down methods

7.3.1 Mechanical compression method

Alternatively, a mechanical compression method may be used to produce 2D metal nanomaterials by growing the dimension of bulk metals. With its outstanding ductility, metals can be quickly compacted and pliant at room temperature.

In a study performed by Wu et al., Fe, Au, and Ag foils with a thickness of $<70\,\mu m$ were stacked and the stacked binary metal foil regularly folded and deposited at room temperature on a rolling frame. The foil, with a thickness of about $5\,\mu m$, is stacked with a sacrificial metal cover in the conventional cycle. The thickness of each metal sheet is reduced to just a few nanometers after rolling 20 times. This idea was initially used for the development of multiatomic Al nanosheets (Gu et al., 2017). This method can also be used in hard metals at high temperatures. In addition, an ultra-hot pressure process was developed for the manufacturing of Bi nanosheets produced from Bi nanoparticles (Hussain et al., 2017).

7.3.2 Type of exfoliation

Though the majority of metallic materials are frameworks that do not fall in the surroundings, a few metals, including Sb and As, have layered frameworks that are firm in ambient circumstances (Zhang et al., 2015). Using Sb as an example, the grayscale of Sb crystallizes in a twisted hcp form of 10%–20% shorter interatomic distances between the intralayer (Vainshtein et al., 2013).

Several 2D metal nanostructures can be made from 2D metallic exfoliated compounds. Matsumoto et al. have shown that monolayer Pt nanosheets can be produced through the reduction of exfoliated Pt oxide monolayers from platinum oxide (Funatsu et al., 2014).

7.3.3 Nanolithography

Nanolithography can be used for the development of well-defined 2D metal arrays on a substrate employing electron beam lithography (EBL), which is widely used to create metal nanostructures through scaling down the existing "parent-focus" layer via an electron beam (Hatzor and Weiss, 2001; Liu et al., 2011). Using this method, nanolithography is a useful method that could also produce nanoscale configurations of precisely organized shapes, sizes, and intervals. Another conventional example is the NSL methodology, which is a multiphase approach for designing surface-restricted nanostructure arrays (Chan et al., 2007, 2008). To store different nanostructures, NSL utilizes monolayer self-assembly of controlled-diameter polystyrene spheres. This technique has been used successfully to produce organized arrays of Ag, Al, and Cu triangulated nanocrystals (Haes et al., 2005).

8. Conclusions

As the nanoparticle has a much greater surface area, its reactivity is improved almost 1000-fold. A particle with a diameter of 60 nm has a ratio of surface area to volume or mass 1000 times higher compared to a particle with a diameter of 60 m. Nanoparticles have different shapes in a liquid, and such shape (i.e., cones, prisms, rods, circles, etc.) and size may be regulated by changing precipitation conditions. There are several synthetic methods for nanoparticle synthesis including bottom-up, hydro/solvothermal, and top-down. In the bottom-up methods, ligands have more than one function. The structure of metal-ligand can contribute in three ways to the creation of the 2D morphology. This involves: (i) lowering the surface energy of metal base; (ii) impulsive 2D packing contributing to a "soft templating" effect that facilitates the forming of 2D types of metal-ligand complexes; and (iii) the long-chain organic ligands providing excellent steric stabilization. The hydrothermal and solvothermal methods are commonly used for the synthesis of different nanomaterials due to their easy set-up and fast operation. Natural compounds or biomimetic hybrid methods are used in biological techniques to generate metal nanostructures.

Acknowledgment

The authors gratefully acknowledge the departments at their respective universities for providing state-of-the-art research facilities.

References

Ah, C.S., Yun, Y.J., Park, H.J., Kim, W.-J., Ha, D.H., Yun, W.S., 2005. Size-controlled synthesis of machinable single crystalline gold nanoplates. Chem. Mater. 17 (22), 5558–5561.

Aherne, D., Ledwith, D.M., Gara, M., Kelly, J.M., 2008. Optical properties and growth aspects of silver nanoprisms produced by a highly reproducible and rapid synthesis at room temperature. Adv. Funct. Mater. 18 (14), 2005–2016.

Arnold, D.E., 2005. Maya Blue and palygorskite: a second possible pre-Columbian source. Anc. Mesoam. 16 (1), 51–62.

Bai, J., Xu, G.-R., Xing, S.-H., Zeng, J.-H., Jiang, J.-X., Chen, Y., 2016. Hydrothermal synthesis and catalytic application of ultrathin rhodium nanosheet nanoassemblies. ACS Appl. Mater. Interfaces 8 (49), 33635–33641.

Bai, X., Zheng, L., Li, N., Dong, B., Liu, H., 2008. Synthesis and characterization of microscale gold nanoplates using Langmuir monolayers of long-chain ionic liquid. Cryst. Growth Des. 8 (10), 3840–3846.

Barber, D., Freestone, I.C., 1990. An investigation of the origin of the colour of the Lycurgus Cup by analytical transmission electron microscopy. Archaeometry 32 (1), 33–45.

Bigioni, T.P., Lin, X.-M., Nguyen, T.T., Corwin, E.I., Witten, T.A., Jaeger, H.M., 2006. Kinetically driven self assembly of highly ordered nanoparticle monolayers. Nat. Mater. 5 (4), 265–270.

Böker, A., He, J., Emrick, T., Russell, T.P., 2007. Self-assembly of nanoparticles at interfaces. Soft Matter 3 (10), 1231–1248.

Bolisetty, S., Vallooran, J.J., Adamcik, J., Handschin, S., Gramm, F., Mezzenga, R., 2011. Amyloid-mediated synthesis of giant, fluorescent, gold single crystals and their hybrid sandwiched composites driven by liquid crystalline interactions. J. Colloid Interface Sci. 361 (1), 90–96.

Cha, S.-H., Kim, J.-U., Kim, K.-H., Lee, J.-C., 2007. Preparation of gold nanosheets using poly(ethylene oxide)-poly(propylene oxide)-poly(ethylene oxide) block copolymers via photoreduction. Mater. Sci. Eng. B 140 (3), 182–186.

Chan, G.H., Zhao, J., Hicks, E.M., Schatz, G.C., Van Duyne, R.P., 2007. Plasmonic properties of copper nanoparticles fabricated by nanosphere lithography. Nano Lett. 7 (7), 1947–1952.

Chan, G.H., Zhao, J., Schatz, G.C., Van Duyne, R.P., 2008. Localized surface plasmon resonance spectroscopy of triangular aluminum nanoparticles. J. Phys. Chem. C 112 (36), 13958–13963.

Cheng, W., Campolongo, M.J., Cha, J.J., Tan, S.J., Umbach, C.C., Muller, D.A., Luo, D., 2009. Free-standing nanoparticle superlattice sheets controlled by DNA. Nat. Mater. 8 (6), 519–525.

Ciesielski, A., Palma, C.A., Bonini, M., Samorì, P., 2010. Towards supramolecular engineering of functional nanomaterials: pre-programming multi-component 2D self-assembly at solid-liquid interfaces. Adv. Mater. 22 (32), 3506–3520.

Duan, H., Wang, D., Kurth, D.G., Möhwald, H., 2004. Directing self-assembly of nanoparticles at water/oil interfaces. Angew. Chem. Int. Ed. 43 (42), 5639–5642.

DuChene, J.S., Niu, W., Abendroth, J.M., Sun, Q., Zhao, W., Huo, F., Wei, W.D., 2013. Halide anions as shape-directing agents for obtaining high-quality anisotropic gold nanostructures. Chem. Mater. 25 (8), 1392–1399.

Fan, Z., Huang, X., Han, Y., Bosman, M., Wang, Q., Zhu, Y., Liu, Q., Li, B., Zeng, Z., Wu, J., 2015. Surface modification-induced phase transformation of hexagonal close-packed gold square sheets. Nat. Commun. 6 (1), 1–9.

Feng, S., Xu, R., 2001. New materials in hydrothermal synthesis. Acc. Chem. Res. 34 (3), 239–247.

Feynman, R.P., 1960. There's Plenty of Room at the Bottom. California Institute of Technology, Engineering and Science Magazine.

Funatsu, A., Tateishi, H., Hatakeyama, K., Fukunaga, Y., Taniguchi, T., Koinuma, M., Matsuura, H., Matsumoto, Y., 2014. Synthesis of monolayer platinum nanosheets. Chem. Commun. 50 (62), 8503–8506.

Gao, S., Lin, Y., Jiao, X., Sun, Y., Luo, Q., Zhang, W., Li, D., Yang, J., Xie, Y., 2016. Partially oxidized atomic cobalt layers for carbon dioxide electroreduction to liquid fuel. Nature 529 (7584), 68–71.

Gao, X., Lu, F., Dong, B., Zhou, T., Tian, W., Zheng, L., 2014. Zwitterionic vesicles with $AuCl_4^-$ counterions as soft templates for the synthesis of gold nanoplates and nanospheres. Chem. Commun. 50 (63), 8783–8786.

Goebl, J., Zhang, Q., He, L., Yin, Y., 2012. Monitoring the shape evolution of silver nanoplates: a marker study. Angew. Chem. Int. Ed. 51 (2), 552–555.

Gu, J., Li, B., Du, Z., Zhang, C., Zhang, D., Yang, S., 2017. Multi-atomic layers of metallic aluminum for ultralong life lithium storage with high volumetric capacity. Adv. Funct. Mater. 27(27), 1700840.

Guo, Z., Zhang, Y., Xu, A., Wang, M., Huang, L., Xu, K., Gu, N., 2008. Layered assemblies of single crystal gold nanoplates: direct room temperature synthesis and mechanistic study. J. Phys. Chem. C 112 (33), 12638–12645.

Haes, A.J., Zhao, J., Zou, S., Own, C.S., Marks, L.D., Schatz, G.C., Van Duyne, R.P., 2005. Solution-phase, triangular Ag nanotriangles fabricated by nanosphere lithography. J. Phys. Chem. B 109 (22), 11158–11162.

Hatzor, A., Weiss, P., 2001. Molecular rulers for scaling down nanostructures. Science 291 (5506), 1019–1020.

Heiligtag, F.J., Niederberger, M., 2013. The fascinating world of nanoparticle research. Mater. Today 16 (7–8), 262–271.

HoáKim, M., KyoungáKwak, S., HyukáIm, S., 2014. Maneuvering the growth of silver nanoplates: use of halide ions to promote vertical growth. J. Mater. Chem. C 2 (30), 6165–6170.

Huang, X., Li, H., Li, S., Wu, S., Boey, F., Ma, J., Zhang, H., **2011**b. Synthesis of gold square-like plates from ultrathin gold square sheets: the evolution of structure phase and shape. Angew. Chem. Int. Ed. 50 (51), 12245–12248.

Huang, X., Li, S., Huang, Y., Wu, S., Zhou, X., Li, S., Gan, C.L., Boey, F., Mirkin, C.A., Zhang, H., **2011**a. Synthesis of hexagonal close-packed gold nanostructures. Nat. Commun. 2 (1), 1–6.

Huang, X., Qi, X., Huang, Y., Li, S., Xue, C., Gan, C.L., Boey, F., Zhang, H., **2010**b. Photochemically controlled synthesis of anisotropic Au nanostructures: platelet-like Au nanorods and six-star Au nanoparticles. ACS Nano 4 (10), 6196–6202.

Huang, X., Zhou, X., Wu, S., Wei, Y., Qi, X., Zhang, J., Boey, F., Zhang, H., **2010**a. Reduced graphene oxide-templated photochemical synthesis and in situ assembly of Au nanodots to orderly patterned Au nanodot chains. Small 6 (4), 513–516.

Hussain, N., Liang, T., Zhang, Q., Anwar, T., Huang, Y., Lang, J., Huang, K., Wu, H., 2017. Ultrathin Bi nanosheets with superior photoluminescence. Small. 13(36), 1701349.

Hwang, S.Y., Zhang, M., Zhang, C., Ma, B., Zheng, J., Peng, Z., 2014. Carbon monoxide in controlling the surface formation of Group VIII metal nanoparticles. Chem. Commun. 50 (90), 14013–14016.

Jang, K., Kim, H.J., Son, S.U., 2010. Low-temperature synthesis of ultrathin rhodium nanoplates via molecular orbital symmetry interaction between rhodium precursors. Chem. Mater. 22 (4), 1273–1275.

Jiang, Y., Su, J., Yang, Y., Jia, Y., Chen, Q., Xie, Z., Zheng, L., **2016**b. A facile surfactant-free synthesis of Rh flower-like nanostructures constructed from ultrathin nanosheets and their enhanced catalytic properties. Nano Res. 9 (3), 849–856.

Jiang, Y., Yan, Y., Chen, W., Khan, Y., Wu, J., Zhang, H., Yang, D., **2016**a. Single-crystalline Pd square nanoplates enclosed by {100} facets on reduced graphene oxide for formic acid electro-oxidation. Chem. Commun. 52 (99), 14204–14207.

Kajimoto, S., Shirasawa, D., Horimoto, N.N., Fukumura, H., 2013. Additive-free size-controlled synthesis of gold square nanoplates using photochemical reaction in dynamic phase-separating media. Langmuir 29 (19), 5889–5895.

Kan, C., Wang, C., Li, H., Qi, J., Zhu, J., Li, Z., Shi, D., 2010. Gold microplates with well-defined shapes. Small 6 (16), 1768–1775.

Kharlamov, A., Skripnichenko, A., Gubareny, N., Bondarenko, M., Kirillova, N., Kharlamova, G., Fomenko, V., 2011. Toxicology of nano-objects: nanoparticles, nanostructures and nanophases. In: Biodefence. Springer, pp. 23–32.

Korgel, B.A., Fullam, S., Connolly, S., Fitzmaurice, D., 1998. Assembly and self-organization of silver nanocrystal superlattices: ordered "soft spheres" J. Phys. Chem. B 102 (43), 8379–8388.

Kortan, A., Hull, R., Opila, R.L., Bawendi, M.G., Steigerwald, M.L., Carroll, P., Brus, L.E., 1990. Nucleation and growth of CdSe on ZnS quantum crystallite seeds, and vice versa, in inverse micelle media. J. Am. Chem. Soc. 112 (4), 1327–1332.

Kouwenhoven, L.P., Austing, D., Tarucha, S., 2001. Few-electron quantum dots. Rep. Prog. Phys. 64 (6), 701.

Lai, J., Niu, W., Luque, R., Xu, G., 2015. Solvothermal synthesis of metal nanocrystals and their applications. Nano Today 10 (2), 240–267.

Lee, C., Josephs, E.A., Shao, J., Ye, T., 2012. Nanoscale chemical patterns on gold microplates. J. Phys. Chem. C 116 (33), 17625–17632.

Lee, Y.H., Shi, W., Lee, H.K., Jiang, R., Phang, I.Y., Cui, Y., Isa, L., Yang, Y., Wang, J., Li, S., 2015. Nanoscale surface chemistry directs the tunable assembly of silver octahedra into three two-dimensional plasmonic superlattices. Nat. Commun. 6 (1), 1–7.

Li, B., Zhou, D., Han, Y., 2016. Assembly and phase transitions of colloidal crystals. Nat. Rev. Mater. 1 (2), 1–13.
Li, N., Zhao, P., Astruc, D., 2014. Anisotropic gold nanoparticles: synthesis, properties, applications, and toxicity. Angew. Chem. Int. Ed. 53 (7), 1756–1789.
Li, Q., Liu, F., Lu, C., Lin, J.-M., 2011. Aminothiols sensing based on fluorosurfactant-mediated triangular gold nanoparticle-catalyzed luminol chemiluminescence. J. Phys. Chem. C 115 (22), 10964–10970.
Li, W., Ma, H., Zhang, J., Liu, X., Feng, X., 2009. Fabrication of gold nanoprism thin films and their applications in designing high activity electrocatalysts. J. Phys. Chem. C 113 (5), 1738–1745.
Li, Z., Liu, Z., Zhang, J., Han, B., Du, J., Gao, Y., Jiang, T., 2005. Synthesis of single-crystal gold nanosheets of large size in ionic liquids. J. Phys. Chem. B 109 (30), 14445–14448.
Liu, N., Tang, M.L., Hentschel, M., Giessen, H., Alivisatos, A.P., 2011. Nanoantenna-enhanced gas sensing in a single tailored nanofocus. Nat. Mater. 10 (8), 631–636.
Liu, X., Li, L., Yang, Y., Yin, Y., Gao, C., 2014. One-step growth of triangular silver nanoplates with predictable sizes on a large scale. Nanoscale 6 (9), 4513–4516.
Lohse, S.E., Burrows, N.D., Scarabelli, L., Liz-Marzán, L.M., Murphy, C.J., 2014. Anisotropic noble metal nanocrystal growth: the role of halides. Chem. Mater. 26 (1), 34–43.
Lou, X., Pan, H., Zhu, S., Zhu, C., Liao, Y., Li, Y., Zhang, D., Chen, Z., 2015. Synthesis of silver nanoprisms on reduced graphene oxide for high-performance catalyst. Catal. Commun. 69, 43–47.
Martin, J.E., Wilcoxon, J.P., Odinek, J., Provencio, P., 2000. Control of the interparticle spacing in gold nanoparticle superlattices. J. Phys. Chem. B 104 (40), 9475–9486.
Mason, R.B., 1997. Early mediaeval Iraqi lustre painted and associated wares: typology in a multidisciplinary study. Iraq 59, 15–61.
Millstone, J.E., Métraux, G.S., Mirkin, C.A., 2006. Controlling the edge length of gold nanoprisms via a seed-mediated approach. Adv. Funct. Mater. 16 (9), 1209–1214.
Millstone, J.E., Wei, W., Jones, M.R., Yoo, H., Mirkin, C.A., 2008. Iodide ions control seed-mediated growth of anisotropic gold nanoparticles. Nano Lett. 8 (8), 2526–2529.
Niu, W., Zhang, L., Xu, G., 2012. Seed-mediated growth method for high-quality noble metal nanocrystals. Sci. China Chem. 55 (11), 2311–2317.
Ortiz, N., Skrabalak, S.E., 2014. On the dual roles of ligands in the synthesis of colloidal metal nanostructures. Langmuir 30 (23), 6649–6659.
Park, S.-W., Lee, J.-W., Choi, B.-S., Lee, J.-W., 2006. Absorption of carbon dioxide into non-aqueous solutions of N-methyldiethanolamine. Korean J. Chem. Eng. 23 (5), 806–811.
Pérez-Arantegui, J., Molera, J., Larrea, A., Pradell, T., Vendrell-Saz, M., Borgia, I., Brunetti, B.G., Cariati, F., Fermo, P., Mellini, M., 2001. Luster pottery from the thirteenth century to the sixteenth century: a nanostructured thin metallic film. J. Am. Ceram. Soc. 84 (2), 442–446.
Redjala, T., Apostolecu, G., Beaunier, P., Mostafavi, M., Etcheberry, A., Uzio, D., Thomazeau, C., Remita, H., 2008. Palladium nanostructures synthesized by radiolysis or by photoreduction. New J. Chem. 32 (8), 1403–1408.
Reincke, F., Hickey, S.G., Kegel, W.K., Vanmaekelbergh, D., 2004. Spontaneous assembly of a monolayer of charged gold nanocrystals at the water/oil interface. Angew. Chem. Int. Ed. 43 (4), 458–462.
Roduner, E., 2006. Size matters: why nanomaterials are different. Chem. Soc. Rev. 35 (7), 583–592.
Santhanam, V., Liu, J., Agarwal, R., Andres, R.P., 2003. Self-assembly of uniform monolayer arrays of nanoparticles. Langmuir 19 (19), 7881–7887.
Sattler, K.D., 2019. 21st Century Nanoscience–A Handbook: Nanophysics Sourcebook (Volume One). CRC Press.
Shaik, F., Zhang, W., Niu, W., 2017. A novel photochemical method for the synthesis of Au triangular nanoplates inside nanocavity of mesoporous silica shells. J. Phys. Chem. C 121 (17), 9572–9578.
Shirai, M., Igeta, K., Arai, M., 2000. Formation of platinum nanosheets between graphite layers. Chem. Commun. 7, 623–624.
Sun, Y., Lei, C., Gosztola, D., Haasch, R., 2008. Formation of oxides and their role in the growth of Ag nanoplates on GaAs substrates. Langmuir 24 (20), 11928–11934.
Tang, B., Xu, S., An, J., Zhao, B., Xu, W., Lombardi, J.R., 2009. Kinetic effects of halide ions on the morphological evolution of silver nanoplates. PCCP 11 (44), 10286–10292.

Tsuji, M., Gomi, S., Maeda, Y., Matsunaga, M., Hikino, S., Uto, K., Tsuji, T., Kawazumi, H., 2012. Rapid transformation from spherical nanoparticles, nanorods, cubes, or bipyramids to triangular prisms of silver with PVP, citrate, and H_2O_2. Langmuir 28 (24), 8845–8861.

Vainshtein, B.K., Friedkin, V.M., Indenbom, V.L., 2013. Structure of Crystals. Springer Science & Business Media.

Walter, J., 2000. Template-assisted growth of hexagonal poly-or single-crystalline quasi-2D palladium nanoparticles. Adv. Mater. 12 (1), 31–33.

Walter, J., Heiermann, J., Dyker, G., Hara, S., Shioyama, H., 2000. Hexagonal or quasi two-dimensional palladium nanoparticles—tested at the Heck Reaction. J. Catal. 189 (2), 449–455.

Wu, L., Willis, J.J., McKay, I.S., Diroll, B.T., Qin, J., Cargnello, M., Tassone, C.J., 2017. High-temperature crystallization of nanocrystals into three-dimensional superlattices. Nature 548 (7666), 197–201.

Wu, Z., Dong, C., Li, Y., Hao, H., Zhang, H., Lu, Z., Yang, B., 2013. Self-assembly of Au15 into single-cluster-thick sheets at the interface of two miscible high-boiling solvents. Angew. Chem. Int. Ed. 52 (38), 9952–9955.

Wu, Z., Li, Y., Liu, J., Lu, Z., Zhang, H., Yang, B., 2014. Colloidal self-assembly of catalytic copper nanoclusters into ultrathin ribbons. Angew. Chem. Int. Ed. 53 (45), 12196–12200.

Wu, Z., Liu, J., Li, Y., Cheng, Z., Li, T., Zhang, H., Lu, Z., Yang, B., 2015. Self-assembly of nanoclusters into mono-, few-, and multilayered sheets via dipole-induced asymmetric van der waals attraction. ACS Nano 9 (6), 6315–6323.

Xia, Y., Gilroy, K.D., Peng, H.C., Xia, X., 2017. Seed-mediated growth of colloidal metal nanocrystals. Angew. Chem. Int. Ed. 56 (1), 60–95.

Xu, B., Wang, X., 2012. Solvothermal synthesis of monodisperse nanocrystals. Dalton Trans. 41 (16), 4719–4725.

Xu, D., Liu, Y., Zhao, S., Lu, Y., Han, M., Bao, J., 2017. Novel surfactant-directed synthesis of ultra-thin palladium nanosheets as efficient electrocatalysts for glycerol oxidation. Chem. Commun. 53 (10), 1642–1645.

Xu, R., Xie, T., Zhao, Y., Li, Y., 2007. Single-crystal metal nanoplatelets: cobalt, nickel, copper, and silver. Cryst. Growth Des. 7 (9), 1904–1911.

Yang, J., Lu, L., Wang, H., Shi, W., Zhang, H., 2006. Glycyl glycine templating synthesis of single-crystal silver nanoplates. Cryst. Growth Des. 6 (9), 2155–2158.

Yang, S., Qiu, P., Yang, G., 2014a. Graphene induced formation of single crystal Pt nanosheets through 2-dimensional aggregation and sintering of nanoparticles in molten salt medium. Carbon 77, 1123–1131.

Yang, Y., Zhong, X.L., Zhang, Q., Blackstad, L.G., Fu, Z.W., Li, Z.Y., Qin, D., 2014b. The role of etching in the formation of Ag nanoplates with straight, curved and wavy edges and comparison of their SERS properties. Small 10 (7), 1430–1437.

Zhang, Q., Hu, Y., Guo, S., Goebl, J., Yin, Y., 2010. Seeded growth of uniform Ag nanoplates with high aspect ratio and widely tunable surface plasmon bands. Nano Lett. 10 (12), 5037–5042.

Zhang, S., Yan, Z., Li, Y., Chen, Z., Zeng, H., 2015. Atomically thin arsenene and antimonene: semimetal-semiconductor and indirect-direct band-gap transitions. Angew. Chem. Int. Ed. 54 (10), 3112–3115.

Zhao, L., Xu, C., Su, H., Liang, J., Lin, S., Gu, L., Wang, X., Chen, M., Zheng, N., 2015. Single-crystalline rhodium nanosheets with atomic thickness. Adv. Sci. 2(6), 1500100.

Zhou, J., Saha, A., Adamcik, J., Hu, H., Kong, Q., Li, C., Mezzenga, R., 2015. Macroscopic single-crystal gold microflakes and their devices. Adv. Mater. 27 (11), 1945–1950.

CHAPTER 2

Sources of nanomaterials

Muneeb ur Rahman
Department of Physics, Islamia College Peshawar (Public Sector University), Peshawar, Pakistan

Nanoparticles or nanomaterials are produced in nature (in events in the skies and via microorganisms in the atmosphere and ecosystem), and can also be synthesized in the laboratory. In this chapter, different sorts of natural and artificial sources of nanoparticles or nanomaterials are highlighted.

1. Natural sources

Engineered nanoparticles and their impacts on the environment are well studied, whereas natural nanoparticles (NNPs) and their influence are often ignored. However, nature itself is an outstanding nanotechnologist. Nanoparticles (NPs) are also formed via natural processes in the universe including the earth. Their formation regions include (but are not limited to) the biosphere, hydrosphere, lithosphere, and atmosphere. Numerous reactions are involved in the natural synthesis mechanisms of these NNPs, which include biological, photochemical, chemical, thermal, volcanic eruptions, winds, mechanical, geothermal, and biogeochemical. These processes either operate separately or work in combination in the synthesis process of these naturally occurring nanoparticles. In the extraterrestrial environment, NPs are formed in the violent explosion mechanism of supernovae and hypernovae (occurring at enormous temperature and pressure), which scatter the synthesized materials into space in the form of cosmic dust with gigantic speed. This mechanism of formation of nanoparticles after the death of stars is shown in Fig. 1. In this whole process, initially the progenitor stars burn their fuels to form heavier and heavier iron group nuclei over millions of years. When the star's inner core, consisted of iron group nuclei, exceeds the appropriate Chandrasekhar's mass limit, the electron degeneracy pressure is unable to stop it from imploding. When the core reaches nuclear density it becomes stiffened and rebounds the in-falling matter at enormous speed. The shock waves thus produced during rebound convert the heavier elements into lighter elements and nuclei, and scatters these lighter elements into space as star dust, which consequently leads to the formation of micro and nanomaterials, as shown in Fig. 1. Similarly, these NPs are produced naturally in soil, deserts, aerosols, waters, microbial systems, deep sea hydrothermal vents, volcanic activity under deep sea water, and natural ore, etc. The schematic diagram of the various sources of these NNPs is shown in Fig. 2.

Nanomaterials: Synthesis, Characterization, Hazards and Safety
https://doi.org/10.1016/B978-0-12-823823-3.00007-0

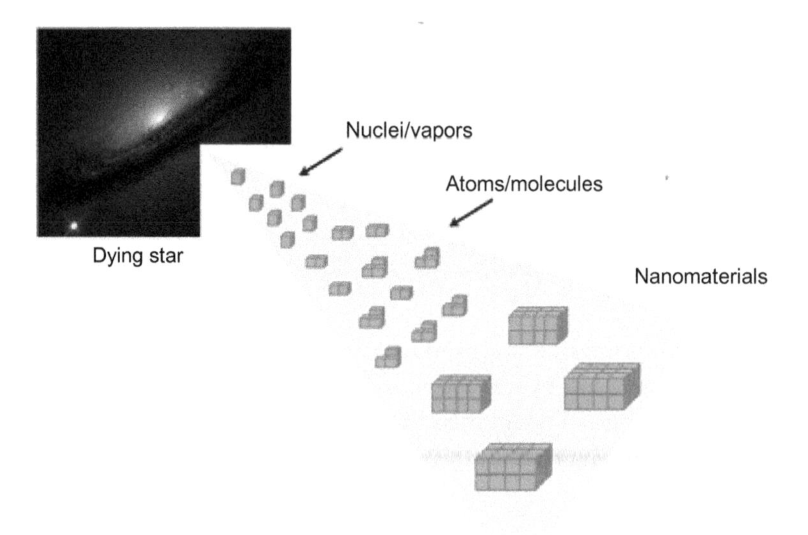

Fig. 1 Dying star in supernova explosion and formation of nanometric-scale materials from the stellar dust.

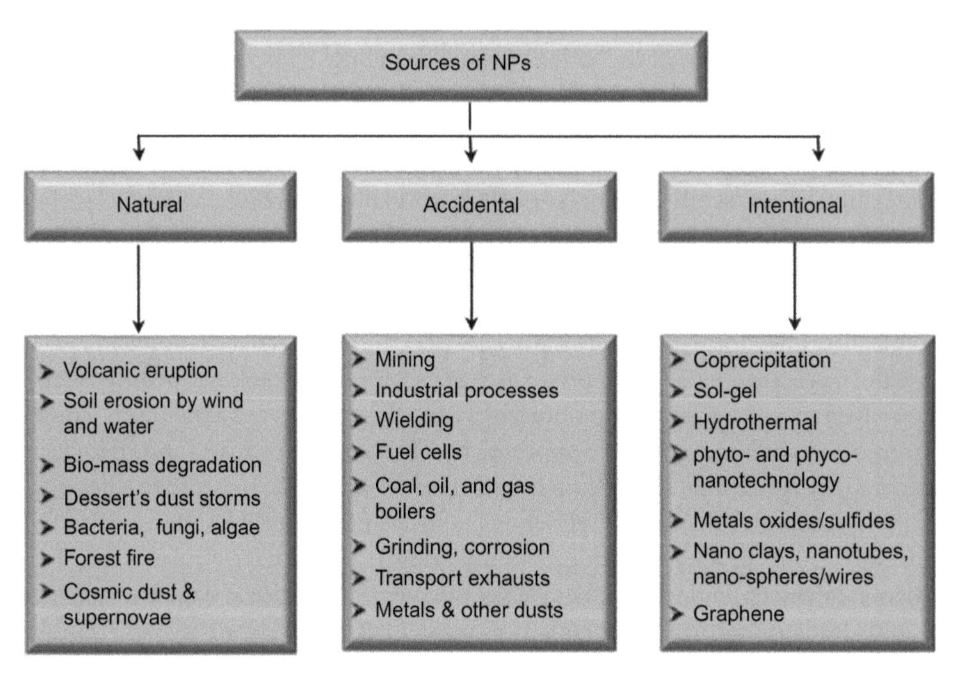

Fig. 2 Different natural, accidental, and intentional sources involved in the production of nanoparticles.

NPs are also spontaneously produced during various industrial processes, natural decay of items, and mining of minerals and elements. Metal oxides/sulfides are examples of naturally occurring nanoparticles in the ecosystem. However, the synthesis mechanisms of these naturally occurring nanoparticles are very complex and are yet to be decoded. These natural sources of nanoparticles are discussed below.

2. Environmental nanoparticles and colloids

Colloid is specific state of matter between suspensions and solutions. Natural colloids may arise due to chemical weathering and other biological decays and contain oxyhydroxides/oxides of Mn, Al, Fe elements, etc., and aluminosilicates (Hartland et al., 2013). The size and distribution of the colloidal particles, their concentration, and composition all depend on the geochemistry, rainfall in a particular area, industrial waste, environmental fates including fates in water, soil, waste water, and other related factors. The important physico-chemical properties of natural colloids are size, shapes and morphologies, surface charge, porosity, surface coating, interaction forces, and binding behavior of contaminants with the natural colloids. The study of these colloidal dynamic systems is very complex and multifaceted. The size of the particles in colloids is usually in the range from a few nanometers to 1 μm; while suspensions are heterogeneous, mixtures containing particles of a size larger than 1 μm are vulnerable to sedimentation. The solvent and the solute could be solids, liquids, and/or gases. The importance of natural colloids in transportation, biogeochemical cycle, fate, and bioavailability of trace pollutants (e.g., trace metals) has been long acknowledged, and thus has fueled the interest in natural colloids and their roles in environmental systems. Colloids coexist with NPs in various compositions. The important consequences, owing to the small particle size of the colloids, are their enormous surface area and energies, which enable these colloids to interact strongly with contaminant elements or species, and their properties are totally different from those of materials having larger particle size and smaller atoms/molecules than those of colloidal particles (Wigginton et al., 2007; Wilkinson et al., 1999, 2007). Natural colloids can be classified into inorganic and organic colloids (Buffle et al., 1998; Wilkinson and Lead, 2017; Baalousha et al., 2009):

i) Inorganic natural colloids

The colloidal particles that can be found in aquatic and oxygenated terrestrial environments are aluminum phyllosilicates (e.g., clay, mica, chlorite) and may be present in the form oxides/hydrous oxides of Fe (magnetite, hematite, etc.), silicon (silicates), and manganese (pyrolusite). Minor colloidal components of FeS, FeS_2, and MnS are also found in anoxic water, while $CaCO_3$ in particulate form is abundant in fresh water lakes (Sigg, 1994).

ii) Humic substances and biopolymer-based natural colloids

Humic substances and biopolymers based colloids are both organic-type natural colloids. The humic substance includes the humic and fulvic acids while nonhumic substances are nucleic acids, sugars, extracellular polymeric substances (polysaccharides, proteins), amino acids, and other small molecules (Mannino and Harvey, 2000).

2.1 Humic-assisted synthesis of nanoparticles

It is very hard to define humus exactly. However, humus or humic substances are hyperbranched self-assemblies of polyelectrolytes having amorphous molecular structures with key biospheric roles (Xu et al., 2013). The biospheric roles include soil structuring, mediating the transportation of nutrients to plants, the chelating ability of carboxyl and hydroxyl groups with metals, regulation of biogeochemical cycles of elements, and mitigation of the antagonistic effects of contaminants. Humus has no definite structure, shape, or quality. It is amorphous and dark in nature with a jelly-like appearance and spongy character. Microscopic studies show that it contains microorganisms and microbial remains, which are mechanically and chemically degraded (Bernier and Ponge, 1994). As an environmental colloid, humic substances agglomerate due to charge and static stabilization, forming bigger molecular structures in the range of nanometers and beyond. High-tech equipment exploits the HS interactions with metals and their oxides on a micro- or nanoscale. In the field of biomedical sciences, HS acts as a stabilizing agent to prevent the agglomeration process of magnetic NPs of iron oxides in aquatic solutions (Illes and Tombacz, 2006). Other advantages of HSs, in comparison to synthetic modifiers, are their detoxification ability, biocompatibility, and capacity to be used as organic refractory materials due to their stability against biodegradation process in oxic and anoxic conditions of soils and subsurface. HS alleviates the threats of hazard to the environment ensuing from release of NPs. Humic substance-assisted synthesis of engineered Au, Ag, ZnS, and other nanoparticles has also been reported (Dubas and Pimpan, 2008). It has been reported that the AgNPs synthesized from Ag^+ sources under ecological conditions and observed in water sources do not have anthropogenic origins alone, as nature has also contributed to them (Mukherjee et al., 2001). The AgNPs formation under ecological condition is given in Fig. 3. Chlorosis, or iron deficiency, is a well-known plant disease caused by the deficiency of bioavailability of iron in soils and its correction is based upon the use of iron (III) chelates. The bioavailability of iron in the

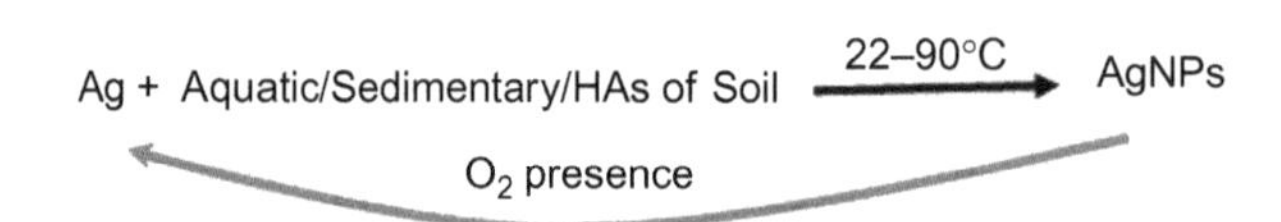

Fig. 3 Humic acid-induced natural synthesis of AgNPs under ecological conditions.

form of humic-based iron composite (iron humates) can be used for controlling iron deficiency and bacterial infections in modern micro-fertilizers.

3. Volcanic eruption and ash

Volcanic eruptions disperse millions of tons of micro and nanoparticles of volcanic ash into the atmosphere and reach the upper troposphere and stratosphere, thus affecting the health of organisms in the ecosystem. The phenomena of eruption of volcanos and erosion are also considered natural sources of nanoparticles production. Volcanic eruption and the ash produced are shown in Fig. 4. The temperature of the ash is above 1400°C during the eruption process. The airborne ash in the atmosphere is carried away by the wind and deposited in the surrounding environment including rocks, soil, plants, and water supplies in the form of a wide range of nanoparticles (100–200 nm) (Griffin et al., 2018). These ashes and lava have negative effects on human health, plants, and ecological systems. These airborne nanoparticles lead to serious respiratory illnesses due to their deposition in the upper tracheobronchial and alveolar sections of the respiratory tract. It has been reported that concentrations of the elements Zn, Cd, Tl, Ni, Se, Cd, Ag, Bi, Te, Sn, Pb, and Hg nanoparticles (with diameter less than 100 nm) were found in volcanic ash. The concentration of these elemental nanoparticles was found to be 10–500 times higher when compared to their total contents in bulk samples. This was found to be valid for the ash of volcanoes in the far east of Russia, Kamchatka, Tolbachik, Klyuchevskoi, Shiveluch-Kamchatka, the Puyehue-Cordón Caulle volcanic complex, and in Chile (Dubas and Pimpan, 2008). It was demonstrated that these NNPs of volcanic ash are important carriers of toxic metals and metalloids on a global scale. Similarly, the composition of the major elements in the ash samples of the Pichincha volcanic complexes (PVC's) eruption in 1999 consisted of Na, Mg, Al, Si, P, S, K, Ca, Ti, Mn, Fe, etc. (Akaighe et al., 2011).

Fig. 4 Volcanic eruption and cloud of volcanic ash as sources of NNPs.

4. Desert sources of nanoparticles

Desert sand is mainly composed of silicon with traces of Al, Ca, Fe elements, etc. Sometimes carbon and nitrogen particles along with ammonium, nitrates, sulfates, and other ions are also found in desert sandstorms. The storms in the deserts and high/low temperature during days/nights play a significant role in grinding sand dust to the nanometric scale. Desert sandstorms are responsible for airborne nanoparticles in the atmosphere. The troposphere's aerosols contain almost 50% of minerals with their origin from the deserts.

5. Biological nanoparticles

Natural and biological process involved in reducing macro-particles to nanoparticles are beneficial. For instance, the biologically derived/synthesized NNPs from microorganism fungi, bacteria, and plants are cost-effective and eco-friendly when compared with nanoparticles synthesized via chemical routes. These biologically evolved nanoparticles have a pivotal role in the amputation process of heavy organic/inorganic pollutants and filtration of other environmental toxins and impurities. The micro-organisms lactobacillus and shewanella bacterial species are involved in the fermentation of proteins in dairy milk and reduction of selenite to elemental selenium nanoparticles [15]. Similarly, the innocuous bacterial species of Staphylococcus carnosus and Saccharomyces cerevisiae are involved in the fabrication of selenium nanoparticles. These nanoparticles are used as antimicrobial agents as well as in food supplements in the agriculture sector [15]. This selenium enrichment of the soil invigorates food production and simultaneously enhances the defense system of plants against hazardous pathogens. When bacteria like Thiobacillus, Stenotrophomonas, Pseudomonas aeruginosa, and Serratia species are exposed to inorganic salts (containing S^{2-}, Au^{3+}, SeO_3^{2-}, Ag^+ ions), these bacteria employ either oxidation or reduction mechanisms of detoxification, which consequently lead to the synthesis of elemental NNPs (Griffin et al., 2018; Ermolin et al., 2018; Aguilera et al., 2018; Janssen et al., 1996; Mishra et al., 2017; Mishra et al., 2014).

6. Synthetically produced nanoparticles

The high surface to volume ratio of nanoparticles, compared to their bulk counterparts, leads to the enhancement of electrical, optical, magnetic, catalytic, mechanical properties, and other functionalities of the nanomaterials. Materials scientists and engineers are developing new tools and methods to improve further the functionalities and cost-effectiveness of the nanomaterials via controlling the sizes and distributions of nanoparticles. A number of production processes have been developed to meet the desired shapes, compositions, and size distributions. Numerous chemical and physical routes are

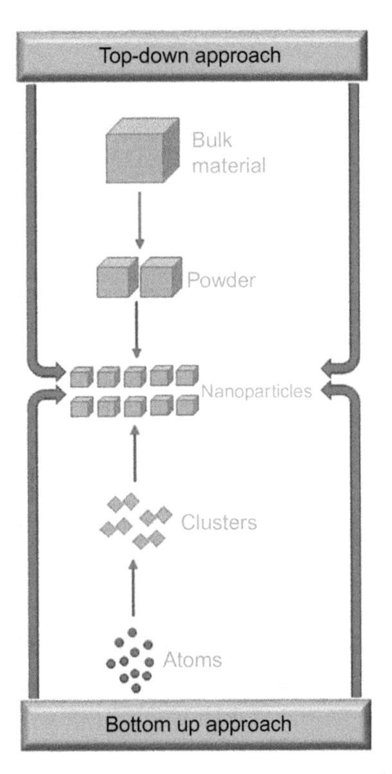

Fig. 5 The top-down and bottom-up approaches for the synthesis of nanoparticles.

available for the synthesis of nanoparticles and nanocomposites. Particle size, chemical composition, crystallinity, and shape can be controlled by temperature, pH-value, concentration, chemical composition, surface modifications, and process control.

In general, there are two main approaches for the synthesis of nanoparticles: top-down (a mechanico-physical process) and bottom-up (a chemo-physical process), as shown in Fig. 5. In the top-down approach, the bulk material is ground to reduce the larger particles to nanometric-scale particles using physical processes such as grinding, mechanical crushing, and ball-milling. The top-down approach is a typical solid-state process. The particles thus obtained are not of uniform sizes and the material obtained has surface imperfections (crystallographic damage). These surface imperfections consequently deteriorate the surface chemistry and other physical properties of the materials. In the bottom-up approach, the nanomaterials are synthesized or built-up in a chemical process via the addition of atoms, molecules, and clusters. These nanoparticles or nanomaterials are of uniform shape, size, and distribution. Nanoparticle growth can be controlled via different reaction parameters such as pH, temperature, concentration, catalyst, etc.

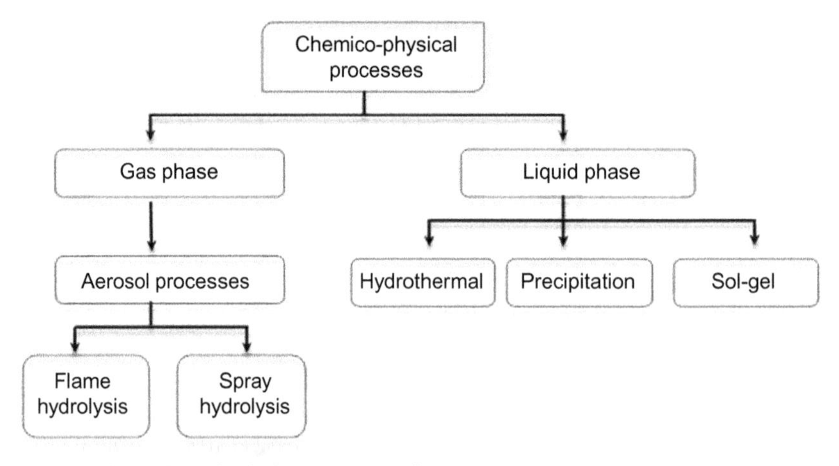

Fig. 6 Chemico-physical methods for synthesis of nanoparticles.

Different techniques are now available for the synthesis of nanoparticles and nanocomposite materials. These techniques have their own pros and cons, and include gas phase (flame and spray hydrolysis), liquid phase (hydrothermal, precipitation, sol-gel), spark discharge, laser ablation, template synthesis, microwave, microemulsion, ultrasound, sputtering, biological synthesis, PVD, CVD, powder process, etc. Some of the common techniques are discussed below. The chemico-physical processes of nanoparticles synthesis include the gas phase and liquid phase methods, as shown in Fig. 6.

7. Gas phase methods

Gas phase processes (aerosol processes) are commonly used at industrial scale for the production of thin films or nanomaterials in powder form. The gas phase process is carried out either physically or chemically. The chemical vapor deposition (CVD) technique is widely used at industrial scale for the coating of large areas. The nanoparticles are synthesized from the vapors of the material in the gas phase via homogeneous nucleation.

Nanoparticles are created from the gas phase by producing a vapor of the product material using chemical or physical means. The production of the initial nanoparticles, which can be in a liquid or solid state, takes place via homogeneous nucleation. The processes involved during the synthesis of nanoparticles are condensation (transition from vapors to liquid state), coagulation (two or more particles adhesion), and coalescence (particles fusion). Sometimes the particles interact chemically with each other. Different sorts of reactors such as laser/plasma/flame and hot wall are used for the synthesis of nanoparticles. A high-power laser is focused on the source material to ablate and decompose its surface layers for yielding the nanoparticles. In plasma reactors, high-energy plasma is used to vaporize and decompose the source material into its nanoparticles. In flame reactors, the source material's molecules are decomposed in a relatively high temperature flame to yield soot, pigment of TiO_2, and SiO_2 NPs. The hot wall reactor is used to produce nickel and iron nanopowders. The source material is converted into vapors at low pressure (~1 mbar) in the presence of inert gas. These vapors are rapidly cooled and are collected in the form of nanoparticles on filters. In the chemical gas phase method, the source material is evaporated under vacuum and condensed/adsorbed chemically on a hot surface (substrate). Similarly, sonic waves, compressed air, ultrasound, centrifugation, electrostatic, spraying pyrolysis, and other techniques can be implemented for the production of nanoparticles.

8. Liquid phase methods

The liquid phase method has many advantages over the gas phase and solid phase synthesis techniques. The liquid phase or wet chemical synthesis technique usually requires a low temperature to control the shape and size of the nanoparticles in a very short time when compared with the gas phase synthesis method. It is the most common synthesis technique used for the synthesis of NPs and is more cost effective than the solid phase synthesis technique. It gives high yield and surface functionalization can be easily done to control physical properties of the nanoparticles. No chromatography or HPLC are required for purification and size separation purposes. Postsynthesis methods are available for the formation of clusters and 2D-superlattices from the as-synthesized NPs and nanostructures (Malhotra et al., 2013; Singh and Kundu, 2014). The liquid phase synthesis techniques are hydrothermal, precipitation, and sol-gel.

8.1 Hydrothermal synthesis method

This method is used to synthesize nanoparticles. Each synthesis method has its own pros and cons. The advantage of using the hydrothermal process is to produce crystalline phases of nanomaterials, which are unstable at elevated temperature, to produce a nanomaterial with high vapor pressure. The hydrothermal synthesis method involves high vapor pressure and temperature of the solutions used inside an autoclave. The autoclave is a specially designed cylindrical steel vessel coated with protective coatings against the corrosion process (resistant to solvents) during chemical reactions and can withstand high temperature and pressure inside the vessel. The protective coating could be Teflon-lined, carbon-free iron, copper, titanium, gold, silver, glass (or quartz), platinum, etc., depending the type of solution, pressure, and temperature involved within the autoclaves. A hermetic (airtight) seal used in the autoclave to tolerate safely the high temperature and pressure for a long time to complete the reactions involved within the autoclave. The different parts of the hydrothermal autoclave are shown in Fig. 7. The outer chamber, the autoclave threaded cap, and lower and upper discs are all made of stainless steel (SS). The hydrothermal autoclave reactors available in the market are in the range

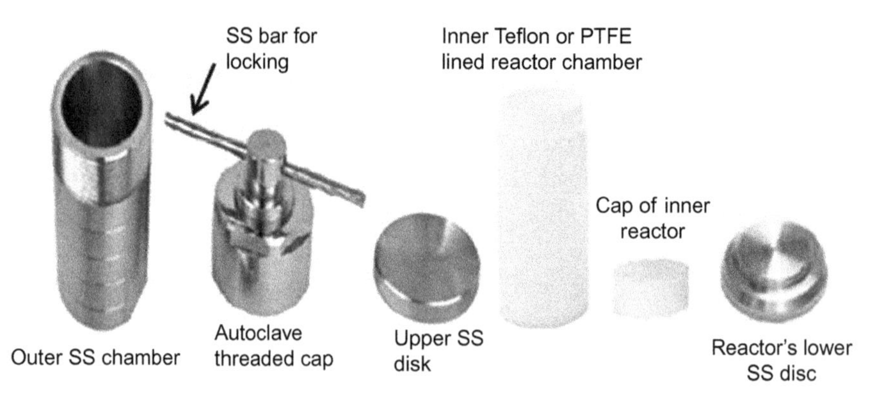

Fig. 7 Different parts of the hydrothermal autoclave reactor.

10–2000 mL capacity. The hydrothermal synthesis process includes the temperature- difference technique, the temperature-reduction technique, and the metastable-phase technique.

8.2 Precipitation or coprecipitation method

The precipitation or coprecipitation method is widely used for the synthesis of metal oxides and metals (nonoxide) nanoparticles. It is an important method used to separate out the product of the reaction or precipitates of the product material from a solution. These precipitates are solid particles settled in a solution due to density and the solution above the precipitates is known as supernatant. In the precipitation method, the reaction kinetics play an important role in controlling particle shapes, sizes and size distribution, crystallinity, and surface morphologies. These properties can also be managed with the help of pH, reactants concentration, and temperature of the solutions, precipitates solubility, rate of reactants mixing, and sequence of mixing mechanism of materials or dopants. In this method, the salts are reacted in the solution used and a precipitating agents are usually added to the solution to yield the precipitates of the desired nanoparticles. These precipitates are later washed or filtered out. The liquid vapors or moisture contents are frequently removed by heat treatment. The various steps involved in the coprecipitation technique are shown in Fig. 8. The precipitating agent's slow addition to the solution of analyte results in supersaturation and thus leads to fast reactions. Precipitates formation involves a two-step process of nucleation and particles growth. Initially the few ions, atoms, and/or molecules interact with each other to form small solids, generally known as seeds or nuclei, in the solution. The formation of these nuclei occurs on the surface of the solid contaminants in the solution. The supersaturated solution leads to increase in the rate of nucleation and, consequently, a precipitate results.

There are three major types of mechanisms involved in the coprecipitation process: (1) surface adsorption, (2) inclusion or mixed-crystal formation, and (3) occlusion. The

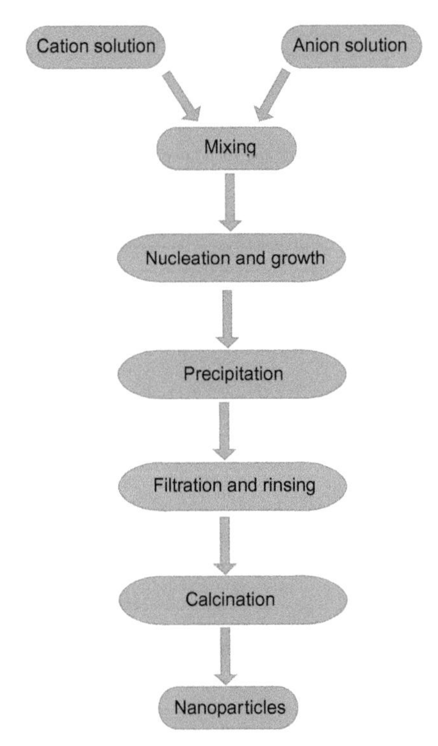

Fig. 8 Different steps involved in coprecipitation technique for the synthesis of nanomaterials.

surface adsorption occurs for precipitates having larger surface area. The adsorbates are weakly bound on the precipitate surface. This method is used to contaminate specially coagulated colloids. In inclusion mechanism, one of the ions (host) in the lattice is replaced by another ion (impurity) and this results in crystallographic defects. This may happen when both the ionic radii and charge of the host and impurity ions are matching. Adsorption and inclusion mechanisms are equilibrium processes, while the occlusion process is a kinetic phenomenon in which the adsorbed impurities are physically trapped inside the lattice during crystal growth.

8.3 Sol-gel method

Sol-gel is a wet-chemical technique used for the synthesis of solid nanomaterials from small molecules, and is also used for glassy and oxide materials (Stefess et al., 1996). Uniform and small-sized powders can also be produced using this technique. The small molecules or monomers in the precursor solution are converted into colloidal solution or gel by continuous stirring. The sol thus produced is converted into an integrated network or gel composed of discrete particles or network polymers. This gel is diphasic,

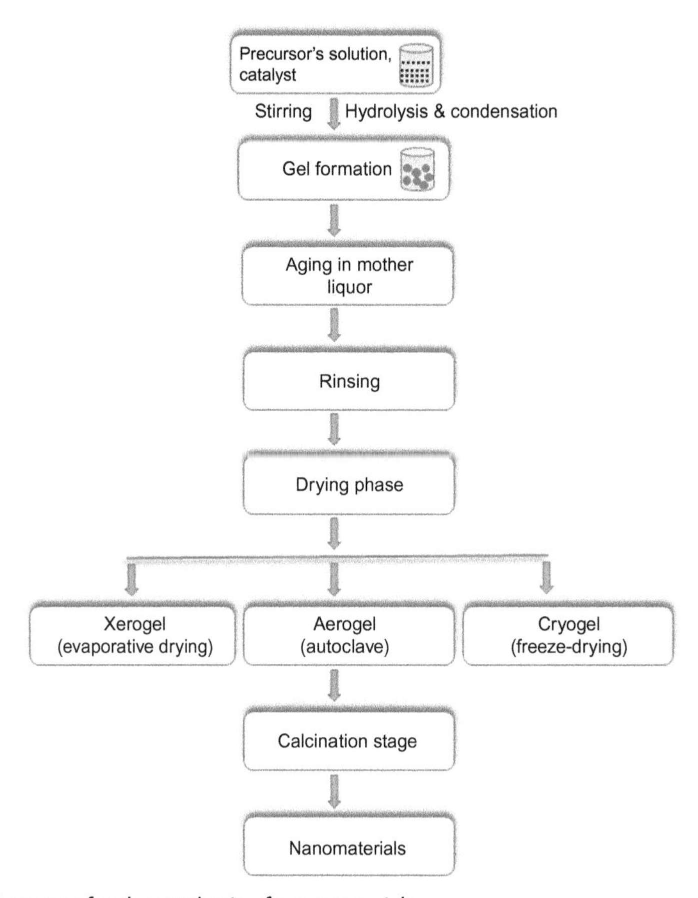

Fig. 9 Sol-gel process for the synthesis of nanomaterials.

containing both solid and liquid phases, with morphologies in the range of discrete particles to continuous chain-like polymeric networks (Lu and Yin, 2012; Eah, 2011). The conventional sol-gel technique involves four stages: hydrolysis, polymerization of monomers or condensation, growth of particles, and gel formation. The complete procedure involved in the sol-gel technique is shown in Fig. 9. Typical precursors used in the sol-gel process are metal alkoxides and metal chlorides, which undergo hydrolysis and polycondensation/polymerization reactions to form a colloid (Fernandez-Garcia et al., 2004). Precursor alkoxysilanes such as tetraethyl orthosilicate (TEOS) and tetramethyl orthosilicate (TMOS) are also in use. However, efforts have been made to find precursors that are less toxic and more environmentally friendly (Klein and Garvey, 1980). In colloid, the particles density is low and significant amounts of liquid need to be removed via sedimentation and centrifugation mechanisms (phase separation).

The time required between gel formation and its drying is called aging. The wet gel is aged either in its mother liquor or in another solvent. During the aging process, the gel is not static and experiences hydrolysis and condensation (Brinker et al., 1986). The remaining liquid phase removal from the gel requires a drying process, which is accompanied by densification and structural shrinkage. The rate of solvent removal is determined by the distribution of porosity in the gel. The three drying processes are xerogel, aerogel, and cryogel (freeze-drying). Thermal treatment or a firing process is frequently used for further polycondensation. Sintering, densification (achieved at comparatively low temperature), and grain growth lead to structural stability and enhancement of mechanical properties. The growth and morphologies of particles/materials can be affected during the experiment via pH, temperature of the solution, concentration of the reactants, types of precursors and solvent, water contents, acid or base contents, additive/catalyst in the solution, and aging. All these sol-gel experimental parameters affect the gel's structure, and consequently the properties the product material. The advantages of using the sol-gel technique are its cost-effectiveness, molecular-scale mixing, good control over doping process by allowing impregnation or coprecipitation, low temperature requirements, control of texture, chemical, and morphological properties of the solid, and synthesis of metal oxides with a high degree of purity and homogeneity (Pierre, 1998; Muresan, 2015; Chen and Mao, 2007; Kolen'ko et al., 2005).

References

Aguilera, C., Viteri, M., Seqqat, R., Navarrette, L.A., Toulkeridis, T., Ruano, A., Arias, M.T., 2018. Biological impact of exposure to extremely fine-grained volcanic ash. J. Nanotechnol. 7543859, 12 pp, https://doi.org/10.1155/2018/7543859.

Akaighe, N., MacCuspie, R.I., Navarro, D.A., Aga, D.S., Banerjee, S., Sohn, M., Sharma, V.K., 2011. Humic acid-induced silver nanoparticle formation under environmentally relevant conditions. Environ. Sci. Technol. 45 (9), 3895–3901.

Baalousha, M., Lead, J.R., Von der Kammer, F., Hofmann, T., 2009. Natural colloids and nanoparticles in aquatic and terrestrial environs. In: Lead, J.R., Smith, E. (Eds.), Environmental and Human Health Effects of Nanoparticles. John Wiley & Sons, Ltd. https://doi.org/10.1002/9781444307504.ch4 (Chapter 4).

Bernier, N., Ponge, J.-F., 1994. Humus form dynamics during the sylvogenetic cycle in a mountain spruce forest. Soil Biol. Biochem. 26 (2), 183–220.

Brinker, C.J., Tallant, D.R., Roth, E.P., Ashley, C.S., 1986. Sol-gel transition in simple silicates. III. Structural studies during densification. J. Non-Crystal. Solids 82 (1–3), 117–126. https://doi.org/10.1016/0022-3093(86)90119-5.

Buffle, J., Wilkinson, K.J., Stoll, S., Filella, M., Zhang, J., 1998. A generalized description of aquatic colloidal interactions: the three-colloidal component approach. Environ. Sci. Technol. 32 (19), 2887–2899.

Chen, X., Mao, S.S., 2007. Titanium dioxide nanomaterials: synthesis properties, modifications, and applications. Chem. Rev. 107, 2891.

Dubas, S.T., Pimpan, V., 2008. Humic acid assisted synthesis of silver nanoparticles and its application to herbicide detection. Mater. Lett. 62, 2661–2663.

Eah, S.-K., 2011. A very large two-dimensional superlattice domain of monodisperse gold nanoparticles by self-assembly. J. Mater. Chem. 21, 16866.

Ermolin, M.S., Fedotov, P.S., Malik, N.A., Karandashev, V.K., 2018. Nanoparticles of volcanic ash as a carrier for toxic elements on the global scale. Chemosphere 200, 16–22. https://doi.org/10.1016/j.chemosphere.2018.02.089.

Fernandez-Garcia, M., Martinez-Arias, A., Hanson, J.C., Rodriguez, J.A., 2004. Nanostructured oxides in chemistry: characterization and properties. Chem. Rev. 104 (9), 4063 4104.

Griffin, S., Masood, M.I., Nasim, M.J., Sarfraz, M., Ebokaiwe, A.P., Schafer, K., Keck, C.M., Jacob, C., 2018. Natural nanoparticles: a particular matter inspired by nature. Antioxidants 7 (1), 29. https://doi.org/10.3390/antiox7010003.

Hartland, A., Lead, J.R., Slaveykova, V.I., O'Carroll, D., Valsami-Jones, E., 2013. The environmental significance of natural nanoparticles. Nat. Educ. Knowl. 4 (8), 7.

Illes, E., Tombacz, E., 2006. The effect of humic acid adsorption on pH-dependent surface charging and aggregation of magnetite nanoparticles. J. Colloid Interface Sci. 295, 115–123.

Janssen, A., de Keizer, A., van Aelst, A., Fokkink, R., Yangling, H., Lettinga, G., 1996. Surface characteristics and aggregation of microbiologically produced sulphur particles in relation to the process conditions. Colloid Surf. B 6, 115–129.

Klein, L.C., Garvey, G.J., 1980. Kinetics of the sol-gel transition. J. Non-Cryst. Solids 38, 45.

Kolen'ko, Y.V., Kovnir, K.A., Gavrilov, A.I., Garshev, A.V., Meskin, P.E., Churagulov, B.R., Bouchard, M., Colbeau Justin, C., Lebedev, O.I., Van Tendeloo, G., Yoshimura, M., 2005. Structural, textural, electronic properties of a nanosized mesoporous $Zn_xTi_{1-x}O_{2-x}$ solid solution prepared by a supercritical drying route. J. Phys. Chem. B 109 (43), 20303–20309.

Lu, Z., Yin, Y., 2012. Colloidal nanoparticle clusters: functional materials by design. Chem. Soc. Rev. 41, 6874–6887.

Malhotra, A., Dolma, K., Kaur, N., Rathore, Y.S., Ashish, Mayilraj, S., Choudhury, A.R., 2013. Biosynthesis of gold and silver nanoparticles using a novel marine strain of Stenotrophomonas. Bioresour. Technol. 142, 727–731.

Mannino, A., Harvey, H.R., 2000. Biochemical composition of particles and dissolved organic matter along an estuarine gradient: sources and implications for DOM reactivity. Limnol. Oceanogr. 45, 775–788.

Mishra, S., Singh, B.R., Naqvi, A.H., Singh, H.B., 2017. Potential of biosynthesized silver nanoparticles using *Stenotrophomonas* sp. BHU-S7 (MTCC 5978) for management of soil-borne and foliar phytopathogens. Sci. Rep. 27 (7), 45154. https://doi.org/10.1038/srep45154.

Mishra, S., Singh, B.R., Singh, A., Keswani, C., Naqvi, A.H., Singh, H.B., 2014. Biofabricated silver nanoparticles act as a strong fungicide against *Bipolaris sorokiniana* causing spot blotch disease in wheat. PLoS One. 9(5), e97881. https://doi.org/10.1371/journal.pone.0097881.

Mukherjee, P., Ahmad, A., Mandal, D., Senapati, S., Sainkar, S.R., Khan, M.I., Parishcha, R., Ajaykumar, P.V., Alam, M., Kumar, R., Sastry, M., 2001. Fungus-mediated synthesis of silver nanoparticles and their immobilization in the mycelial matrix: a novel biological approach to nanoparticle synthesis. Nano Lett. 1 (10), 515–519. https://doi.org/10.1021/nl0155274.

Muresan, L.M., 2015. Corrosion protective coatings for Ti and Ti alloys used for biomedical implants. Intell. Coat. Corros. Control 585–602. https://doi.org/10.1016/b978-0-12-411467-8.00017-9.

Pierre, A.C., 1998. Introduction to Sol-Gel Processing. Kluwer Academic Publishers, Boston, p. 394.

Sigg, L., 1994. Regulation of trace elements in lakes. In: Buffle, J., de Vitre, R. (Eds.), Chemical and Biological Regulation of Aquatic Processes. Lewis Pub., Boca Raton, FL, pp. 177–197

Singh, P.K., Kundu, S., 2014. Biosynthesis of gold nanoparticles using bacteria. Proc. Natl. Acad. Sci. India Sect. B 84, 331–336.

Stefess, G.C., Torremans, R.A.M., DeSchrijver, R., Robertson, L.A., Kuenen, J.G., 1996. Quantitative measurement of sulphur formation by steady state and transient state continuous cultures of autotrophic thiobacillus species. Appl. Microbiol. Biotechnol. 45, 169–175.

Wigginton, N.S., Haus, K.L., Hochella, M.F., 2007. Aquatic environmental nanoparticles. J. Environ. Monit. 9, 1306–1316.

Wilkinson, K.J., Lead, J.R. (Eds.), 2017. Environmental Colloids and Particles: Behaviour, Separation and Characterisation. In: IUPAS Series of Analytical and Physical Chemistry of Environmental Systems10, John Wiley and Sons, New York n.d.

Wilkinson, K., Lead, J.R., Wilkinson, K., Lead, J.R., 2007. Environmental Colloids: Behavior, Separation and Characterization. John Wiley and Sons, New York.
Wilkinson, K.J., Balnois, E., Leppard, G.G., Buffle, J., 1999. Characteristic features of the major components of freshwater colloidal organic matter revealed by transmission electron and atomic force microscopy. Colloids Surf. A: Physicochem. Eng. Aspects 155 (2–3), 287–310.
Xu, J., Wu, J., He, Y. (Eds.), 2013. Functions of Natural Organic Matter in Changing Environment. Jointly Published by Zhejiang University Press and Springer. https://doi.org/10.1007/978-94-007-5634-2_133.

CHAPTER 3

Types and classification of nanomaterials

M. Rizwan[a], Aleena Shoukat[a], Asma Ayub[a], Bakhtawar Razzaq[a], and Muhammad Bilal Tahir[b]
[a]Department of Physics, University of Gujrat, Gujrat, Pakistan
[b]Department of Physics, Khwaja Fareed University of Engineering and Information Technology, RYK, Pakistan

1. Introduction

Materials that possess structures and size in the range of nanometers (from 1 to 100 nm) and in one or more further dimensions are known as nanomaterials, whereas a nanoparticle (Xie et al., 2007) is any particle that has at least one external dimension in the nano range. These materials play a vital role in technological revolutions, due to their exclusive mechanical, thermal, electrical, and biological characteristics, which are absent in ordinary materials. This transition in behavior mainly occurs due to two main factors, which are quantum effect and surface effect, which occur due to the decrement in size. The surface effect increases the material's chemical reactivity and lowers its melting point, while the quantum effect confines electrons in a very small NP (Buzea and Pacheco, 2017). Nanoparticles inimitable characteristics and abilities make them highly viable in various technological applications of several fields of chemistry, physics, biology, and engineering. Most of the characteristics of these materials depend upon the size, geometry, and surface morphologies like area, smoothness, energy, and electrons allocations, in addition to their low weight and great mechanical strength (Scida et al., 2011).

The arrangement and classification of NMs in accordance with their chemical and physical characteristics is essential to understand the diversity of nanomaterials. Nanomaterials are fabricated by different approaches like ball milling, cutting, chipping, and extruding, and create different kinds of structures with variant kinds of coatings on the surface, thus leading to their classifications. Materials that differ in dimensions, morphology, sizes, compositions, and uniformity are categorized discretely. There are various types of nanomaterials on the basis of their composition. The main types include organic-based NMs, inorganic-based NMs, and composite-based or hybrid NMs (Makhlouf and Barhoum, 2018). Organic NMs mainly consist of carbon; thus a special class of nanomaterials is carbon-based nanomaterials, which include fullerene, carbon nanotubes, and graphene, which are briefly discussed in the following sections. Inorganic NMs include metals and metal oxide-based nanomaterials like Al, Ag, and Cu, iron oxide, titanium oxide, and aluminum oxide. Metalloid NMs like CdSe, ZnO, and ZnS, which are

Nanomaterials: Synthesis, Characterization, Hazards and Safety
https://doi.org/10.1016/B978-0-12-823823-3.00001-X

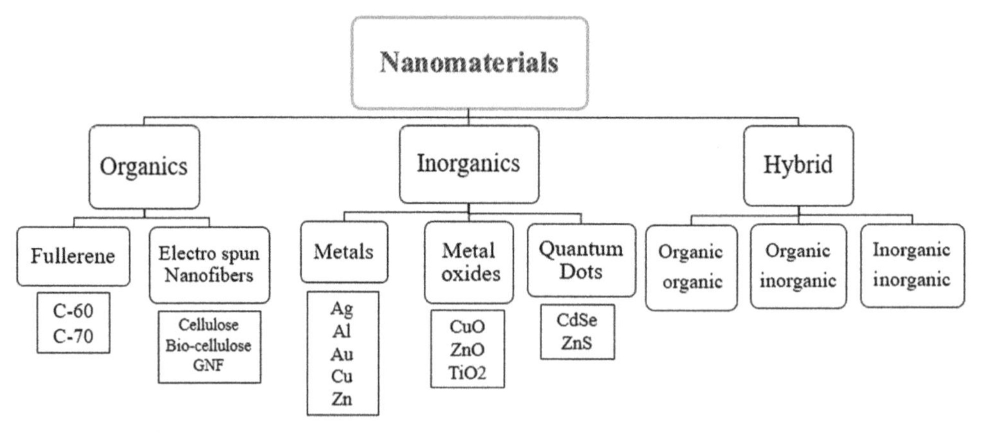

Fig. 1 Types of nanomaterials on the basis of composition (Zayed et al., 2019).

known as quantum dots, are also included in the class of inorganic NMs. Hybrid NMs or composite NMs, which are made by the combination of organic and inorganic nanomaterials, are considered to be the third class of nanomaterials. Fig. 1 shows the types of nanomaterials (Al-Kayiem et al., 2013).

1.1 Organic-based nanomaterials

As the name suggests, this class of NMs contains materials that are mainly composed of organic compounds, such as carbohydrates, lipids, or polymers that are in the range of 10 nm to 1 μm. Polymeric nanoparticles have greater structural stability, integrity, and control, thus they have attracted great attention from researchers (Hadinoto et al., 2013). These polymer NPs, along with several liposomes, dendrimers, and micelles, are widely implemented in drug delivery systems (Kumar and Lal, 2014). These organic-based nanoparticles are shown in Fig. 2. Polymer NPs form branched units

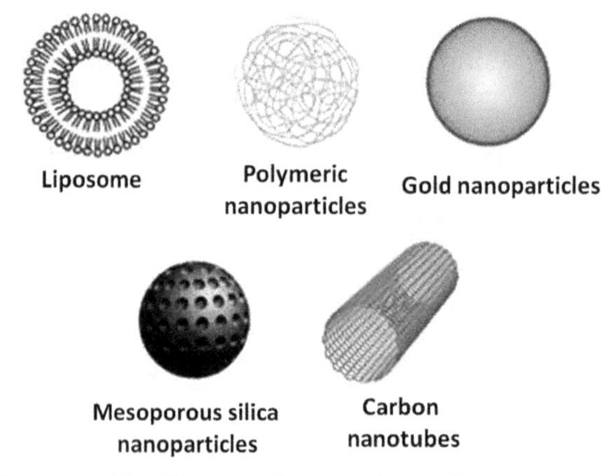

Fig. 2 Organic-based nanoparticles (Chattopadhyay et al., 2017).

and compose a nanomaterial. The dendrimer surface has chains in the end, which are tailored and used as catalysts to perform various chemical functions. The three-dimensional (3D) version of dendrimer accumulates other molecules inside its cavity, and is thus implemented in drug delivery systems. Organic-based NMs can be synthesized by various approaches and techniques such as laser ablation, ion association, and mechanical grinding (Makhlouf and Barhoum, 2018).

As discussed earlier, materials that include carbon are also categorized as organic nanomaterials. There is a wide range of nanomaterials that include carbon, thus forming a special class of nanomaterials known as "carbon-based nanomaterials" (CBNs). These CBNs are discussed below in detail.

1.2 Carbon-based nanomaterials

In these nanomaterials, carbon is the main element and they exist in the form of ellipsoids, hollow spheres, and tubes. Graphite and diamond are allotropes of the most abundant element, carbon. The element carbon has a certain uniqueness and diversity in its structure, with structures like 3D diamond and graphite, 2D graphite sheets, nanotubes in 1D, and fullerene in zero-dimension (Makhlouf and Barhoum, 2018). The structural configuration and hybridization states of carbon strongly influence the electronic, physical, and chemical behavior of the nanomaterial. Carbon has six electrons and its ground state configuration is $1s^2 2s^2 2p^2$. The low gap among 2s and 2p allows electron transition, thus supporting hybridization into sp, sp^2, or sp^3. The covalent bonding with neighbor atoms at higher levels provides energy to compensate for this configuration, which is the same for sp^2 and sp^3 hybridization. The unhybrid p-orbitals consider π-bonding among themselves (Hu et al., 2007; Al-Kayiem et al., 2013).

Thus diversity among the various organic compounds mainly occurs owing to the variable hybridization states. Substantial differences can be observed among configurations of carbon's bulk, which is presented in Fig. 3. The promising trigonometric sp^3 configuration exists for diamond at high pressures and temperatures. With the decrease

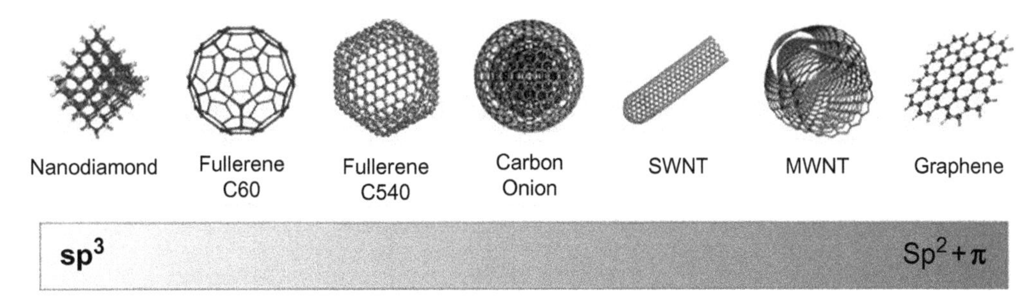

Fig. 3 Carbon-based nanomaterials and their hybridization states (Mauter and Elimelech, 2008).

in the heat formation, the planar sp^2 configuration adds up, thus forming a single-layered sheet structure with single π-bond and three sigma covalent bonding. Slight shear forces and chemical-physical separation induce weak interplanar forces among graphene sheets, and induces slip-up among them (Mauter and Elimelech, 2008).

1.2.1 Graphene

One of the carbon-based nanomaterials is graphene—including graphite sheets—which is considered to be a material with superior characteristics. Carbon atoms with sp^2 hybridization in an excellent crystal structure form a quasitwo-dimensional graphene nanomaterial. Graphite sheets are more stable compared to 3D structures at nanoscale. Numerous relativistic phenomena of quantum electrodynamics are made possible attributable to strange electronic behavior of graphene. Cooperative electrons' behavior in graphene gives rise to its exclusive properties and honeycomb atomic structure with two dimensionalities (Meyer et al., 2008). Electrons travel in graphene at actual speed of light as compared to in vacuum and allow the observation of relativistic effects in quantum electrodynamics, making it viable for various fields like optics, electronics, spintronics, composite and thermal materials, and thus hydrogen storage is based on graphene because of its extraordinary properties (Makhlouf and Barhoum, 2018). The nanostructures of graphene and graphene sheets are displayed in Fig. 4.

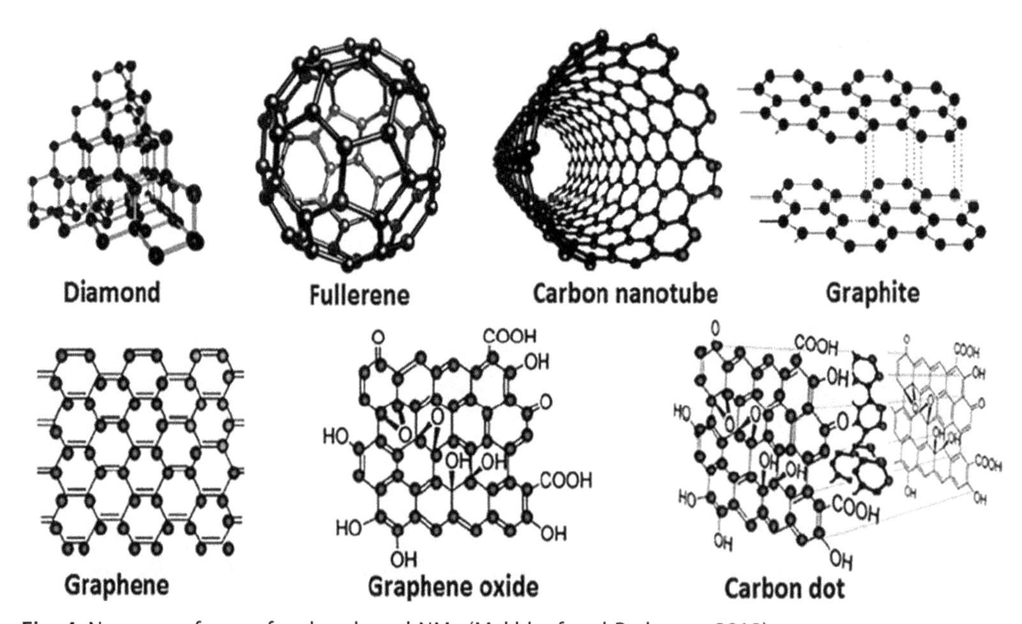

Fig. 4 Numerous forms of carbon-based NMs (Makhlouf and Barhoum, 2018).

1.2.2 Fullerenes

Ellipsoid and spherical carbon nanomaterials are usually known as fullerene, which retains exclusive characteristics. The sp^2 bonding exists in it to form a hexagonal ground structure with exceptionally symmetry, making it a prominent feature of fullerene. Rotation about the axis and reflection form a plane along with 120 symmetry operations exist in fullerene's molecular structure (Müller et al., 2004). Thus the C60 molecule of fullerene, as shown in Fig. 4, is the utmost symmetrical molecule with prevalent symmetry operations. The fullerences are mostly considered as anisotropic semimetal in 2D. The buckminsterfullerene ("buckyballs") structure of fullerene consists of 20 hexagon rings with 12 pentagonal, which form a spherical shape with 60 vertices of 60 carbons (Sun et al., 2002). It is an exceptionally strong material, which can resist high pressure of about 3000 atm. and retain its original shape. Occurrence of delocalized π-electrons induces nonlinear optical response in it. In addition to the 60 carbon atoms, the fullerene molecule may consist of 30–980 carbon atoms to form variant structures with diverse properties that can be implemented in numerous applications. The soccer ball-like structure can be lost by the molecule by the addition or removal of the hexagons in it. With 70 carbon atoms, a rugby ball-like structure is formed by C70 with 25 hexagons. Maasive fullerences occur in a pentagonal shape, while asteroids shape occurs for smaller fullerene. The density of fullerene is 1.65 g/cc, which is lower than that of diamond at 3.51 g/cc (Makhlouf and Barhoum, 2018; Mauter and Elimelech, 2008).

1.2.3 Nanotubes

Nanomaterials in spherical or ellipsoid form are known as fullerene, while those which exist in cylindrical form are referred to as carbon nanotubes (CNTs). These tubes are the elongated version of the buckyball structure that exists in one dimension, compared to fullerene, which is zero-dimensional. Theoretically, a micro-sized sheet of graphene is trundled in cylindrical form in nano range, then topped with spherical fullerene. Carbon nanotubes have extraordinary structural aspect ratio due to pairing of nano-diameter with the length in cm. Carbon nanotubes' sidewalls are sp^2 hybridized with the graphene hexagon, while the sp^3 bonding character is stronger, which reduces strain on carbons at the sidewalls, thus making CNTs less susceptible to chemical alterations than fullerene molecules are. The nanostructure of a CNT is shown in Fig. 4 (Mauter and Elimelech, 2008).

Interaction of CNTs with organic molecules is through noncovalent means, such as electrostatic or van der Waals forces, hydrogen bonding, or hydrophobic interactions. Their higher strength, rigidity, conductivity, and elasticity make them promising materials to be employed in various applications. Laser ablation, the arc discharge method, and chemical vapor deposition (CVD) can be used to synthesize CNTs (Makhlouf and Barhoum, 2018). CNTs can exist in single-walled (SWCNT), double-walled (DWCNT), and multiwalled (MWCNT) structures. The SWCNTs are of smaller diameter and exhibit semiconducting, metallic, or semimetallic character, which depends upon hexagonal lattice orientation. DWNTs and MWNTs are "carbon onions" in a single dimension. MWNTs have similar characteristics to the bulk SWNTs, and the coupling along layers is weak in both NTs. But MWNTs display semiconducting behavior instead of metallic character as SWNTs do (Mauter and Elimelech, 2008).

1.2.4 Nanodiamond

A new kind of diamond, the size of which is in the range of 4–10 nm, is known as a nanodiamond, and has held the attention of scientists for several decades. Adamantane was the first diamond discovered from crude oil back in 1933, which has the form $C_{10}H_{16}$, then followed by tetra-$C_{22}H_{28}$ adamantine and so on (Landa and Macháček, 1933). These diamond NPs have a sphere-like shape, and are also known as detonation diamonds, which is a special class of carbon-based NP having unique magnetic and optical behavior with larger surface and unique lattice structure. More smooth carbons in tetrahedral amorphous form are obtained when a diamond is converted into nano range. These have proved to be chemically stable and can be employed as abrasives, protective coatings, and semiconductors. These diamond NPs possess a high absorption rate, and they are widely utilized in skincare products due to their deep layer penetration along with the active ingredients (Makhlouf and Barhoum, 2018).

1.3 Inorganic-based nanomaterials

Metals and metal oxide-based NMs, along with ceramic-based NMs, are part of the class of inorganic-based NMs, and exhibit unique electronic and optical features on the nanoscale. Clay is a natural inorganic NM that has evolved from the Earth's crust due to variant chemical circumstances. Similarly, cement, pigments, and fumed silica are obtained from volcanic eruptions. On the other hand, metal and metal oxide-based NMs are engineered nanoparticles, which have attracted significant attention. Inorganic quantum dots are usually semiconductors in the range of 2–10 nm and exhibit unique features like brightness and photo stability, which are widely implemented in diagnostics and therapeutic gears (Bailey et al., 2004; Makhlouf and Barhoum, 2018).

1.3.1 Metallic nanostructures

There are various nanostructures that are metallic or metal alloys and which include silver, copper, iron, gold, and many more. Silver nanoparticles are utilized in medicinal diagnosis processes, conductive and optical solicitations, in antimicrobial and in antiseptic activities. Copper nanomaterial is widely used as a conductor, catalyst, lubricant, and for sintering, whereas gold nanoparticles are widely implemented as a drug delivery system, in medication, detection of cancer, as alloys, and in fuel cells. Iron nanoparticles are

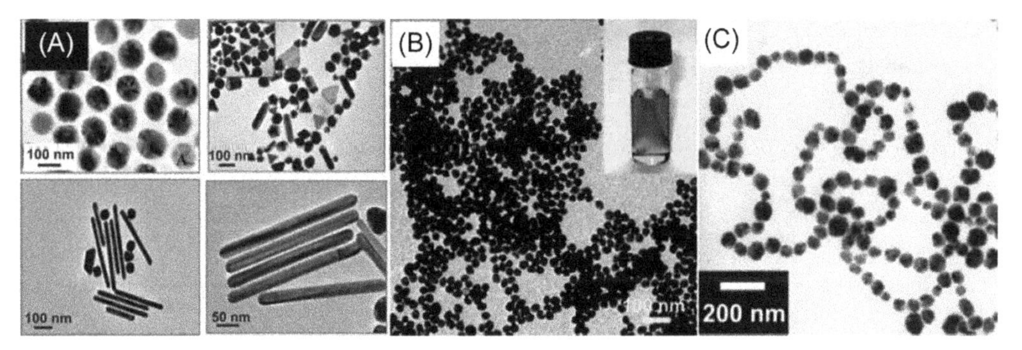

Fig. 5 TEM images of metallic nanostructures: (A) Au NPs (Kundu et al., 2008), (B) Ag NPs (Hu et al., 2008), and (C) Ni NPs.

employed as contaminant treatment due to reactivity, in super paramagnetic and magnetic detectors, and as a data recorder. Various alloys like Ti-Al and Al-Mg alloys, owing to their strength, are widely utilized in aerospace and in applications which require high temperature, while Fe-Si-Bo alloy nanomaterial is implemented in electronics, due to its magnetic character (Dolez, 2015; Bilecka and Niederberger, 2010). Fig. 5 shows various metal NMs, like gold, silver, and nickel NPs, obtained through TEM.

1.3.2 Metal oxides

Metal oxide-based inorganic NMs are another category of nanomaterials, which comprise of titanium dioxide, zinc oxide, and many more. Transparent UV filters, catalysts for cleaning products, and solar cells utilize TiO_2 nanoparticles, whereas zinc oxide also has numerous applications in UV, and as antimicrobial and antistatic agents. Iron oxides have several electrical and biomedical applications owing to their extraordinary magnetic behavior, and can be utilized for purification of water, as ion exchangers, and pigments. Cerium oxide, nanosilica, and nanoalumina are essential NPs for abrasion resistance coatings, as polishing agents, UV absorbers in lacquers, and many more. Metal chalcogenides, metal phosphates, metal pnictides, and zeolites are also inorganic-based nanomaterials that have found numerous applications in various fields of life (Dolez, 2015; Bilecka

Fig. 6 SEM images of various metal oxide NMs (Polshettiwar et al., 2009).

and Niederberger, 2010). Fig. 6 displays images of several metal oxide-based NMs like iron, cobalt, and manganese and cerium oxide, achieved through SEM.

1.4 Composite-based nanomaterials

The next class of nanomaterials is composite nanoparticles, which are made by the combination of two or more different materials, mixed together in order to merge their best properties. This fusion of features leads to overcome the detriments of individual materials and increase the benefits of the fused compound formed. This nanocomposite is a solid material, which may have many phases, but one dimension of any of the phases must be in the nanoscale. Such as a many nanoparticles made of multiple materials combine in order to form a composite nanomaterial with unique features (Makhlouf and Barhoum, 2018). These composite-based nanomaterials are either the combination of organic-organic NMs, inorganic-inorganic NMs, or organic-inorganic NMs, also known as hybrid NMs. The organic component or network of mainly organic polymers fuse together with the inorganic parts, like metal oxides or metal oxo polymers at the nanoscopic level, coupled via covalent or noncovalent bonding (Sanchez et al., 2011). Many natural nanocomposites like eggshell, abalone shell, and bones exist. Many kinds of nanocomposites like metal-metal, ceramic-metal, ceramic-ceramic, polymer/nonpolymer-based, polymer-ceramic, carbon-metal, and carbon-polymer nanocomposites, have been formed with improved physio-chemical, mechanical, and biological characteristics, with extraordinary flexibility. Recently, the most popular nanocomposites have been of polymer, layered silica, and metal-organic frameworks (MOFs) (Yuan and Müller, 2010; Taylor-Pashow et al., 2010).

2. Classification of nanomaterials

Materials having size equal to 100 nm or less are said to be nanomaterials. There are a variety of materials, either naturally occurring or synthesized by different methods, currently known, and it is expected that more will be discovered. The concept of calcification of nanomaterials was first proposed by Gleiter in 1995 (Gleiter, 2000). This classification of NMs is done on the basis of crystalline form as well as chemical composition. The classification scheme of Gleiter was not fully complete because it did not include the dimensionality of nanostructures. In 2007, a new classification scheme of NMs was introduced by Pokropivny and Skorokhod, which includes dimensional classification (0D, 1D, 2D, and 3D) as well as components (Pokropivny and Skorokhod, 2007). NMs can exist in any form such as single, cluster, spherical, fused, and irregular, fibers, tubes, etc. Classifications of NMs according to their composition, dimension, morphology, uniformity, and agglomeration state are given in Fig. 7 (Buzea et al., 2007). Nanomaterials have different characteristics compared to normal materials and are used in a variety of fields, such as nanotechnology (Lahann, 2008), air purification

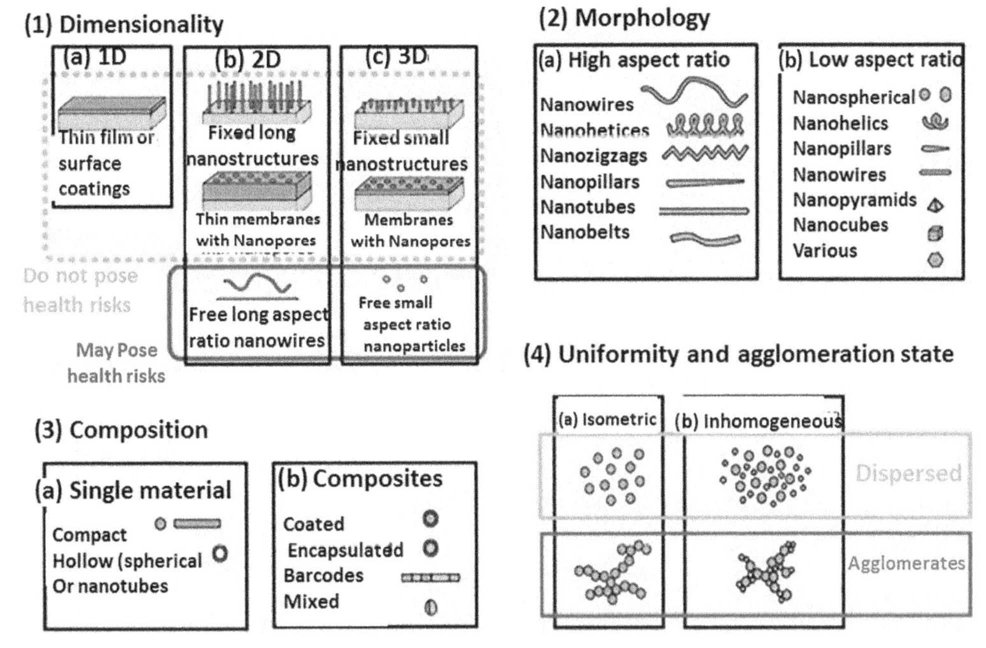

Fig. 7 Classification of nanomaterials (Buzea et al., 2007).

(Granqvist et al., 2007), water purification, medical devices, reactive membrane (Fryxell and Cao, 2012), etc.

2.1 Types of nanomaterials according to dimension

According to dimensions, nanomaterials are divided into four types:

- zero-dimensional (0D)
- one-dimensional (1D)
- two-dimensional (2D)
- three-dimensional (3D)

2.1.1 Zero-dimensional nanomaterials

Materials in which electron motion is confined along all directions within the system are said to be 0D NMs. The simplest example of 0D nanomaterials is nanoparticles. Other than nanoparticles, atomic clusters, filaments, etc. are also 0D NMs (Capsal et al., 2011). It contains materials that have all their properties within size less than 100 nm and in these materials length and width both are equal. The 0D NMs can be of any form, such as given below:

- crystalline or polycrystalline
- amorphous or crystalline
- single or multichemical elements
- contains many shapes and forms (An et al., 2006)

2.1.1.1 Quantum dots

Quantum dots (QDs) are an example of a zero dimensional NM, having size within 2 and 20 nm (Kluson et al., 2007). According to the literature, their size must be less than 10 nm (Kluson et al., 2007; Ferancová and Labuda, 2008; Kral et al., 2006). QDs' diameter depends on the material being used. QDs exhibit quantum size effect. The nature of the QD may be metallic (cobalt, gold, nickel, etc.) (Ghanem et al., 2004) or semiconductor, and they have been a subject of research due to their novel properties such as electronic, optical, and catalytic. The number of atoms occupied by the QD, ranges between 1000 and 100,000 atoms, due to which these materials can never possess single molecule entity and neither can be extended to a solid structure. QDs are also used as silicon QDs in their metalloid form (Fujioka et al., 2008).

Due to their wide range of properties, QDs can be used in a variety of fields such as biology (Jamieson et al., 2007), chemistry, materials science, etc., and different synthesis methods have been developed for creating QDs, such as the wet chemical colloidal process and plasma synthesis. The advantage of QDs is that the light they emit can be tuned to a desired suitable wavelength, and for this purpose, particle size is altered by carefully controlling the growth factor. It contains discrete energy states and in semiconductor QDs, emitted light wavelength depends upon the energy difference of valance and conduction band. The emitted light wavelength also depends upon the size of QDs (2–3 nm), larger the size of QDs (5–6 nm diameter), larger the wavelength and smaller the size smaller the wavelength emitted (Nath, 2018). QDs are also referred to as "artificial atoms" (Jortner and Ratner, 1997). The potential applications of QDs include quantum computing, inkjet printing, solar cells, medical imaging, etc. (Medintz et al., 2008).

2.1.2 One-dimensional nanomaterials

Nanomaterials such as nanofibers (Choi et al., 2003), nanowires (Wang and Song, 2006), and nanotubes (She et al., 2007) are 1D NMs. In these, the electrons are confined in two dimensions (*x*&*y* or *y*&*z* or *z*&*x*) with undefined surface boundaries and allowed to move only in 1D. They are used in various fields such as thin films in electronic engineering, reflectors and computer chips, etc. These NMs contain all properties greater than 100 nm, and in this, length is much greater than width. Deposition of thin film of 1D nanomaterials is done in a controlled manner; this deposited layer is known as a monolayer. For synthesizing 1D NMs, different techniques are used, such as vapor deposition, electro deposition, etc. Different materials such as nanowhiskers, nanocables, and nanofibers are also included in one-dimensional NMs. However, each material has different diameter ranges; for example, whiskers and fibers range from several nm to a hundred microns (Gong and Cheng, 2017). ZnO nanowires are an example of a 1D nanomaterial (Zhang et al., 2012).

2.1.2.1 Nanotubes

Carbon nanotubes (CNTs) are tubes made of carbon with diameters typically measured in nanometers. The carbon nanotubes were discovered by Ijma and Ichihashi in 1991 by using the arc-discharge method in the form of new carbon molecules. These tubes are basically originated in the form of nanoparticles. These structures are matched with spherical C60 molecules (Ebbesen, 1996), which are also known as "buckyballs," as discovered in 1980 (Dresselhaus et al., 2000). These are elongated to form long, hollow, tube-like structures with diameters of 1–2 nm. They can be generated in large amounts and their length is greater than 1 mm. Put simply, nanotubes comprise of one single layer of carbon molecules, which are arranged in a cylindrical pattern. These tubes also termed as single-walled carbon nano-tubes having diameter between 0.6 and 5 nm. These tubes are also converted into multiple concentric tubes Nanotubes having inner and outer diameter of 1.5–15 and 2.5–50 nm respectively (Kaneto et al., 1999). Carbon nanotubes are considered as the special form of fullerenes (Dresselhaus et al., 2000) or can be constructed by joining the concentric layers of graphite. CNTs cannot break under the action of tension and are considered 100 times stronger than steel. CNT properties such as aspect ratio, length, etc. depend on the techniques which are being used to synthesize them (Asmatulu et al., 2009). CNTs possess a variety of properties such as high surface area, molecular adsorption capacity, and also have high conductivity (Dresselhaus and Riichiro, 1998); due to these properties, they are employed in various applications such as lithium ion batteries (Landi et al., 2009), polymer composition (Bhatt, 2005), etc. Some other applications of nanotubes are given below:

1. CNTS field emission (Kim et al., 2000).
2. CNTS thermal conductivity (Berber et al., 2000).
3. CNTS energy storage (Zhang et al., 2011).
4. CNTS thermal materials (Wen and Ding, 2004).
5. CNTS conductive adhesive (Yim and Kim, 2010).
6. Molecular electronics based on CNTS (Rueckes et al., 2000; Avouris, 2002).
7. CNTS structured application (Salunkhe et al., 2015).

2.1.2.2 Nanowires

Nanowires can be organized from conductive and nonconductive materials by various methods including deposition and suspension. These materials have diameters of a few tens of nanometers (Li et al., 2012). Nanowires are also known as "quantum wires." The basic function of nanowires is to transport electrons in nanoelectronic devices (Dresselhaus et al., 2001). Different types of metals are used to create nanowires such as cobalt, gold (Mishra, 2013), and copper. Nanowires of silicon have also been produced (Li et al., 2003). The most appropriate methods used in the production of nanowires are derived from methods used in the semiconductor industry (Markussen et al., 2008) for the formation of microchips. One of the most successful linear nanostructures is carbon nanostructure, which are available in different forms such as (single- or multiwalled, filled or surface modified) carbon modified, and are the most wide field in nanoelectronics for (data storage or wiring) (Hashim and Sidek, 2012).

2.1.3 Two-dimensional nanomaterials

The materials such as nanosheets, nanoribbons (Yamano et al., 2012), graphene, nanofilms, nanolyers, and nanocoating are included in 2D NMs (Qi et al., 2010). In two dimensional nanomaterials, the electrons are confined in one dimension only and cannot move freely in this direction. They also have large aspect ratio (length and width relation). These materials possess high mechanical flexibility as well as optical transparency. The two-dimensional NMs are synthesized by using different synthesis techniques such as bottom-up and top-down lithography. These materials have gained attention due to their applications in a variety of fields such as electronics, biomedicines, sensors, photodetectors, battery electrodes, topological insulators, etc. (Lee et al., 2016; Vaughn et al., 2011). Due to their size characteristics and specific geometry, 2D NMs can be utilized in a variety of nanodevices.

2.1.3.1 Graphene and hexagonal boron nitride

Graphene was first discovered as a 2D NM in 2004 by Novoselov et al. (2004). There are many examples of 2D NMs (Mounet et al., 2018) due to their extensive range of properties. A 2D nanomaterial, graphene is a lattice of carbon atoms, which are covalently bounded and have a hexagonal arrangement. In this 2D NM, the thickness of one atom is about 0.14 nm. The graphene is considered to be a semimetal material (valance and conduction band touch each other). The electrons of graphene have an extremely high speed, at 1/3000th of the speed of light. Due to this, they have fascinating properties including matchless thermal conductivity. Graphene is optically transparent and has the capability to absorb 2% of light which is incidental on it. Graphene possesses the highest tensile strength compared to any other material. A single layer of graphene, with a thickness of about 0.3 nm, has the strength to bear the weight of a football.

The crystallographic appearance of hexagonal boron nitride (h-BN) is the same as that of graphene. But within this, carbon atoms are exchanged by the atoms of boron and nitrogen. h-BN has a wide band gap compared to graphene and is known to be an insulator (Lee et al., 2008). Due to the ecofriendly nature of h-BN nanosheets, they are widely used in industries for different purposes such as for making composites, surface coating, etc.

Graphene is a well-known example of a nanosheet. The absorption coefficient of graphene is 2.3%. It possesses a number of properties including electron transfer capabilities, high mechanical strength, electrical conduction, optical transmittance, etc. Graphene has a large surface area of 2630 m^2/g, which is larger than the surface area of carbon nanotubes. A single graphene layer is used as a substrate for carbon nanotubes, and furthermore can produce 3D graphite. Single-layer graphene can be converted into 0D or 1D materials by the rotation around its own axis to form nanotubes or by rapping spherically to form fullerene (Sinitskii and Tour, 2010).

2.1.4 Three-dimensional nanomaterials

The nanomaterials in which electrons are not confined to any dimension—termed as delocalized electrons—are said to be 3D NMs. Three dimensional NMs are also known as bulk materials and contain arbitrary directions beyond 100 nm. The materials containing equiaxed nm-sized grains are included in the 3D NMs category (Liu and Bashir, 2015; Zhang et al., 2014). These materials have high absorption capability and large surface area, due to which they are taken into account in research fields and applications. Due to these properties, they are used for transportation of molecules or drugs. 3D NMs can be found in various structural forms such as nanocoils, nanoflowers, nanoballs, nanocones, nanopillars, etc. (Tiwari et al., 2012).

2.1.4.1 Dendrimers

Dendrimers are nanoscale materials and considered as 3D macromolecules. They have a star-like appearance and a highly branched architectural structure. The first dendrimer was synthesized in 1978. Dendrimers basically consist of three components; one is the inner core, the second is the interior structure, and the third is the exterior surface having surface groups. Using these components, different shapes and sizes of the dendrimers can be produced. Their structure has a great influence on their physical and chemical properties, due to which dendrimers have many applications in biology, chemistry, materials science, and engineering (Niu and Crooks, 2003).

The diversity of nanoparticles is colossal and cannot be realized under one class; thus, classification of nanoparticles is required. Nanoparticles can be classified on the basis of their diameter, size, surface area, morphology, dimensions, and fabrication techniques (Buzea et al., 2007). However, generally nanoparticles are classified on the following basis:

- Morphology
- Composition
- Uniformity and agglomeration
- Dimensionality

All the classifications are described in detail in the following sections.

3. Morphology

In morphology-based classification of nanoparticles, aspect ratio, sphericity, and flatness are considered. However, most studies classify nanoparticles on the basis of their aspect

Table 1 Classification of aspect ratio and related nanoparticles (Bray et al., 2013).

Type	Area fraction	Aspect ratio	Nanoparticles
Low	$A_f=0.05$	$f<18$	Aluminum borate whisker, nanoclay, rod-like wollastonite, ceramic, plate-like
	$A_f=0.1$	$f<9$	Rod-like wollastonite, nanoclay, ceramic
Moderate	$A_f=0.2$	$f<4$	Ceramic, core-shell, rubber, plate-like
	$A_f=0.1$	$9<f<36$	Aluminum borate whisker
	$A_f=0.2$	$4<f<18$	Aluminum borate whisker, nanoclay
	$A_f=0.05$	$f>72$	Multi/single-wall nanotube, fibers, nanowire
High	$A_f=0.1$	$f>36$	Multi/single-wall nanotube, nanowire
	$A_f=0.2$	$f>18$	Aluminum borate whisker, nanotube, nanowire

ratio (Sekunowo et al., 2015). On the basis of aspect ratio, nanoparticles are divided into two categories:

- high aspect ratio nanoparticles
- low aspect ratio nanoparticles

On the basis of low aspect ratio, nanoparticles are categorized as follows:

- Oval
- Cubical
- prism-shaped
- spherical
- pillar-like
- colloids

The concept of high and low and moderate aspect ratio classification is explained in Table 1. Low aspect ratio nanoparticles are those whose ratio is less than 18 or 9 whereas moderate aspect ratio have different criterion as shown in Table 1. High aspect ratio nanoparticles have aspect ratio greater than 18 or 36. Based on different criteria of low, moderate, and high aspect ratio, examples of nanoparticles are given in Table 1 (Bray et al., 2013).

3.1 High aspect ratio nanoparticles

On the basis of morphological characterization nanoparticles can be divided into

- nanorods
- nanohooks
- nanohelices
- nanostars
- nanospring
- nanoplates

Some of the above examples are shown in Fig. 8.

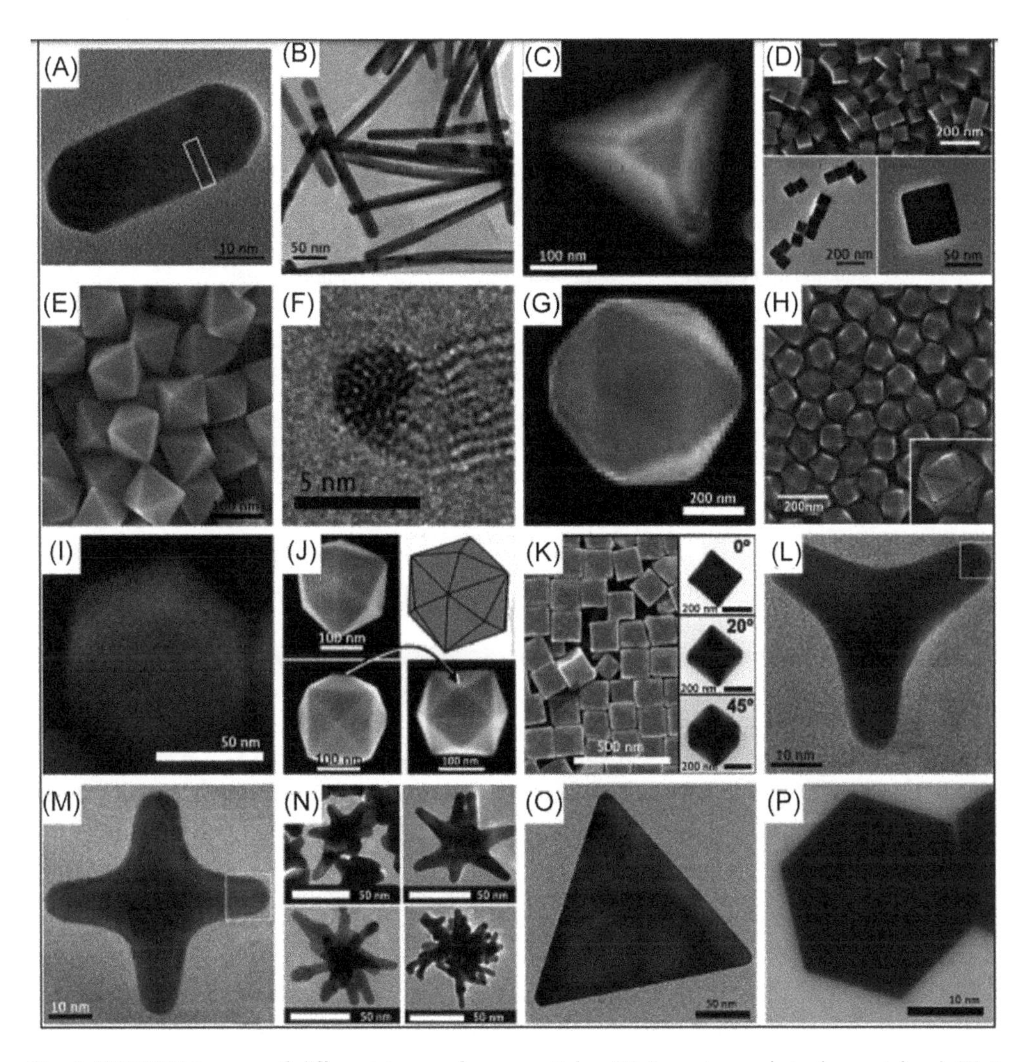

Fig. 8 SEM/TEM images of different types of nanoparticles. (A) Au octagonal single-crystal rod, (B) Au pentagonally twinned rods, (C) Au tetrahedron NP, (D) Pd hexahedron (i.e., cube) NPs, (E) Au octahedron NPs, (F) decahedron, (G) Au icosahedron NP, (H) Au trisoctahedron NPs, (I) Au rhombic dodecahedron NP, (J) Pt tetrahexahedron NPs, (K) Au concave hexahedron NPs, (L) Au tripod NP, (M) Au tetrapod NP, (N) Au star NPs, (O) Au triangular plate/prism NP, and (P) Au hexagonal plate/prism NP (Lohse et al., 2014).

As discussed earlier, on the basis of morphology, nanoparticles can also classified based on aspect ratio. High or long aspect ratio nanoparticles include carbon and palladium nanotubes or silicon nanowires and silver nanowires (Cheng et al., 2010; Leonard et al., 2014; Liu et al., 2011). Complex morphologies such as core-shell and spherical core-shell are also possible; these include CdSe-CdS and Pd-Cu core-shell nanoparticles

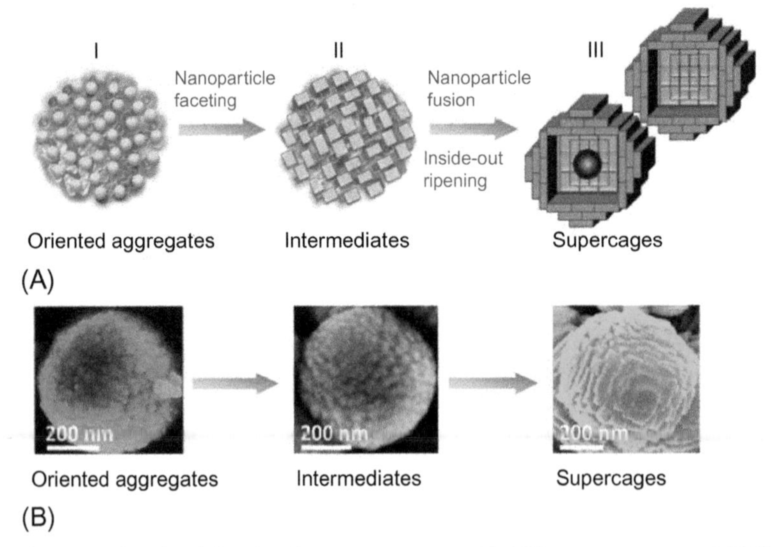

Fig. 9 (A) Schematic sketch of the development process of $BaTiO_3$ supercages and (B) SEM images showing the morphology transition process (Li et al., 2015).

(Mark et al., 2013; Xia and Tang, 2012; Jin et al., 2012). Fig. 9 explains the complex morphologies of barium titanate nanoparticles (Li et al., 2015).

There is a large body of data available on the subject of nanomaterials or nanoparticles due to the colossal importance it holds in fields like physics, chemistry, and the medical industry (Buzea et al., 2007).

3.1.1 Nanotubes

In nanotechnology, nanotubes are mostly referred to as carbon nanotubes. Nanotubes have been the most discussed morphology of nanoparticles for a long time because of their intriguing applications. Nanotubes can be made from boron nitride in the case of inorganic nanoparticles and from peptide proteins in the case of organic nanoparticles. Carbon nanotubes, however, have the most interesting applications and properties (Holister et al., 2003). Carbon nanotubes are shown in Fig. 10.

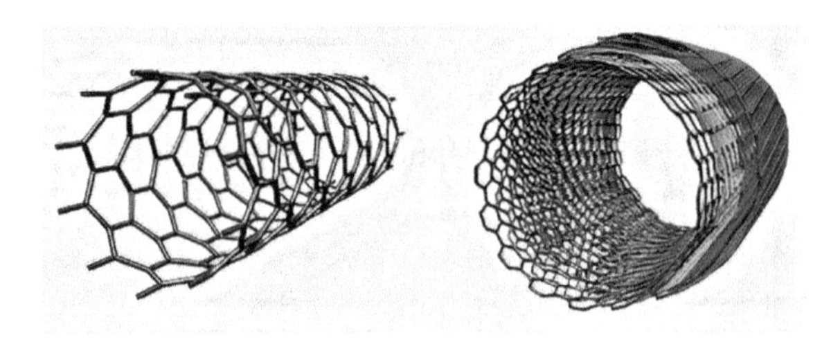

Fig. 10 An image of carbon nanotubes (Gupta et al., 2018).

Nanotubes are one form of nanoparticle that has sparked great interest. Carbon nanotubes were discovered back in 1991. The diameter of the nanotubes is 1–2 nm. Carbon nanotubes are classified as high aspect ratio nanoparticles and can be made as long as 1 mm (Kroto et al., 1985). Carbon nanotubes can further be divided into:

- single-walled nanotubes
- multiwalled nanotubes (diameter = 20 nm, length = 1 mm)

Nanotubes are produced from carbon, silicon, and germanium, but carbon nanotubes have the most applications (Njuguna et al., 2014).

3.1.2 Nanowires

Nanowires can be made using different techniques using metals or semiconductors (Gleixner, 2000). Nanowires have high aspect ratio, a diameter of 10 nm, and a single crystal structure. Nanowires are extensively used in nanoelectronic devices as connectors for the transportation of electrons. Cobalt, copper, silicon, and gold have been utilized to make nanowires. Chemical vapor deposition is used for the production of nanowires (Njuguna et al., 2014). Nanotubes and nanowires can then be categorized on the basis of their shape and length as follows:

- zigzags
- helices
- belts

Nanowires are classified as long aspect ratio nanoparticles.

4. Techniques used for morphological characterization

Various techniques are utilized to assess the physical and chemical properties of nanoparticles in order to understand their morphology and composition. These techniques are X-ray diffraction (XRD), infrared, scanning electron microscopy (SEM), tunneling electron microscopy (TEM), X-ray photoelectron spectroscopy (XPS), and Brunauer-Emmett-Teller theory (BET).

Morphology plays a pivotal role in the basic properties of nanoparticles. For morphological characterization, different techniques are utilized such as SEM, polarized optical microscopy (POM), and TEM. For nanoscale information regarding nanoparticles, SEM is utilized, since it involves the mechanism of scanning of electrons. SEM is not only useful to study the morphology of nanoparticles, but is also implemented to study the dispersion of nanoparticles in bulk materials or matrices. TEM is another technique for morphological characterization of nanoparticles that is based on transmission of electrons. TEM provides information about nanoparticles in bulk with different magnifications. Tunneling electron microscopy is also essential in providing useful information regarding layered nanoparticles (Saeed and Khan, 2014, 2016; Khlebtsov and Dykman, 2011; Khan et al., 2019; Vickers, 2017).

5. Composition

Nanoparticles can also be classified based on their composition. Some naturally existing nanoparticles are composed of single materials, and some are composed of more than one material and are called composites. Naturally occurring nanoparticles are usually composites and are agglomerates of many constituting materials (Buzea and Pacheco, 2017). Single and pure nanoparticles are easy to synthesize with today's technology. Nanomaterials can be either composed of a single material such as antimony or bismuth, or made of composite materials such as hybrid nanozigzags made of nickel, silicon, and titanium oxide (Njuguna et al., 2014).

Nanoparticles undergo oxidation in air and form thin films. Currently, chemical vapor deposition is utilized to make composite nanoparticles. The composite particles' motion and temperature are utilized to make different nanoparticles. Chemical vapor deposition can also be employed to alter the shape of single nanomaterials (Buzea and Pacheco, 2017). Table 2 explains the classification of nanoparticles based on composition. Table 2 gives the classification based on composition such as single phase nanoparticles are crystalline polymers, amorphous particles and layers where the multiphase nanoparticles are matrix composites, coated particles colloids, aerogels, fluids (Njuguna et al., 2014).

6. Uniformity and agglomeration state

Composition is not the only parameter that is used to distinguish and classify nanoparticles; based on the uniformity and agglomeration state, nanoparticles can exist in

- dispersed aerosols
- agglomerate state

In dispersion aerosols, they can exist in the form of colloids or suspensions. This classification is based on their electromagnetic properties such as magnetism, surface charge, and chemistry (Tyagi et al., 2013).

Table 2 Classification of nanomaterials based on phase composition (Njuguna et al., 2014).

Phase composition	Example
Single-phase solids	Crystalline polymers, amorphous particles and layers
Multiphase solids	Matrix composites, coated particles
Multiphase systems	Colloids, aerogels, fluids

Table 3 Classification of nanomaterials based on agglomeration state (Njuguna et al., 2014).

	Homogeneous	Heterogeneous
Dispersed		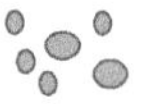
Agglomerated	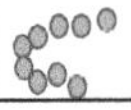	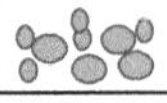

6.1 Agglomerate state

Nanoparticles that are present in agglomerate state behave as colossal particles that are dependent on their size and dimension of their agglomerate. For example, nanoparticles that are magnetic exist in agglomerate in the form of clusters, unless there is some kind of nonmagnetic coating on them. Size, dimensionality, surface area, surface reactivity, and agglomerate size are all essential for the better comprehension of nanoparticles (Buzea and Pacheco, 2017). Surface morphology plays a role in the agglomeration of liquid nanoparticles, which in turn tells us about the hydrophobicity or hydrophilicity of the NPs.

Hydrophobic nanoparticles repel water and hydrophilic nanoparticles are soluble in water. Self-engineered nanoparticles are generally hydrophobic because they are composed mostly of metals. In order to improve their performance, they are coated with different particles or other nanoparticles (Podila and Brown, 2013). Hence size, shape, diameter, crystal structure, morphology, and reactivity, as well as agglomeration of nanoparticles, are fundamental when regulating nanoparticles, as explained in Table 3. Nanoparticles can be classified in terms of their agglomeration state and dispersion; the dispersed and agglomerate states for homogeneous and heterogeneous systems are different.

7. Conclusion

Nanotechnology is pushing the boundaries of technology day by day as the fields of nanoscience and materials are making tremendous progress very rapidly. Nanomaterials are at the heart of nanoscience and technology. This chapter gave an overview of all types of nanomaterials, including organic, inorganic, carbon, and composite-based nanomaterials. Various organic-based nanomaterials such as carbon nanotubes, fullerenes, graphite, and metal nanotubes were discussed. Organic nanomaterials are found to have strong stability and control, and are employed in drug delivery systems. Inorganic nanomaterials are thought to be photostable, have unique properties to be implementable in therapeutic systems, and are metal- and metal oxide-based. Composite nanomaterials are amalgamations of organic and inorganic nanomaterials. Classification of nanomaterials based on dimensionality, morphology, composition, uniformity, and agglomeration state was also explained, with examples and scanning electron microscope and tunneling electron microscope images. Size, shape, and dimensions of nanoparticles are of colossal importance since they change the characteristics of nanomaterials such as optical properties. Morphology characterization was taken into account in the context of aspect ratio. Better comprehension of types and classification will enable the discovery of novel nanomater ials that can revolutionize the nanotechnology industry.

References

Al-Kayiem, H.H., Lin, S.C., Lukmon, A., 2013. Review on nanomaterials for thermal energy storage technologies. Nanosci. Nanotechnol. 3, 60–71.

An, K., Lee, N., Park, J., Kim, S.C., Hwang, Y., Park, J.-G., Kim, J.-Y., Park, J.-H., Han, M.J., Yu, J., 2006. Synthesis, characterization, and self-assembly of pencil-shaped CoO nanorods. J. Am. Chem. Soc. 128, 9753–9760.

Asmatulu, R., Asmatulu, E., Yourdkhani, A., 2009. Toxicity of nanomaterials and recent developments in the protection methods. In: SAMPE Fall Technical Conference, Wichitapp. 1–12.

Avouris, P., 2002. Molecular electronics with carbon nanotubes. Acc. Chem. Res. 35, 1026–1034.

Bailey, R.E., Smith, A.M., Nie, S., 2004. Quantum dots in biology and medicine. Physica E Low Dimens. Syst. Nanostruct. 25, 1–12.

Berber, S., Kwon, Y.-K., Tománek, D., 2000. Unusually high thermal conductivity of carbon nanotubes. Phys. Rev. Lett. 84, 4613.

Bhatt, S., 2005. Compositions Comprising Carbon Nanotubes and Articles Formed Therefrom. Google Patents.

Bilecka, I., Niederberger, M., 2010. Microwave chemistry for inorganic nanomaterials synthesis. Nanoscale 2, 1358–1374.

Bray, D., Gilmour, S., Guild, F., Taylor, A., 2013. The effects of particle morphology on the analysis of discrete particle dispersion using Delaunay tessellation. Compos. A: Appl. Sci. Manuf. 54, 37–45.

Buzea, C., Pacheco, I., 2017. Nanomaterials and their classification. In: EMR/ESR/EPR Spectroscopy for Characterization of Nanomaterials. Springer, New Delhi.

Buzea, C., Pacheco, I.I., Robbie, K., 2007. Nanomaterials and nanoparticles: sources and toxicity. Biointerphases 2, MR17–MR71.

Capsal, J.-F., Dantras, E., Dandurand, J., Lacabanne, C., 2011. Molecular mobility in piezoelectric hybrid nanocomposites with 0–3 connectivity: volume fraction influence. J. Non-Cryst. Solids 357, 3410–3415.

Chattopadhyay, S., Chen, J.-Y., Chen, H.-W., Hu, C.-M.J., 2017. Nanoparticle vaccines adopting virus-like features for enhanced immune potentiation. Nanotheranostics 1, 244.

Cheng, B., Le, Y., Yu, J., 2010. Preparation and enhanced photocatalytic activity of Ag@ TiO2 core–shell nanocomposite nanowires. J. Hazard. Mater. 177, 971–977.

Choi, S.-S., Lee, S.G., Im, S.S., Kim, S.H., Joo, Y.L., 2003. Silica nanofibers from electrospinning/sol-gel process. J. Mater. Sci. Lett. 22, 891–893.

Dolez, P.I., 2015. Nanomaterials definitions, classifications, and applications. In: Nanoengineering. Elsevier, Amsterdam.

Dresselhaus, G., Riichiro, S., 1998. Physical Properties of Carbon Nanotubes. World Scientific, London.

Dresselhaus, M.S., Dresselhaus, G., Eklund, P., Rao, A., 2000. Carbon nanotubes. In: The Physics of Fullerene-Based and Fullerene-Related Materials. Springer, Dordrecht.

Dresselhaus, M., Lin, Y., Koga, T., Cronin, S., Rabin, O., Black, M., Dresselhaus, G., 2001. Quantum wells and quantum wires for potential thermoelectric applications. In: Recent Trends in Thermoelectric Materials Research III.vol. 71. .

Ebbesen, T.W., 1996. Carbon Nanotubes: Preparation and Properties. CRC Press, Boca Raton, FL; London.

Ferancová, A., Labuda, J., 2008. DNA biosensors based on nanostructured materials. In: Nanostructured Materials in Electrochemistry. Wiley-VCH, Weinheim, pp. 409–434.

Fryxell, G.E., Cao, G., 2012. Environmental Applications of Nanomaterials: Synthesis, Sorbents and Sensors. World Scientific, London.
Fujioka, K., Hiruoka, M., Sato, K., Manabe, N., Miyasaka, R., Hanada, S., Hoshino, A., Tilley, R.D., Manome, Y., Hirakuri, K., 2008. Luminescent passive-oxidized silicon quantum dots as biological staining labels and their cytotoxicity effects at high concentration. Nanotechnology 19, 415102.
Ghanem, M.A., Bartlett, P.N., De Groot, P., Zhukov, A., 2004. A double templated electrodeposition method for the fabrication of arrays of metal nanodots. Electrochem. Commun. 6, 447–453.
Gleiter, H., 2000. Nanostructured materials: basic concepts and microstructure. Acta Mater. 48, 1–29.
Gleixner, S.H., 2000. The edible microchip: a hands on overview of the semiconductor manufacturing process. Mater. Res. Soc. Symp. Proc. 632.
Gong, S., Cheng, W., 2017. One-dimensional nanomaterials for soft electronics. Adv. Electron. Mater. 3, 1600314.
Granqvist, C.G., Azens, A., Heszler, P., Kish, L., Österlund, L., 2007. Nanomaterials for benign indoor environments: electrochromics for "smart windows", sensors for air quality, and photo-catalysts for air cleaning. Sol. Energy Mater. Sol. Cells 91, 355–365.
Gupta, S., Murthy, C., Prabha, C.R., 2018. Recent advances in carbon nanotube based electrochemical biosensors. Int. J. Biol. Macromol. 108, 687–703.
Hadinoto, K., Sundaresan, A., Cheow, W.S., 2013. Lipid–polymer hybrid nanoparticles as a new generation therapeutic delivery platform: a review. Eur. J. Pharm. Biopharm. 85, 427–443.
Hashim, Y., Sidek, O., 2012. Effect of temperature on the characteristics of silicon nanowire transistor. J. Nanosci. Nanotechnol. 12, 7849–7852.
Holister, P., Harper, T.E., Vas, C.R., 2003. Nanotubes White Paper. CMP Científica January.
Hu, Y., Shenderova, O.A., Brenner, D.W., 2007. Carbon nanostructures: morphologies and properties. J. Comput. Theor. Nanosci. 4, 199–221.
Hu, B., Wang, S.-B., Wang, K., Zhang, M., Yu, S.-H., 2008. Microwave-assisted rapid facile "green" synthesis of uniform silver nanoparticles: self-assembly into multilayered films and their optical properties. J. Phys. Chem. C 112, 11169–11174.
Jamieson, T., Bakhshi, R., Petrova, D., Pocock, R., Imani, M., Seifalian, A.M., 2007. Biological applications of quantum dots. Biomaterials 28, 4717–4732.
Jin, M., Zhang, H., Wang, J., Zhong, X., Lu, N., Li, Z., Xie, Z., Kim, M.J., Xia, Y., 2012. Copper can still be epitaxially deposited on palladium nanocrystals to generate core–shell nanocubes despite their large lattice mismatch. ACS Nano 6, 2566–2573.
Jortner, J., Ratner, M., 1997. Molecular Electronics. Blackwell Science, Oxford.
Kaneto, K., Tsuruta, M., Sakai, G., Cho, W., Ando, Y., 1999. Electrical conductivities of multi-wall carbon nano tubes. Synth. Met. 103, 2543–2546.
Khan, I., Saeed, K., Khan, I., 2019. Nanoparticles: properties, applications and toxicities. Arab. J. Chem. 12, 908–931.
Khlebtsov, N., Dykman, L., 2011. Biodistribution and toxicity of engineered gold nanoparticles: a review of in vitro and in vivo studies. Chem. Soc. Rev. 40, 1647–1671.
Kim, J.M., Choi, W.B., Lee, N.S., Jung, J.E., 2000. Field emission from carbon nanotubes for displays. Diam. Relat. Mater. 9, 1184–1189.
Klusoň, P., Drobek, M., Bartkova, H., Budil, I., 2007. Welcome in the nanoworld. Chem. List. 101, 262–272.
Kral, V., Šotola, J., Neuwirth, P., Kejik, Z., Zaruba, K., Martasek, P., 2006. Nanomedicine-current status and perspectives: a big potential or just a catchword? Chem. List. 100, 4–9.
Kroto, H.W., Heath, J.R., O'Brien, S.C., Curl, R.F., Smalley, R.E., 1985. C60: Buckminsterfullerene. Nature 318, 162–163.
Kumar, R., Lal, S., 2014. Synthesis of organic nanoparticles and their applications in drug delivery and food nanotechnology: a review. J. Nanomater. Mol. Nanotechnol. 3 (4), 2–11.
Kundu, S., Peng, L., Liang, H., 2008. A new route to obtain high-yield multiple-shaped gold nanoparticles in aqueous solution using microwave irradiation. Inorg. Chem. 47, 6344–6352.
Lahann, J., 2008. Environmental nanotechnology: nanomaterials clean up. Nat. Nanotechnol. 3, 320.

Landa, S., Macháček, V., 1933. Sur l'adamantane, nouvel hydrocarbure extrait du naphte. Collect. Czechoslov. Chem. Commun. 5, 1–5.

Landi, B.J., Ganter, M.J., Cress, C.D., Dileo, R.A., Raffaelle, R.P., 2009. Carbon nanotubes for lithium ion batteries. Energy Environ. Sci. 2, 638–654.

Lee, C., Wei, X., Kysar, J.W., Hone, J., 2008. Measurement of the elastic properties and intrinsic strength of monolayer graphene. Science 321, 385–388.

Lee, J.Y., Shin, J.-H., Lee, G.-H., Lee, C.-H., 2016. Two-dimensional semiconductor optoelectronics based on van der Waals heterostructures. Nanomaterials 6, 193.

Leonard, S.S., Cohen, G.M., Kenyon, A.J., Schwegler-Berry, D., Fix, N.R., Bangsaruntip, S., Roberts, J.R., 2014. Generation of reactive oxygen species from silicon nanowires. Environ. Health Insights 8, S15261.

Li, D., Wu, Y., Kim, P., Shi, L., Yang, P., Majumdar, A., 2003. Thermal conductivity of individual silicon nanowires. Appl. Phys. Lett. 83, 2934–2936.

Li, Y., Yang, X.-Y., Feng, Y., Yuan, Z.-Y., Su, B.-L., 2012. One-dimensional metal oxide nanotubes, nanowires, nanoribbons, and nanorods: synthesis, characterizations, properties and applications. Crit. Rev. Solid State Mater. Sci. 37, 1–74.

Li, J., Hietala, S., Tian, X., 2015. BaTiO3 supercages: unusual oriented nanoparticle aggregation and continuous ordering transition in morphology. ACS Nano 9, 496–502.

Liu, J.L., Bashir, S., 2015. Advanced Nanomaterials and Their Applications in Renewable Energy. Elsevier, Amsterdam.

Liu, L., Yoo, S.-H., Lee, S.A., Park, S., 2011. Wet-chemical synthesis of palladium nanosprings. Nano Lett. 11, 3979–3982.

Lohse, S.E., Burrows, N.D., Scarabelli, L., Liz-Marzán, L.M., Murphy, C.J., 2014. Anisotropic noble metal nanocrystal growth: the role of halides. Chem. Mater. 26, 34–43.

Makhlouf, A.S.H., Barhoum, A., 2018. Emerging Applications of Nanoparticles and Architectural Nanostructures: Current Prospects and Future Trends. William Andrew, Amsterdam.

Mark, A.G., Gibbs, J.G., Lee, T.-C., Fischer, P., 2013. Hybrid nanocolloids with programmed three-dimensional shape and material composition. Nat. Mater. 12, 802–807.

Markussen, T., Rurali, R., Jauho, A.-P., Brandbyge, M., 2008. Transport in silicon nanowires: role of radial dopant profile. J. Comput. Electron. 7, 324–327.

Mauter, M.S., Elimelech, M., 2008. Environmental applications of carbon-based nanomaterials. Environ. Sci. Technol. 42, 5843–5859.

Medintz, I.L., Mattoussi, H., Clapp, A.R., 2008. Potential clinical applications of quantum dots. Int. J. Nanomedicine 3, 151.

Meyer, J.C., Kisielowski, C., Erni, R., Rossell, M.D., Crommie, M., Zettl, A., 2008. Direct imaging of lattice atoms and topological defects in graphene membranes. Nano Lett. 8, 3582–3586.

Mounet, N., Gibertini, M., Schwaller, P., Campi, D., Merkys, A., Marrazzo, A., Sohier, T., Castelli, I.E., Cepellotti, A., Pizzi, G., 2018. Two-dimensional materials from high-throughput computational exfoliation of experimentally known compounds. Nat. Nanotechnol. 13, 246–252.

Müller, R., Radtke, M., Wissing, S., Nalwa, H., 2004. Nalwa, H.S. (Ed.), Encyclopedia of Nanoscience and Nanotechnology. American Scientific Publishers.

Mishra, P., 2013. Emerging nanotechnology and it's impact on electronic circuit designing. IJECT 4 (5), 64–65.

Nath, D., 2018. Scientoonic representation of nanotechnology. IOSR-JAP 10 (3), 27–38.

Niu, Y., Crooks, R.M., 2003. Dendrimer-encapsulated metal nanoparticles and their applications to catalysis. C.R. Chim. 6, 1049–1059.

Njuguna, J., Ansari, F., Sachse, S., Zhu, H., Rodriguez, V., 2014. Nanomaterials, nanofillers, and nanocomposites: types and properties. In: Health and Environmental Safety of Nanomaterials. Elsevier, United Kingdom.

Novoselov, K.S., Geim, A.K., Morozov, S.V., Jiang, D., Zhang, Y., Dubonos, S.V., Grigorieva, I.V., Firsov, A.A., 2004. Electric field effect in atomically thin carbon films. Science 306, 666–669.

Podila, R., Brown, J.M., 2013. Toxicity of engineered nanomaterials: a physicochemical perspective. J. Biochem. Mol. Toxicol. 27, 50–55.

Pokropivny, V., Skorokhod, V., 2007. Classification of nanostructures by dimensionality and concept of surface forms engineering in nanomaterial science. Mater. Sci. Eng. C 27, 990–993.

Polshettiwar, V., Baruwati, B., Varma, R.S., 2009. Self-assembly of metal oxides into three-dimensional nanostructures: synthesis and application in catalysis. ACS Nano 3, 728–736.

Qi, Y., Jafferis, N.T., Lyons Jr., K., Lee, C.M., Ahmad, H., Mcalpine, M.C., 2010. Piezoelectric ribbons printed onto rubber for flexible energy conversion. Nano Lett. 10, 524–528.

Rueckes, T., Kim, K., Joselevich, E., Tseng, G.Y., Cheung, C.-L., Lieber, C.M., 2000. Carbon nanotube-based nonvolatile random access memory for molecular computing. Science 289, 94–97.

Saeed, K., Khan, I., 2014. Preparation and properties of single-walled carbon nanotubes/poly (butylene terephthalate) nanocomposites. Iran. Polym. J. 23, 53–58.

Saeed, K., Khan, I., 2016. Preparation and characterization of single-walled carbon nanotube/nylon 6, 6 nanocomposites. Instrum. Sci. Technol. 44, 435–444.

Salunkhe, R.R., Ahn, H., Kim, J.H., Yamauchi, Y., 2015. Rational design of coaxial structured carbon nanotube–manganese oxide (CNT–MnO2) for energy storage application. Nanotechnology 26, 204004.

Sanchez, C., Shea, K.J., Kitagawa, S., 2011. Recent progress in hybrid materials science. Chem. Soc. Rev. 40, 471–472.

Scida, K., Stege, P.W., Haby, G., Messina, G.A., García, C.D., 2011. Recent applications of carbon-based nanomaterials in analytical chemistry: critical review. Anal. Chim. Acta 691, 6–17.

Sekunowo, O.I., Durowaye, S.I., Lawal, G.I., 2015. An overview of nano-particles effect on mechanical properties of composites. Int. J. Mech. Aerosp. Ind. Mechatron. Manuf. Eng. 9, 1–7.

She, G., Zhang, X., Shi, W., Fan, X., Chang, J.C., 2007. Electrochemical/chemical synthesis of highly-oriented single-crystal ZnO nanotube arrays on transparent conductive substrates. Electrochem. Commun. 9, 2784–2788.

Sinitskii, A., Tour, J., 2010. Graphene electronics, unzipped. IEEE Spectr. 47, 28–33.

Sun, X.-H., Li, C.-P., Wong, N.-B., Lee, C.-S., Lee, S.-T., Teo, B.-K., 2002. Templating effect of hydrogen-passivated silicon nanowires in the production of hydrocarbon nanotubes and nanoonions via sonochemical reactions with common organic solvents under ambient conditions. J. Am. Chem. Soc. 124, 14856–14857.

Taylor-Pashow, K.M., Della Rocca, J., Huxford, R.C., Lin, W., 2010. Hybrid nanomaterials for biomedical applications. Chem. Commun. 46, 5832–5849.

Tiwari, J.N., Tiwari, R.N., Kim, K.S., 2012. Zero-dimensional, one-dimensional, two-dimensional and three-dimensional nanostructured materials for advanced electrochemical energy devices. Prog. Mater. Sci. 57, 724–803.

Tyagi, P.K., Tyagi, S., Verma, C., Rajpal, A., 2013. Estimation of toxic effects of chemically and biologically synthesized silver nanoparticles on human gut microflora containing Bacillus subtilis. J Toxicol Environ Health Sci 5, 172–177.

Vaughn, D.D., In, S.-I., Schaak, R.E., 2011. A precursor-limited nanoparticle coalescence pathway for tuning the thickness of laterally-uniform colloidal nanosheets: the case of SnSe. ACS Nano 5, 8852–8860.

Vickers, N.J., 2017. Animal communication: when I'm calling you, will you answer too? Curr. Biol. 27, R713–R715.

Wang, Z.L., Song, J., 2006. Piezoelectric nanogenerators based on zinc oxide nanowire arrays. Science 312, 242–246.

Wen, D., Ding, Y., 2004. Effective thermal conductivity of aqueous suspensions of carbon nanotubes (carbon nanotube nanofluids). J. Thermophys. Heat Transf. 18, 481–485.

Xia, Y., Tang, Z., 2012. Monodisperse hollow supraparticles via selective oxidation. Adv. Funct. Mater. 22, 2585–2593.

Xie, J., Lee, J.Y., Wang, D.I., Ting, Y.P., 2007. Identification of active biomolecules in the high-yield synthesis of single-crystalline gold nanoplates in algal solutions. Small 3, 672–682.

Yamano, A., Takata, K., Kozuka, H., 2012. Ferroelectric domain structures of 0.4-μm-thick Pb (Zr, Ti) O3 films prepared by polyvinylpyrrolidone-assisted sol-gel method. J. Appl. Phys. 111, 054109.

Yim, B.-S., Kim, J.-M., 2010. Characteristics of isotropically conductive adhesive (ICA) filled with carbon nanotubes (CNTs) and low-melting-point alloy fillers. Mater. Trans. 51, 2329–2331.

Yuan, J., Müller, A.H., 2010. One-dimensional organic–inorganic hybrid nanomaterials. Polymer 51, 4015–4036.

Zayed, M.E., Zhao, J., Elsheikh, A.H., Du, Y., Hammad, F.A., Ma, L., Kabeel, A., Sadek, S., 2019. Performance augmentation of flat plate solar water collector using phase change materials and nanocomposite phase change materials: a review. Process. Saf. Environ. Prot. 128, 135–157.

Zhang, R., Wen, Q., Qian, W., Su, D.S., Zhang, Q., Wei, F., 2011. Superstrong ultralong carbon nanotubes for mechanical energy storage. Adv. Mater. 23, 3387–3391.

Zhang, Y., Ram, M.K., Stefanakos, E.K., Goswami, D.Y., 2012. Synthesis, characterization, and applications of ZnO nanowires. J. Nanomater. 2012, 1–2.

Zhang, Q., Deng, Y., Hu, Z., Liu, Y., Yao, M., Liu, P., 2014. Seaurchin-like hierarchical NiCo2O4@ NiMoO4 core–shell nanomaterials for high performance supercapacitors. Phys. Chem. Chem. Phys. 16, 23451–23460.

CHAPTER 4

Fabrication strategies for functionalized nanomaterials

Nisar Ali[a], Muhammad Bilal[b], Adnan Khan[c], Farman Ali[d], Hamayun Khan[e], Hassnain Abbas Khan[f], and Hafiz M.N. Iqbal[g]

[a]Key Laboratory for Palygorskite Science and Applied Technology of Jiangsu Province, National & Local Joint Engineering Research Center for Deep Utilization Technology of Rock-salt Resource, Faculty of Chemical Engineering, Huaiyin Institute of Technology, Huaian, China
[b]School of Life Science and Food Engineering, Huaiyin Institute of Technology, Huaian, China
[c]Institute of Chemical Sciences, University of Peshawar, Peshawar, Khyber Pakhtunkhwa, Pakistan
[d]Department of Chemistry, Hazara University, Mansehra, Pakistan
[e]Department of Chemistry, Islamia College University, Peshawar, Pakistan
[f]Clean Combustion Research Centre, Division of Physical Sciences and Engineering, King Abdullah University of Science and Technology (KAUST), Thuwal, Saudi Arabia
[g]Tecnologico de Monterrey, School of Engineering and Sciences, Monterrey, Mexico

1. Introduction

The potential to manipulate and understand functional materials is fundamental in the development of science and technology. In this context, research is underway around the globe to exploit functional materials, and their economic and environmental applications (Joshi et al., 2015). Functional materials are a group of engineered and advanced materials designed and synthesized for some specific function with proper surface morphology and tailored properties. The tailored features can only be achieved by providing graded compositions in-depth, nano-, and microscale material, and properties (Kawasaki and Watanabe, 1997). Naturally functional materials are not new. Many functional natural materials, for example, bamboo, have been used for hundreds of years in construction and decoration (Gottron et al., 2014). Bever et al., in 1970 and 1971, reported the theoretical implications of functionally graded materials (FGMs) (Shen and Bever, 1972; Bever and Duwez, 1970). Yet, at that time, due to the lack of knowledge and very limited available fabrication processes, there was a delay in the development of functional materials (Kieback et al., 2003). After one decade, in 1984 the functional material term was utilized for the first time in Japan (Koizumi, 1997). It is interesting that the understanding of functional materials nowadays has improved due to the enhanced research to design and prepare materials with some tailored properties for applications in several domains such as bioengineering, aerospace, and nuclear industries. Fig. 1 explains the different applications of functional materials.

Herein, we discuss functional materials as hetero nanoparticles containing both inorganic and organic parts. The fascination of functional materials explains the properties and

Nanomaterials: Synthesis, Characterization, Hazards and Safety
https://doi.org/10.1016/B978-0-12-823823-3.00010-0

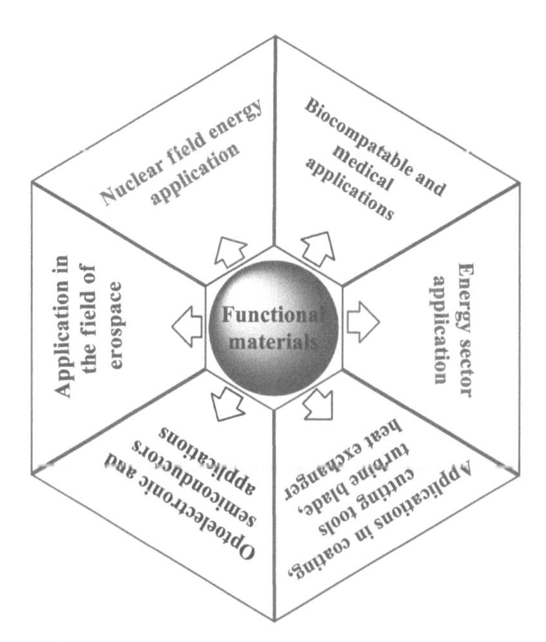

Fig. 1 Simple illustration of functional materials applications.

applications of the individual inorganic and organic parts, with the production of new characteristics, for example, desirable magnetic, electrical and optical (Ruiz-Hitzky et al., 2010). The remarkable characteristic of functional materials fabricated at the nano/micro scale is that they show clearly different characteristics and properties from their original individual components and materials (Nguyen and Zhao, 2015), which are promising in many practices (Sanchez et al., 2005; Chen et al., 2016). Therefore, functional materials integrating organic (polymers), inorganic (metals), and metal oxides may be among the best candidates to fabricate materials for a generation.

2. Construction of functional materials

The fabrication of functionalized nanomaterials, such as polymers and ceramics perceive the change in their amount and composition. To achieve properly tailored surface morphologies and engineered distinctive features, such as mechanical and physical grades in a definite direction, is one of the main advantages of functionalized nanomaterials as compared to other composites (Ma et al., 2011; Cannillo et al., 2007). A simple explanation of functional materials fabrication is given in Fig. 2. There are many organic species that may be suitable choices for the construction of functionalized nanomaterials. In general practice, the surface of inorganic nanoparticles usually needs functionalization to adjust the nanoparticle and organism interactions (Jha et al., 2013; Albanese et al., 2012; Verma et al., 2008). To effectively decrease the toxicity and improve the properties of the

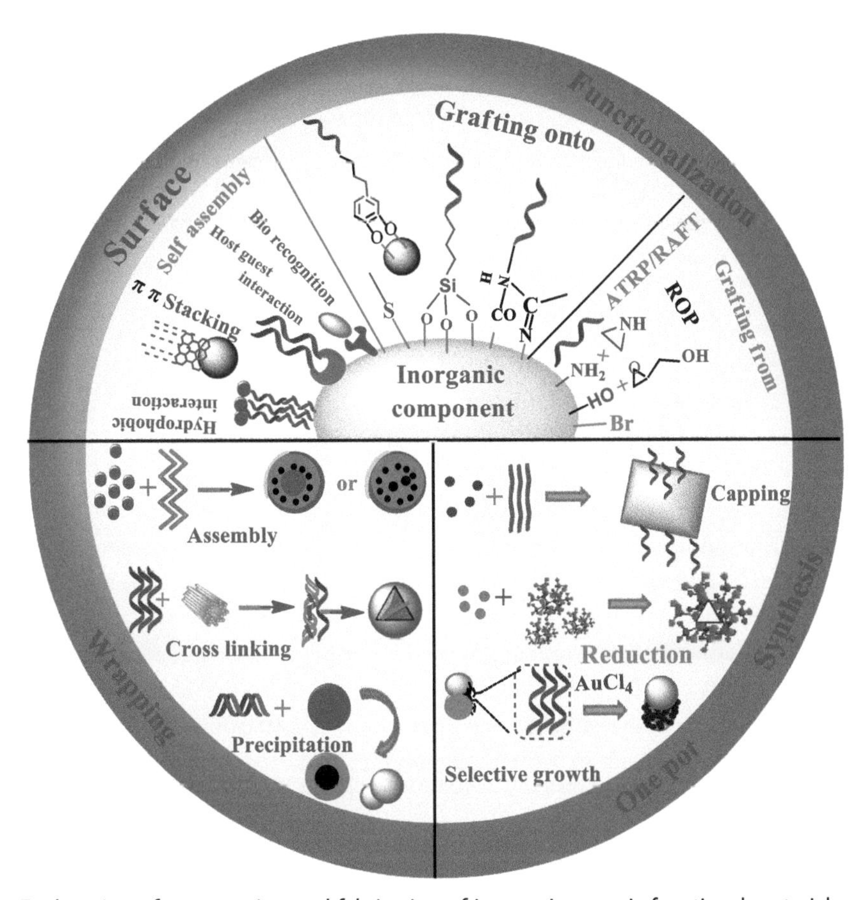

Fig. 2 Explanation of construction and fabrication of inorganic-organic functional materials.

functionalized materials, some organic or inorganic materials could be utilized to make a shell on the surface of the nanoparticle (Kim et al., 2010).

2.1 Organic component

Organic components of the functional material may possess excellent flexibility to properly tailor the functionality of the material and improve its stability, dispersibility, and compatibility. The functionalization of the nanoparticle surface with some suitable organic component can impart nanoparticles with versatile properties and efficient site-specificity, which depend on the properties and nature of the nanoparticles (Burda et al., 2005; Xu et al., 2006).

In the case of polymers for the surface functionalization of synthesized nanoscale hybrid materials, some widely used small organic molecules are 3-aminopropyltriethoxysilane (APTES), 3-aminopropyl-trimethoxysilane (APTMS), and 3-(tri hydroxyl silyl)-propyl methyl phosphonate carboxy ethyl silane triol, to be fastened onto the silica nanoparticles surface having alkoxysilane functional groups. The amino groups of 3-aminopropyltriethoxysilane can be used as a binding source for the introduction of some more functions (Huang et al., 2011a). The amino group at the terminal position can be engaged to link or combine to the polymers having *N*-hydroxy succinimide (NHS) esters or carboxylic groups. Small organic molecules having thiol groups are widely applied for the surface functionalization of gold (Au) nanoparticles or semiconductors through covalent bonding (Sandhu et al., 2002; Joseph et al., 2006; Lee et al., 2002).

Three types of small molecules with the epoxy group—glycidyl methacrylate and 3-mercaptoethanoic acid having a carboxylic group, amin ethane thiol having amino group, and 2-mercaptoethanol containing hydroxyl groups—were widely introduced for the modification of iron oxide and gold (Au) nanoparticles (Li et al., 2015a). After modification, the resultant iron oxide Fe_3O_4 and gold (Au) nanoparticles have different surface morphologies and properties that were used to investigate for future demulsification applications and tissue engineering information. Some small molecules work as a reductant after conjugation with polymers. Dimethyl amino propyl amine was attached to hyaluronic acid (HA) backbone through amide bonds followed by the synthesis of amphiphilic dimethyl aminopropyl amine block copolymers conjugated with (hyaluronic acid-polycaprolactone) (HA-PCL) form micelles. After the reduction of $HAuCl_4$ with dimethyl aminopropyl amine DAA, gold (Au) nanoparticles in proper core-shell morphology were produced, in which the polymer makes a proper shell on the surface of gold (Au) nanoparticles (Han et al., 2016). Polymers and many other small organic components combine to coat the surface of the nanoparticles properly to give some fruitful applications.

Polymeric ligands of different natures, like block copolymers, linear homopolymers, branched polymers, provide more diverse designs and structures to nanoscale hybrid materials. In the demulsification field, some block copolymers show the capacity to make a layer along the surface of hybrid organic and inorganic nanoparticles, which is affected by tunable surface properties, particle size, and amphiphilic properties (Li et al., 2015b). In addition to the surface properties, this particularly influences the composition of amphiphilic polymers and the correlation with nanoparticle interactions, which ultimately determines the nanoparticle transport to the oil-water interface (Lynch and Dawson, 2008; Tenzer et al., 2013). This suggests that the organic portion of the nanoscale hybrid material is more value. Furthermore, in recent years, polymers leading to materials of advanced interfaces with super-wetting properties have emerged and been explored in new research fields and the developing interface sciences (Behzadi et al., 2017). All these views contribute to the synthesis and design of appropriate nanoscale hybrid and assembly materials. The hydrophilicity and excellent solvent properties of polyethylene glycol (PEG) make it a more effective polymer, which is usually employed in the synthesis of surface-active nanoparticles. Hydrophilic polyethylene glycol (PEG) is primarily applied for nanoparticle functionalization to gain high colloidal stability (Walkey et al., 2012). It is proven that gold (Au) nanoparticles functionalized the polymer of carboxy betaine with longer life than polyethylene glycol (PEG) functionalized analogs (Yang et al., 2014a).

Many earlier research investigations demonstrated that inorganic core and polymeric components can be functionalized with the layering of different materials, including zirconium hydrous oxide, yttrium basic carbonate, and silica, either by direct surface reaction on the surface of core particles or controlled surface precipitation. The core-shell surface morphology of the nanoparticles exhibits properties that are substantially different from their template's core, such as increased stability, surface compositions, high surface area, and varying magnetic and optical characteristics, making these materials more attractive from both a technological and scientific point of view. The applications of such tunable nanoscale hybrid materials are very diverse, including drug delivery, coatings, catalysis, and composite materials for environmental applications (Bever and Duwez, 1970; Kieback et al., 2003; Zou et al., 2008; Rosaria et al., 2011). One of the environmental applications of magnetic composite core-shell nanoparticles is to remove zinc from aqueous media (Zhang et al., 2008). Redox-, pH-, enzyme-, or temperature-responsive polymers could be utilized to design and construct some smart nanoscale hybrid materials adapted for the treatment of microemulsion environments. A polymer containing a carboxylic group on its surface bonds can be applied for the synthesis of surface-active polymer, while polymer with a linker of acid-cleavable hydrazone can be used for the preparation of pH-sensitive nanohybrid materials (Naseem and Tahir, 2001; Srivastava et al., 1989).

One of the emerging areas of research is nanocomposites of the block copolymer and both experimental and theoretical research is in progress to address the fundamental problems related to nanocomposites of block copolymer systems. In the literature, many reviewers report various aspects of block copolymers nanocomposites (Bektas et al., 2004; Emadi et al., 2013; Shim and Kwon, 2012; Schadler et al., 2007; Tjong, 2006; Vaia and Maguire, 2007; Balazs et al., 2006; Bockstaller et al., 2005; Forster and Plantenberg, 2002; Balazs, 2007; Darling, 2007; Hamley, 2003; Haryono and Binder, 2006; Lazzari and Lopez-Quintela, 2003; Matsushita, 2007; Smart et al., 2008; Ji et al., 2012; Zhang et al., 2013a,b, 2015a; Wen et al., 2013; Li et al., 2015c,d; Lin et al., 2015a; Ganesan et al., 2010; Pyun, 2007; Haraguchi, 2007). Considerable research work is in progress in controlling the distribution of nanoparticle and their location in the body of block copolymer (Crosby and Lee, 2007; Chen et al., 2015). Despite such valuable progress, the applications of block copolymer nanocomposites may require an economically capable continuous processing of block copolymer nanocomposites materials on a large scale with controlled surface morphologies, structure, and tunable properties

(Jiang et al., 2017; Wu et al., 2012). The incorporation of nanoparticles in the body of block copolymer matrices can be investigated and discussed under various controlling factors; furthermore, some small changes in the design parameters may impose direct effects on the resulting block copolymer nanocomposites. Thus, it is important to understand properly the long-term effects of several components in designing and synthesis of block copolymer nanocomposites.

3. Inorganic component

Inorganic nanoparticles as a component with a particle size of 1–100 nm and a wide range of surface makeups can be designed and prepared with both chemical and physical strategies (Bombalski et al., 2006). As well as nontoxicity being essential, for the biocompatibility of these inorganic nanoparticles it is crucial that the surface should be modified or functionalized to gain proper dispersibility and stability inside body fluids for biomedical applications (Shenhar et al., 2005; Kao et al., 2013). Fortunately, stimuli-responsive or intrinsic hydrophilic inorganic materials exhibit a high degree of stability and efficiency in the destabilization of oil/water emulsions (Topham et al., 2011; Adhikari et al., 2008; Burda et al., 2005). Copper nanoparticles with hydrophilic characters (Liu et al., 2019), $Cu(OH)_2$, palygorskite (Yang et al., 2014b), titanium dioxides, tungsten oxide (Zhu and Chen, 2017), zeolite (Hinde et al., 2017), and NiOOH (Geng et al., 2007) loaded materials were reported in different shapes and structures and found to be more efficient in terms of oil/water separation. Nevertheless, the design and preparation of tunable inorganic nanoscale material need complex operating systems and instruments. Therefore, to combine a universal approach and some cheaper reagents and follow an easy operation for the fabrication of stable hydrophilic nanoscale materials for efficient oil/water separation is a vital area for researchers.

Several types of inorganic nanoparticles, including C, silicon oxide, gold, quantum dots, iron oxide, and rare earth nanoparticles have shown clear advantages in the area of biomedicine, environment, and catalysis. In this part of the review, the synthetic idea of nanoparticles will be introduced to explain the preparation, functionalized synthesis, and variety of applications. Among chemical routes, synthesis using organic solvents at high temperatures, seed-mediated growth, synthetic methods (solution phase), and reaction in poles are most widely used (Chauhan et al., 2011). The final surface engineering and sound structure of nanoparticles will be modified by growth steps and nucleation. Usually, the encapsulation of the nanoparticles surface with some capping and coating ligands, such as polymers, and surface-active agents are required to explain the development of particles at the nano and micro scale. The mutual interactions of a variety of nanoparticles and polymers and ligands will significantly modify their speed of respective growth, producing several sizes and forms, which will heavily affect the penetration and circulation states of nanoparticles (Yang et al., 2012; Li et al., 2012; Tang et al., 2012). The design and synthetic techniques of some inorganic nanoparticles with proper and controlled surface morphologies are briefly discussed below.

3.1 Silica nanoparticles

With their advantages of porous structure, controlled engineering and morphology, biocompatibility, and easy and facile functionalization of surface, silica nanoparticles stand as worthy candidates for a diversity of applications, especially in the environment, because of their limited capability to conduct several types of loads. Porous and amorphous silica particles both at the nanoscale (1–100 nm) and microscale (>100 nm) are highly attractive for demulsification applications (Wu et al., 2013; Dong et al., 2019). Controlled hydrolysis may be the main synthesis route of the organosilane (tetraethyl orthosilicate (TEOS)) precursor particles, at the same time the geometry and surface morphology could be adjusted and controlled by variations in the reaction reagent concentration, reaction time, temperature, and also by the addition of various kinds of surfactants. To prepare multifunctional and monodispersed silica nanoparticles (2–50 μm) a well-known method is the Stöber method (Stober et al., 1968; Laurent et al., 2008; Reddy et al., 2012). Mesoporous silica nanoparticles can be prepared through the Stöber method with some changes by the addition of coping agents and surfactants while using the growth method of ammonia catalyst through condensation surface reaction. Silica nanoparticles (MCM-41) with two-dimensional (2D) ordered hexagonal mesoporous structure were synthesized through the Stöber method (Ling and Hyeon, 2013; Abareshi et al., 2010). Using the same Stöber method, utilizing pluronic F127 and cetyltrimethylammonium bromide (CTAB), surfactants were used for the preparation of silica nanoparticles (MCM-48) in three-dimensional (3D) geometry (Zhou et al., 2009). Silica nanoparticles with a hollow structure could be designed and constructed by the application of soft templates such as vesicle, micelle, and microemulsion (Choi et al., 2007; Karaca et al., 2015; Qiu et al., 2005; Ge et al., 2009).

Although it is crystal clear that the soft template method is more powerful, to obtain silica nanoparticles requires a wide range of size distribution, while the main problem is usually that an aggregation occurs in solution. The same templating technique can be employed for the fabrication of defined hollow silica nanoparticles and the templates can be washed out using solvent or template calcination (Liu et al., 2015a). It is worth mentioning that an etching strategy to scale selectively was designed for silica nanoparticles with an entirely hollow structure. Firstly, multilayer typical core/shell silica nanoparticles were designed and fabricated, obtaining a hollow nanoparticle after the middle layer or core etching of the original structures. In silica, nanoparticles having three layers were prepared, in which the inner core and outer layer constituents were condensed

TEOS silica, and composite siloxane made up the middle part, which was hydrolyzed from *N*-[3-(trimethoxysilyl) propyl] ethylenediamine and TEOS mixture (Yang et al., 2015a). A nanorattle structure was achieved after the less compact middle layer etching. Silicon oxide nanoparticles with a PVP shell on the surface can be switched to hollow structures after treatment with alkali (Hyeon, 2003; Hyeon et al., 2001; Kim et al., 2009). This is called "surface protected etching" and this technique makes the operation of core-shell nanoparticles selective etching easy.

In addition, homogeneous silica nanoparticles were prepared through the Stöber method. This method work based on the inner part or the core, and selective etching without any surface protection, thus, results in nanorattles or hollow nanospheres (Xia et al., 2003; Klokkenburg et al., 2004). Silica nanotubes with one-dimensional (1D) geometries could also be designed and prepared using templating methods (Zhou et al., 2010; Korth et al., 2006). Nanorods of silica with varying ratios were produced by systematically changing the reaction reagent concentration with the assistance of cetyltrimethylammonium bromide (CTAB) (Yang et al., 2015b). It is noted that silica nanoparticles are degradable, although they may take a long time to degrade because of the complex silicate network (Ye et al., 2010). Different routes have been adopted to prepare biodegradable silica and increase the rate of biodegradation (Park et al., 2009; Peiris et al., 2011; Arami et al., 2015; Ruiz et al., 2013; Ma et al., 2015). The entrapment method was used to modify melamine sponges with silica dodecyltrimethoxysilane (SiO_2-DTMS) showing high water absorption capacity (Wang et al., 2014, 2017).

3.2 Iron oxide Fe_3O_4 nanoparticles

Iron oxide Fe_3O_4 nanoparticles with extraordinary magnetic domains have been reported to have great potential due to their external response and separation. It is believed that materials with a strong magnetic domain can be reused and recycled using external magnets and can have potential applications to apply for large-scale cleaning of oil spills. In the current era, there are numerous materials with magnetic properties, like Fe_3O_4, that are of great interest because of their low cost, nontoxic characteristics, and strong magnetic response. There are many synthetic procedures proposed for the effective synthesis of Fe_3O_4 nanoparticles, such as chemical, physical, and microbial (Wu et al., 2015, 2016; Dan et al., 2016). There are various methods to prepare Fe_3O_4 nanoparticles like the coprecipitation method (Ge et al., 2007a; Zhang et al., 2018a), the template method (Chiu et al., 2018), the solution casting technique (Sun et al., 2017), and so on (Song et al., 2013). Among all these, coprecipitation methods is the most simple synthetic method, in which iron oxide Fe_3O_4 nanoparticles are designed and prepared through ferric salt and ferrous in 1:2 stoichiometric ratios in the presence of a base in aqueous solution (Na et al., 2012).

Iron oxide Fe_3O_4 nanoparticles' growth rate can be tailored by varying the experimental conditions, i.e., acidity, temperature, and iron strength, but it is experimentally proved that the size distribution control seems weak. The superiority growth rate control strategy relies the production on a large scale, which may advantageous for promising applications. The solvothermal/hydrothermal route for the preparation of iron oxide Fe_3O_4 nanoparticles is carried out using high temperature and pressure in autoclaves and can produce iron oxide Fe_3O_4 nanoparticles in engineered shape and morphology (Bever and Duwez, 1970; Kieback et al., 2003; Basuki et al., 2013a). The hydrothermal method followed by a reduction reaction can lead to the production of nanorods and nanorings of iron oxide Fe_3O_4 with unique surface morphology. Another process, called the polyol process, was employed to prepare iron oxide Fe_3O_4 nanoparticles, in which polyol was used as the solvent, reducing agent, and stabilizer (Basuki et al., 2014a). One of the main tactics is the thermal decomposition to prepare well-designed and monodispersed iron oxide Fe_3O_4 nanoparticles. In this process, the breakdown of iron and organic precursor particles takes place at about 300°C in suitable solvents and surface-active agents. The control size and surface architecture of iron oxide nanoparticles may be adjusted using proper precursor, temperature, and a solvent with head covering ligands. Using the hydrothermal decomposition method a nanotube of iron oxide Fe_3O_4 from 20 to 160 nm in size can be prepared in which iron acetylacetone is mixed in benzyl ether and oleic acid is used as a surface modifier (Basuki et al., 2013b). The main drawback of the thermal decomposition method is the use of organic solvent at a high temperature, which is not environmentally friendly. One-dimensional iron oxide Fe_3O_4 nanoparticles exhibit unique surface properties and applications (Dong et al., 2011).

One-dimensional iron oxide Fe_3O_4 can be induced through the assembly process of a direct magnetic dipole, in which the ultrasonic irradiation or external magnetic field is usually applied, and many research efforts have been applied to maintain the prepared one-dimensional structure after the removal of external force (Wang et al., 2013a). The nanoscale particle size could be stable after the capping or coating of inorganic iron oxide Fe_3O_4 nanoparticle surface with some suitable polymers (Zhang et al., 2015b). In addition, one-dimensional iron oxide Fe_3O_4 nanoworms can be prepared by coprecipitation with dextran—the surface of iron oxide, and hydroxyl group's hydrogen bonding from dextran played a vital role (Zhao et al., 2015a). Table 1 explains the different fabrication approaches to functional materials.

4. Fabrication art of nanoscale functional materials

For the engineering design and synthesis of nanoscale hybrid material, a variety of processes and methods have been introduced and developed. Among them, there are three prominent types, which include one-pot synthesis, surface functionalization, and wrapping. In this section of the chapter, we present a comprehensive summary of all possible

Table 1 Construction routes of organic/inorganic nanoscale hybrid materials.

Organic component	Inorganic component	Fabrication technique	Morphology	Reference
Poly(glycidyl methacrylate) (PGMA)/ethylene diamine	SiO_2 nanoparticles	Atom transfer radical polymerization (ATRP)	Core-shell	Huang et al. (2011b)
Hydrazide-functionalized poly-GMA	Fe_3O_4	Surface-initiated atom transfer radical polymerization (SI-ATRP)	Core-shell	Cao et al. (2014)
Poly(glycidyl methacrylate-poly(ethylene glycol) methyl ether methacrylate) (SPIONs-P(GMA-PEGMA))	Fe_3O_4 nanoparticles	Atom transfer radical polymerization (ATRP)	Magnetic core nanoparticles	Huang et al. (2012a)
	Fe_3O_4/SiO_2	Precipitation	Core shell	Kokate et al. (2013)
Poly(ethylene glycol) methyl ether methacrylate (PEGMA)/iron oxide magnetite	Fe_3O_4 nanoparticles	Surface-initiated living radical polymerization/ grafting	Magnetic core shell nanoparticles	Ohno et al. (2013)
Poly(poly-(ethylene glycol) mono methacrylate)	Fe_3O_4 nanoparticles	Postatom transfer radical polymerization		Lu et al. (2012)
Poly(poly(ethyleneglycol) monomethacrylate)	Fe_3O_4 nanoparticles	Atom transfer radical polymerization	Core-shell	Fan et al. (2007)
Zwitterionic 3-dimethyl (meth acryloyloxyethyl) ammonium propane sulfonate (DMAPS)		Surface-initiated atom transfer radical polymerization (ATRP)		Hu et al. (2012)
Poly(ethylene glycol) monomethacrylate	Fe_3O_4 nanoparticles	Grafting	Core-shell	Qin et al. (2010)
Polystyrene (PS)	Fe_3O_4 nanoparticles	Reversible addition-fragmentation chain transfer (RAFT) polymerization	Raspberry like	Tian et al. (2011)
Poly(acrylic acid) (PAA)		Reversible addition-fragmentation chain transfer (RAFT)	Open nanoshell	Hong et al. (2009)

Table 1 Construction routes of organic/inorganic nanoscale hybrid materials—cont'd

Organic component	Inorganic component	Fabrication technique	Morphology	Reference
Poly-(4-vinylpyridine) (P4VP)		Surface reversible addition-fragmentation chain transfer (RAFT) polymerization	Core-shell	Liu et al. (2010)
pH-responsive poly diethyl aminoethyl methacrylate	SiO_2 nanoparticles	Surface reversible addition-fragmentation chain transfer (RAFT) polymerization	Core-shell	Huang et al. (2012b)
Poly(2-(diethylamino)ethyl meth acrylate)-poly(oligo (ethylene glycol) methacrylate)-coated (SNT-PDEAEMA-POEGMA)	SiO_2 nanoparticles	Reversible addition fragmentation chain transfer (RAFT) polymerization	Core shell nanorods	Zhou et al. (2015)
Poly(2-diethylaminoethyl methacrylate) (PDEAEMA)/ mesoporous silica nanoparticles		Surface-initiated atom transfer radical polymerization (ATRP)/ reversible addition-fragmentation chain transfer (RAFT) polymerization	Core-shell	Huang et al. (2012c)
Multihydroxy hyperbranched polyglycerol (HPG)	MoS_2 nanoparticles	Surface-initiated ring-opening polymerization	Nanosheets	Huang et al. (2017) and Wang et al. (2016a)
Poly(oligoethylene glycol)/ MoS_2		One-pot solvothermal	Nanosheets	Wang et al. (2015a)
Poly(oligo(ethylene glycol) methacrylate-methacrylic acid) P(OEGMA-MAA)	Fe_3O_4 nanoparticles	Atom-transfer radical polymerization	Core shell	Lutz et al. (2006)

Continued

Table 1 Construction routes of organic/inorganic nanoscale hybrid materials—cont'd

Organic component	Inorganic component	Fabrication technique	Morphology	Reference
Poly(oligoethylene glycol acrylate)/Fe_3O_4	Fe_3O_4 nanoparticles	In situ coprecipitation polymerization	Core shell	Basuki et al. (2014b)
Poly(ethylene glycol)	Calcium phosphate nanoparticles		Core shell	Mi et al. (2016)
Doxorubicin	$CaCO_3$ nanoparticles	In situ mineralization approach	Flower like	Min et al. (2015) and Karagoz et al. (2014)
Graphene oxide anchoring iron oxide (RGI) with a polyethyleneimine (PEI) modification		One pot method	Core shell	Du et al. (2017)
Concentric and eccentric poly(acrylic acid) (PAA)	Fe_3O_4 nanoparticles		Janus, Core shell and rode shape	Li et al. (2013, 2016) and Lin et al. (2015b)
Polycarboxybetaine methacrylate (CBMA), pCBMA-coated Fe_3O_4	Fe_3O_4 nanoparticles	Encapsulation		Zhang et al. (2011)
Polystyrene (FPS)	Fe_3O_4 nanoparticles	Solvothermal/ emulsion polymerization	Core shell	Ge et al. (2007b)
(Sodium alginate and hyaluronic acid, chitosan), graphene oxide (GO), polysaccharides	Fe_3O_4 nanoparticles	Simple layer-by-layer technique	Core shell	Deng et al. (2016)
Hyaluronic acid nanoparticles (HANPs), ingle-walled carbon nanotubes (SWCNTs)		Encapsulation		Wang et al. (2016b)
Ethyl cellulose (EC)	Fe_3O_4 nanoparticles	Grafting	Core shell	Peng et al. (2012)
Methyloxy-poly(ethylene glycol)-*block*-poly [dopamine-2-(dibutylamino) ethylamine-L-glutamate] (mPEG-*b*-P(DPA-DE)LG)		Typical ring-opening polymerization/ grafting	Flower like	Yang et al. (2016)
Poly(lactic-*co*-glycolic acid)	Fe_3O_4 nanoparticles	Fabrication by encapsulation		Ye et al. (2014)

Table 1 Construction routes of organic/inorganic nanoscale hybrid materials—cont'd

Organic component	Inorganic component	Fabrication technique	Morphology	Reference
Heterobifunctional amphiphilic triblock copolymers R (R=methoxy or folate (FA))-PEG(114)-PLA(x)-PEG(46)-acrylate		Double emulsion polymerization	Wormlike vesicles	Yang et al. (2010)
Polystyrene	Fe_3O_4/SiO_2 nanoparticles	Combined process of miniemulsion and sol-gel reaction	Janus like	Wang et al. (2013c)
Polystyrene and silica		Mini emulsion/one-pot colloidal reaction strategy	Janus like	Wang et al. (2011)
Glucose-grafted superparamagnetic (G-g-SNPs)		Supercooling self-assembling method/fabrication	Core shell	Zheng et al. (2016)
Melamine foam/poly (vinylidene fluoride-hexafluoropropylene) (PVDF-HFP)	Fe_3O_4 nanoparticles	Single-step dipping method		Li et al. (2018)
Polyvinylpyrrolidone-coated iron oxide	Fe_3O_4 nanoparticles	One-step modified polyol synthesis	Core shell	Palchoudhury and Lead (2014)
N,N'-methylene bis acrylamide	Fe_3O_4 nanoparticles	Solvothermal/miniemulsion polymerization	Core shell	Xu et al. (2004)
	Fe_3O_4/SiO_2 nanoparticles	Stepwise solution-phase interface deposition approach with sol-gel chemistry/surfactant-involved coassembly process	Core shell	Yue et al. (2015)
Resorcinol formaldehyde	$Fe_3O_4/Si_2/TiO_2$ nanoparticles	Solvothermal/sol-gel coating	Core shell	Wang et al. (2015b)

Continued

Table 1 Construction routes of organic/inorganic nanoscale hybrid materials—cont'd

Organic component	Inorganic component	Fabrication technique	Morphology	Reference
Carbon	Fe_3O_4/Ag nanoparticles	One-pot solvothermal method/*situ* reduction	Core shell	An et al. (2012)
Carbon	$Fe_3O_4/Pd/SiO_2$ nanoparticles	Solvothermal/reduction	Core shell	Sun et al. (2014)
Carbon cloth/manganese dioxide	MnO_2/Fe_2O_3		Sheets	Yang et al. (2014c)
	TiO_2/MnO_2	Coating	Core shell	Lu et al. (2013)
Polydivenyl benzene/polystyrene	SiO_2 nanoparticles	Seed emulsion polymerization	Snowman-like Janus particles	Liu et al. (2017)
Polystyrene poly(4-vinylpyridine) (PS-P4VP)/1,5-diiodopentane, DIP	Au/SiO_2 nanoparticles	Emulsion solvent-evaporation method	Janus	Deng et al. (2015)
Poly(2-methacryloyloxyethyl ferrocenecarboxylate)-*block*-poly(*N*-isopropylacryamide)/Au, PMAEFc-*b*-PNIPAM@Au		Two-step successive in-situ reduction	Core shell	Zhang et al. (2018b)
NaYF4:Yb/Tm/4-cyano-4[(dodecylsulfanylthiocarbonyl)sulfanyl]pentanoic acid (CDTPA), was first grafted onto the silica-coated UCNPs, UCNP@CDTPA		Electron transfer-reversible addition-fragmentation chain transfer polymerization (PET-RAFT)	Core shell	Bagheri et al. (2017)
Bi-functionalize mesoporous silica nanoparticles (MSN)/PHPMA and poly (2-diethylaminoethyl methacrylate) (PDEAEMA)		Surface-initiated atom transferred radical polymerization/reversible addition-fragmentation chain transfer (TRP/RAFT) polymerization	Core shell	Huang et al. (2012d)

Table 1 Construction routes of organic/inorganic nanoscale hybrid materials—cont'd

Organic component	Inorganic component	Fabrication technique	Morphology	Reference
Poly (methymeth acrylate-acrylic acid-divinely benzene)	Fe_3O_4 nanoparticles	Solvothermal/ soap-free emulsion polymerization/ precipitation polymerization	Janus/ Raspberry/ core shell	Ali et al. (2014, 2015a, b,c)
Poly(glycidylmethacrylate-methyl methacrylate-divinyl benzene)	Fe_3O_4 nanoparticles	Solvothermal/ precipitation polymerization	Core shell	Ali et al. (2015d)

design and synthetic strategies for nanoscale hybrid materials with some useful information and data about their performance and applications. To tailor properties and applications of hybrid materials, the surface functionalization of inorganic nanoparticles is an important tool in the field of nanotechnology. Not only can the surface of inorganic nanoparticles be functionalized with some organic or polymers but the polymer surface can also be fabricated with inorganic materials, which can give a versatile type of hybrid material with promising applications. In this part of the chapter, we explain the construction of functional materials through the surface functionalization of inorganic nanoparticles using different techniques such as grafting from, grafting onto, self-assembly strategies, wrapping, and one-pot synthesis.

4.1 Grafting from technique

"Grafting from" is a bottom-up technique in which the polymer grows through surface-initiated polymerization from the surface of the nanoparticles and produces a high level of grafting density. This is one of the foremost techniques in the fabrication of functional materials. In the grafting from technique, controlled surface-initiated polymerization is enabled by different initiators to fix firmly the tethered brushes of the polymer. This process leads to a controlled, versatile, and chemically robust polymer (Boyer et al., 2016; Matyjaszewski and Tsarevsky, 2009). The following subsection introduces different examples of functional materials constructed by the typical grafting from technique.

To fabricate functional materials effectively from polymer, surface-initiated controlled radical polymerization is more suitable for the controllable composition and architecture of the resulting material (Dong et al., 2011). In particular, surface-initiated atom transfers radical polymerization (ATRP) is a suitable method for the immobilization of

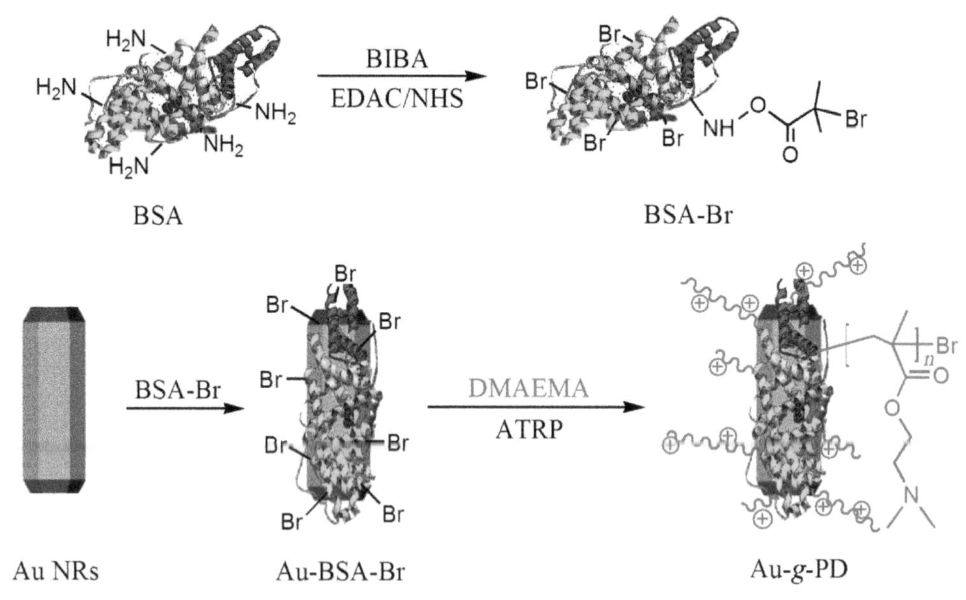

Fig. 3 Illustration of the fabrication processes of Au@PDs via ATRP. *(Reprinted from Yan, P., Zhao, N., Hu, H., Lin, X., Liu, F., Xu, F.J., 2014. A facile strategy to functionalize gold nanorods with polycation brushes for biomedical applications. Acta Biomater. 10, 3786–3794 with permission from Elsevier. Copyright © 2014 Acta Materialia Inc. Published by Elsevier Ltd.)*

initiators on the surface of the nanoparticles. The reaction processes such as monomers, catalyst, and solvent also affect the polymerization process. The effective immobilization of initiators on the nanoparticles' surface, the binding interaction, and the surface properties of the nanoparticles will also be considered. Ring-opening metathesis polymerization (ROMP) and reversible addition-fragmentation chain transfer polymerization (RAFT) are also effective processes for the controlled polymerization techniques to modify polymer.

The grafting from technique can be effectively used for the fabrication of PDMAEMA/Au functional material in different shapes and morphologies Fig. 3 (Yan et al., 2014). Silane chemistry helps in the conjugation of APTES on the silica nanoparticle surface by siloxane bonds. Fig. 4 explains the grafting of PDMAEMA on silica nanoparticles surface in different shapes like nanorods and spheres in different aspect ratios (Lin et al., 2015c) (Fig. 4). A two-step procedure was used for the proper coating of Fe_3O_4 magnetic nanoparticles with poly(PEGMEA). In the first step, which was tri thiocarbonate-based, a well-known chain transfer agent (CTA) having catechol functionality, was successfully used to firmly fix the surface of Fe_3O_4 magnetic nanoparticles (Fig. 5). The catechol group is very effective to anchor functional groups on metal oxides

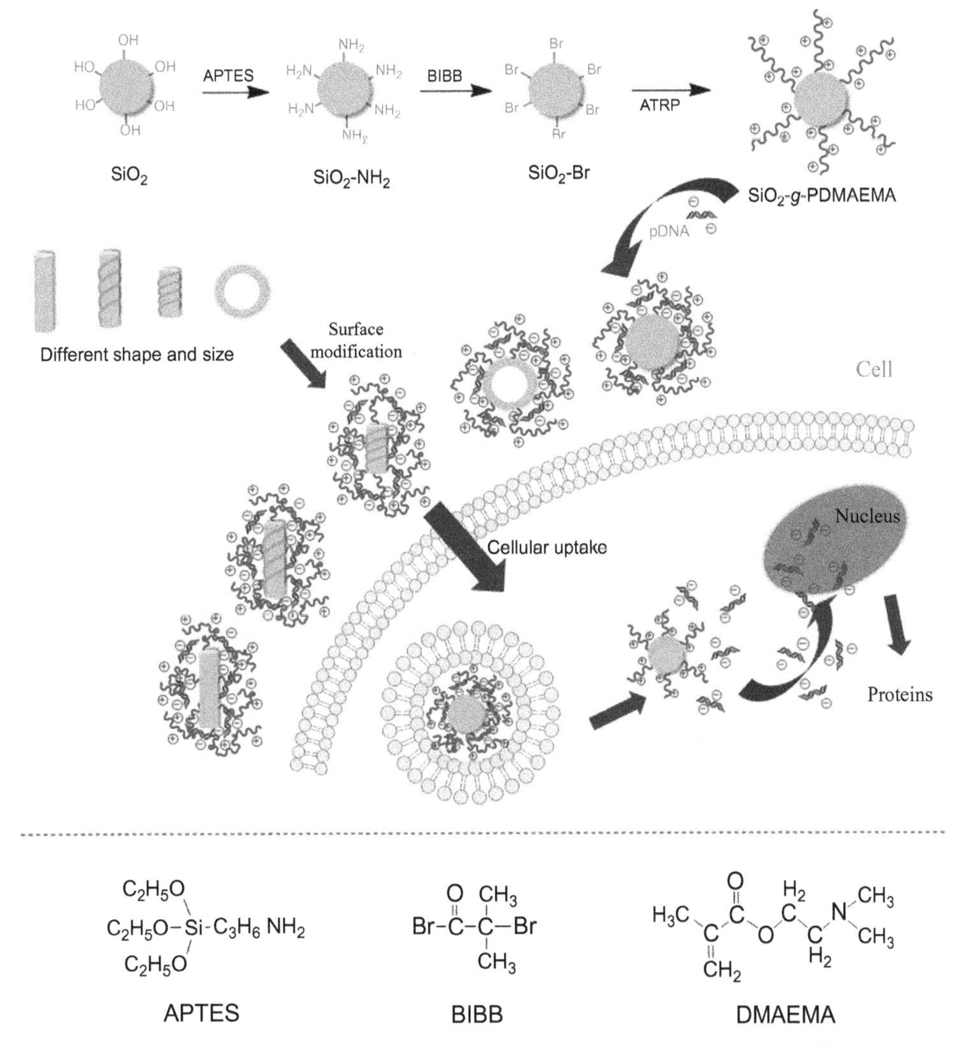

Fig. 4 Synthetic route of PDMAEMA/SiO_2 nanocomposites. *(Reprinted from Lin, X., Zhao, N., Yan, P., Hu, H., Xu, F.J., 2015c. The shape and size effects of polycation functionalized silica nanoparticles on gene transfection. Acta Biomater. 11, 381–392 with permission from Elsevier. Copyright © 2014 Acta Materialia Inc. Published by Elsevier Ltd.)*

and metal surfaces. Furthermore, the catechol group, having strong effects on different surfaces, has been investigated in the modification of planar surfaces and the surface of Fe_3O_4 magnetic nanoparticles with polymers. Surface-initiated RAFT polymerization was used in the next step, to obtain polymer shell on Fe_3O_4 magnetic nanoparticles surface (Oz et al., 2016).

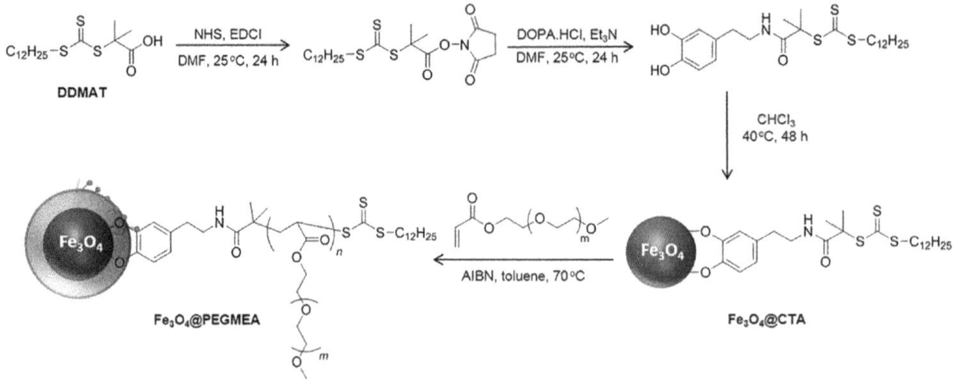

Fig. 5 Synthetic route of p(PEGMEA)@ Fe_3O_4 magnetic nanoparticles. *(Reprinted from Oz, Y., Arslan, M., Gevrek, T.N., Sanyal, R., Sanyal, A., 2016. Modular fabrication of polymer brush coated magnetic nanoparticles: engineering the interface for targeted cellular imaging. ACS Appl. Mater. Interfaces 8, 30, 19813–19826 with permission from American Chemical Society. Copyright © 2016 American Chemical Society.)*

The surface functionalization of inorganic nanoparticles using polymers leads to the preparation of a versatile type of nanoscale hybrid materials with compatible hydrophilic, hydrophobic, and amphiphilic surface properties. This kind of hybrid material may have excellent properties in the migration to the oil/water interface, and the separation of oil/water emulsions. Ali et al. described the encapsulation of iron oxide Fe_3O_4 with poly(glycidyl methacrylate-methyl methacrylate-divinyl benzene) block copolymer to prepare a proper core shell-like (Fe_3O_4@P(GMA-MMA-DVB)) iron oxide@poly(glycidyl methacrylate-methyl methacrylate-divinyl benzene) magnetic nanohybrid materials. The well-designed core-shell surface morphology was achieved using a solvothermal process followed by precipitation polymerization, in which the surface of nanoscale hybrid material contains the epoxy group as a main functional group. To get an interfacially active and amphiphilic surface, the epoxy group of Fe_3O_4@P (GMA-MMA-DVB) was successfully modified with an amino group by reaction with ethylenediamine (EDA) to finally prepared Fe_3O_4@P(GMA-MMA-DVB)/NH_2 nanoscale magnetic hybrid materials (Ali et al., 2015d) (Fig. 6). It is important to mention that some nanoscale hybrid materials with an asymmetric structure can offer unique properties, which are not in the scope of homogeneous nanomaterials.

4.2 Grafting onto technique

In the case of the "grafting onto" technique for the integration of inorganic nanoparticles with polymers, the simple process is grafting or coating of the polymers required onto the surface of the inorganic nanoparticles. Herein, the molecular weight of the

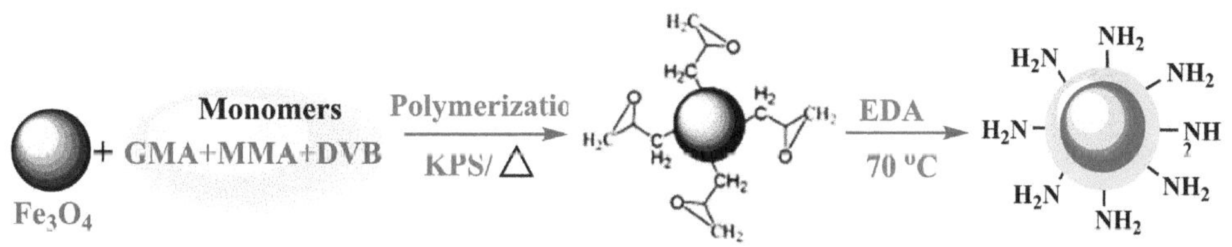

Fig. 6 Synthetic approach of amino modified Fe_3O_4@P(GMA-MMA-DVB)/NH_2 magnetic composite core shell microspheres. *(Reprinted from Ali, N., Zhang, B., Zhang, H., Zaman, W., Ali, S., Ali, Z., Li, W., Zhang, Q., 2015d. Monodispers and multifunctional magnetic composite core shell microspheres for demulsification applications. J. Chin. Chem. Soc. 62, 695–702 with permission from John Wiley and Sons. Copyright © 2015 The Chemical Society Located in Taipei & Wiley-VCH Verlag GmbH & Co. KGaA, Weinheim, Germany.)*

polymer should be fixed. Polymers that contain carboxyl groups as the main functional group can be easily grafted onto inorganic nanoparticles and also on amino groups-functionalized nanoparticles' surfaces. Recently, it has been noted that designing a desirable procedure for the reproducible synthesis of core shell-like morphology with the magnetic core can be started with accessible and inexpensive material. The solvothermal method is one of the easy and approachable routes for the synthesis of magnetic nanoparticles with tunable size monodispersity. The modified solvothermal processes were used for the synthesis of these magnetite particles by the reduction of iron chloride $FeCl_3$ in ethylene glycol (EG) at 200°C, in which tri-sodium citrate (Na_3Cit) was used as an electro-static stabilizer and sodium acetate as an alkali source. Na_3Cit contains strong carboxylate groups, which show a strong affinity for Fe^{+3} ions, which help in the desirable citrate groups attachment on the magnetic nanoparticle surface and also prevents them from agglomeration into a big particle. The uniform and tunable particle size can easily be obtained in the range of 10–400nm by changing the amount of Na_3Cit, or $FeCl_3$, or both of them. Furthermore, the most important property of these magnetic nanoparticles is their electrostatic repulsion effect of the ligand tri citrate, the prepared magnetic nanoparticles may be easily dispersed in alcohols, water, and tetrahydrofuran to form a dispersion leading to the subsequent surface modification or coating by polymers or other oxides. This strategy could be helpful in the coating and fabrication of magnetic nanoparticles with a protective shell, for example a layer of polymer (Ali et al., 2015c,d).

The same method was used with iron oxide Fe_3O_4 magnetite nanoparticles prepared using the solvothermal process. To obtain properly interfacially active material, the prepared Fe_3O_4 magnetite nanoparticles were encapsulated by poly(methylmethacrylate-acrylic acid-divinylbenzene) block copolymer using the precipitation polymerization method to get (Fe_3O_4@P(MMA-AA-DVB)) magnetic hybrid microspheres with proper

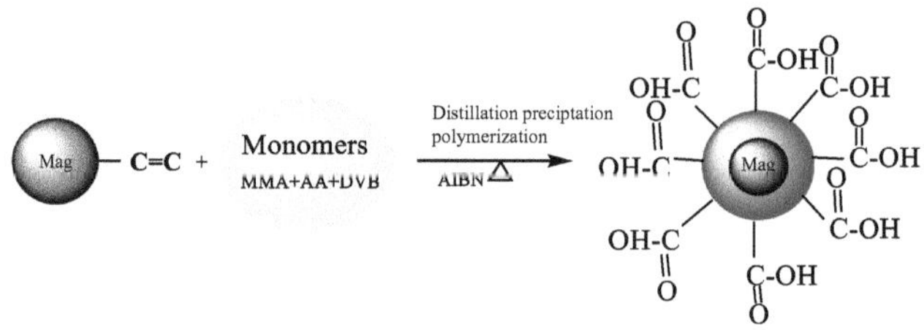

Fig. 7 Synthetic approach of Fe_3O_4@P(MMA-AA-DVB) magnetic core shell nano-scale hybrid material.

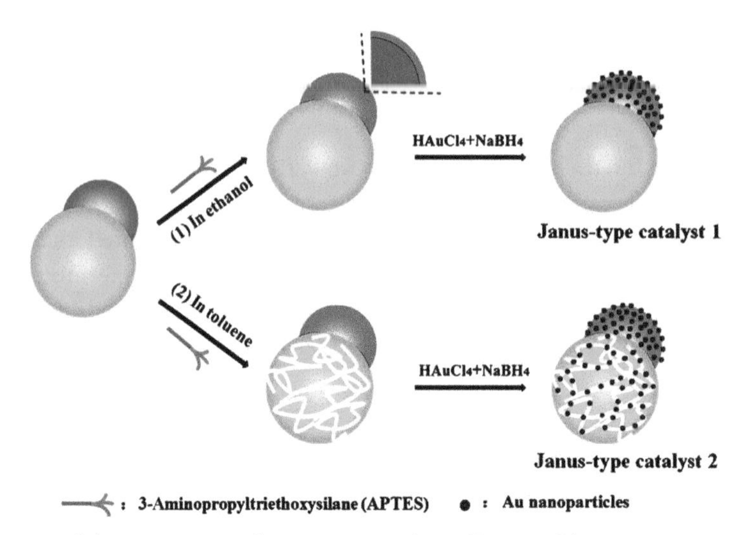

Fig. 8 Schematic of the preparation of Janus type catalysts. *(Reprinted from Liu, Y., Hu, J., Yu, X., Xu, X., Gao, Y., Li, H., Liang, F., 2017. Preparation of Janus-type catalysts and their catalytic performance at emulsion interface. J. Colloid Interface Sci. 490, 357–364 with permission from Elsevier. Copyright © 2016 Elsevier Inc.)*

core-shell morphology (Ali et al., 2015c) (Fig. 7). Liu et al. in 2017 reported seed emulsion polymerization for the preparation of two catalysts with Janus-type surface morphology, which contained SiO_2@PDVB/PS with snowman-like Janus particles. Then the surface of SiO_2@PDVB/PS was modified with 3-aminopropyltriethoxysilane (APTES) and newly NH_2-modified silica@PDVB/PS Janus particles were produced (Fig. 8) (Liu et al., 2017). Deng et al. (2015) presented the method of solvent/emulsion for the self-assembly of polystyrene-*b*-poly(4-vinyl pyridine) (PS-*b*-P4VP) diblock copolymer for the preparation of patchy Janus-type particles (Fig. 9), which bears amphiphilic characters having the capability of self-organization into superstructures.

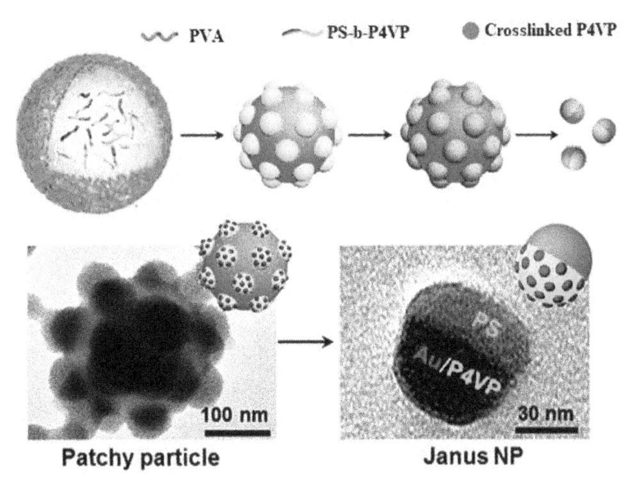

Fig. 9 Diblock copolymer polystyrene-poly(4-vinylpyridine) (PS-*b*-P4VP) self-assembly by an emulsion evaporation for the preparation of Janus nanoparticles. *(Reprinted from Deng, R., Liang, F., Qu, X., Wang, Q., Zhu, J., Yang, Z., 2015. Diblock copolymer based janus nanoparticles. Macromolecules 48, 750–755 with permission from American Chemical Society. Copyright © 2015 American Chemical Society.)*

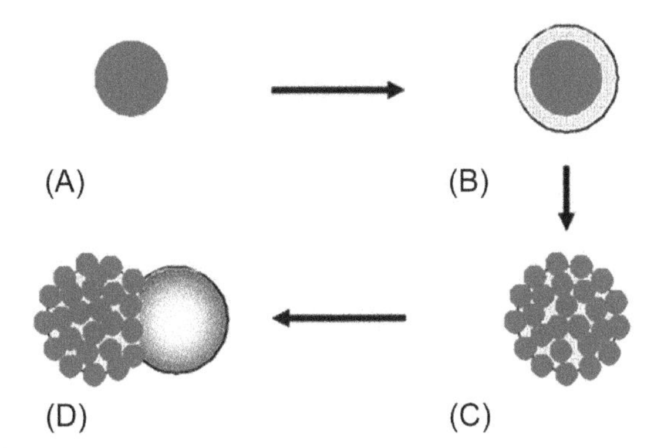

Fig. 10 Schematic procedure for preparation of asymmetrically nanoparticle-supported composite dumbbells. (A) Inorganic sphere. (B) Inorganic sphere covered with cross-linked polymer (core-shell particles). (C) Core-shell particles supporting nanoparticles. (D) Asymmetrically nanoparticle-supported composite dumbbells. *(Reprinted from Nagao, D., Goto, K., Ishii, H., Konno, M., 2011. Preparation of asymmetrically nanoparticle-supported, monodisperse composite dumbbells by protruding a smooth polymer bulge from rugged spheres. Langmuir 27(21) 13302–13307 with permission from American Chemical Society. Copyright © 2011 American Chemical Society.)*

Fig. 10 explains the preparation method of hybrid and functional Janus nanoparticles prepared using a soap-free emulsion polymerization method. In this process, the asymmetry of nanoparticle was supported and prepared in proper monodispersed and dumbbell-shaped architectures (Nagao et al., 2011). The method can be completed in three main steps, having two soap-free emulsion polymerizations (before and after the hetero coagulation). Initially, silica core was covered by poly(methyl methacrylate) (PMMA) cross-linked polymer via soap-free emulsion polymerization, followed by the hetero coagulation of core shell silica@PMMA nanoparticles. Finally, at different pH values, again soap-free polymerizations were conducted to generate a protuberance or bulge of polystyrene from the core shell charged silica nanoparticles. This method is very effective in the composition and dumbbell shape of the particles.

4.3 One-pot synthetic technique

One-pot synthesis is an efficient and simple approach for the construction of nanoscale organic/inorganic hybrid functional materials. In this process, a direct reaction in one step is used for the preparation of inorganic parts while the organic component works as surface capping material or template. Generally, the organic component plays a capping role and does not participate in any reaction. For example, the preparation of polymeric Janus-type hybrid materials is phase separation because of their tunable composition, size, and shape. Soap-free emulsion polymerization, selective cross-linking, and self-assembly are the main examples of phase separation. Ali et al. (2014, 2015a,b) successfully reported the preparation of (P(MMA-AA-DVB)/Fe_3O_4, Poly(methylmethacrylate-acrylic acid-divinylbenzene)/iron oxide) magnetic composite microspheres, which were designed and prepared with Janus and raspberry-like surface morphology, strong magnetic domains, and unique surface functionalities. The P(MMA-AA-DVB) microspheres were prepared via soap-free emulsion polymerization and used as a precursor for preparing Janus and raspberry-like composite microspheres via the solvothermal process (Figs. 11 and 12).

Fig. 13 shows the complete scheme and representation of the one-pot synthesis of 3-mercaptopropyltrimethoxysilane-functionalized mesoporous fibrous silica (SH-MFS) nanospheres followed by the subsequent adsorption of crystal violet (CV)

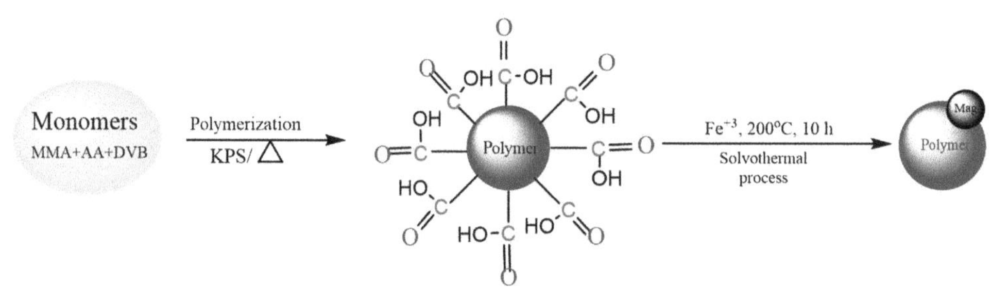

Fig. 11 Synthetic procedures of Janus type P(MMA-AA-DVB)/Fe_3O_4 magnetic composite microspheres. *(Reprinted from Ali, N., Zhang, B., Zhang, H., Zaman, W., Li, W., Zhang, Q., 2014. Key synthesis of magnetic janus nanoparticles using a modified facile method. Particuology 17, 59–65 with permission from Elsevier. Copyright © 2014 Elsevier B.V.)*

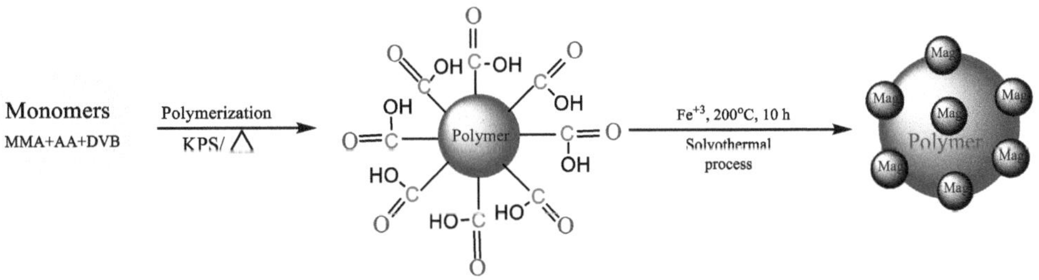

Fig. 12 Preparation of raspberry-like P(MMA-AA-DVB)/Fe_3O_4 magnetic composite microspheres.

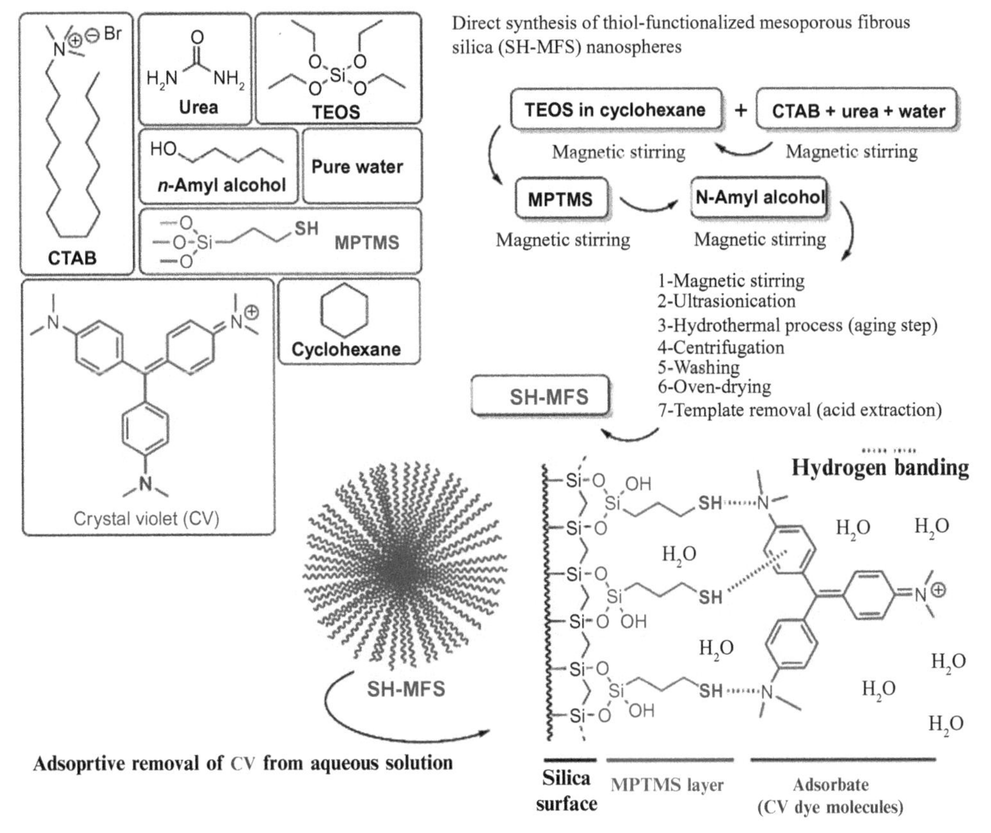

Fig. 13 A schematic representation of the synthesis of SH-MFS and its capability for removal of CV dye by its surface functional groups. *(Reprinted from Roghanizad, A., Abdolmaleki, M.K., Ghoreishi, S.M., Dinaric, M., 2020. One-pot synthesis of functionalized mesoporous fibrous silica nanospheres for dye adsorption: isotherm, kinetic, and thermodynamic studies. J. Mol. Liq. 300, 112367 with permission from Elsevier. Copyright © 2019 Elsevier B.V.)*

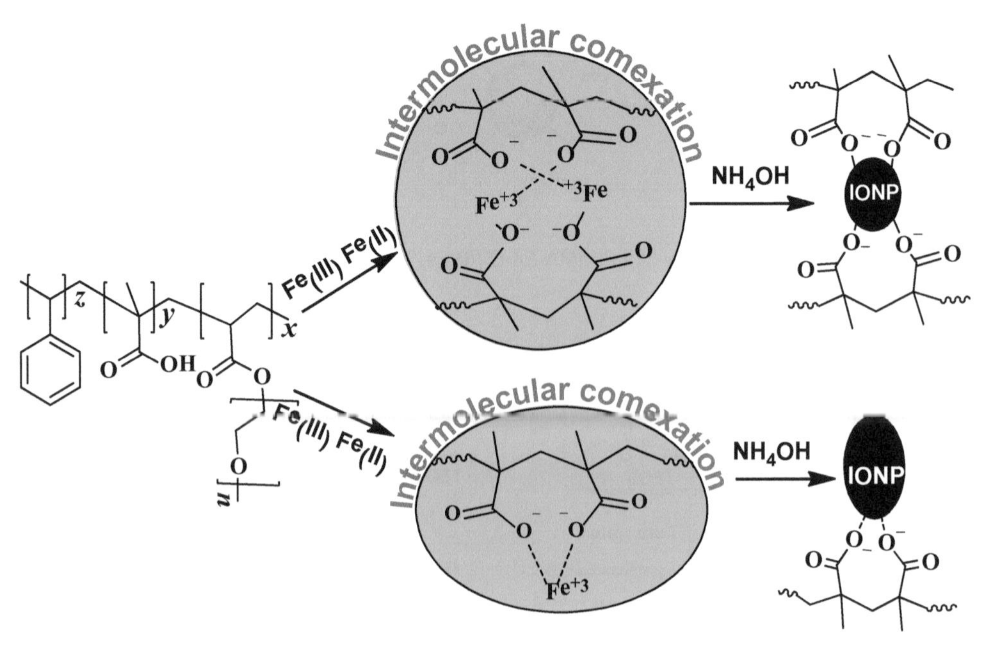

Fig. 14 Schematic explanation of iron complexations and iron oxide formation in POEGMA-PMAA-PST triblock copolymer aqueous dispersion. *(Reprinted from Karagoz, B., Yeow, J., Esser, L., Prakash, S.M., Kuchel, R.P., Davis, T.P., Boyer, C., 2014. An efficient and highly versatile synthetic route to prepare iron oxide nanoparticles/nanocomposites with tunable morphologies. Langmuir 30, 10493–10502 with permission from American Chemical Society. Copyright © 2014 American Chemical Society.)*

dye from aqueous media. This is a simple, short, and nonlaborious synthesis process, and the SH-MFS nanospheres could an economic fibrous silica-based adsorbent compared to the other fibrous silica. In addition, the organic/inorganic hybrid silica-based adsorbent shows benefits as a functional material (mercapto grafted at the surface having adsorption sites) with porosity (a micro and mesoporous network) (Roghanizad et al., 2020).

Recently, a useful report on a very simple and versatile in situ self-assembly synthetic method for magnetic/polymeric functionalized in nano morphologies, having rod-like, spherical micelles, vesicles, and worm-like micelles. Simultaneously, triblock copolymer chains (POEGMA-PMAA-PST) poly(oligo ethylene glycol methacrylate-methacrylic acid-styrene) were self-assembled and propagated using the approach of polymerization-induced self-assembly (PISA) (Fig. 14). Afterwards, the carboxylic acid groups of the copolymer were successfully used to make a complex mixture of iron ions (Fe^{+2} and Fe^{+3}). As a result, iron oxide Fe_3O_4 nanoparticles were produced within the main block of polymer, through the coprecipitation process of the Fe^{+2} and Fe^{+3} salts in an alkaline environment (Karagoz et al., 2014).

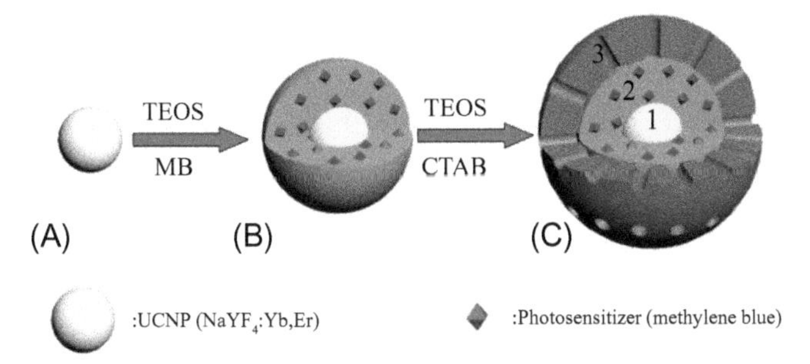

Fig. 15 Fabrication and NIR-triggered activation of one core plus two shells UCNP@SiO_2@mSiO_2 nanoparticles. *(Reprinted from Wang, H., Han, R.L., Yang, L.M., Shi, J.H., Liu, Z.J., Hu, Y., Wang, Y., Liu, S.J., Gan, Y., 2016c. Design and synthesis of core-shell-shell upconversion nanoparticles for NIR-induced drug release, photo dynamic therapy, and cell imaging. ACS Appl. Mater. Interfaces 8, 4416–4423 with permission from American Chemical Society. Copyright © 2016 American Chemical Society.)*

4.4 Self-assembly

Self-assembly is an effective, universally and spontaneously occurring bottom-up process (Elacqua et al., 2017). Therefore, self-assembly is regarded as a practical and promising approach to fabricate functional materials (Cui et al., 2016; Levchenko et al., 2018). Gan et al. presented the synthetic process of the double shell (one core plus two shells), i.e., core-shell/shell UCNP@SiO_2@mSiO_2 nanoparticles to conjugate properly the guest Ad via O_2-sensitive linkers (Wang et al., 2016c). Furthermore, a complete β-CD host-guest interaction resulting in photo responsive nanocomposites is shown in Fig. 15. Qu et al. (2016) presented a quadruple supramolecular polymer (SP) having a [c2]daisy chain structure in its central body (Xin et al., 2016). Photo-triggered polymerization of [c2]daisy chain monomer was used for the effective fabrication of quadruple supramolecular polymer (QSP) (Fig. 16). The structure of monomer 7 is like a daisy chain rotaxane which contains two (UPy) ureido pyrimidinone terminals protected by coumarin at both ends.

Light irradiation was used for supramolecular polymerization, which can remove coumarin components and yield chain 8 at both terminals of ureido pyrimidinone (UPy) groups. Further, linear supramolecular polymer (SP) can then be self-assembled through quadruple hydrogen bonding (Fig. 16). This technique of photo-triggered supramolecular polymerization is an excellent treatment for the construction of stimuli-responsive supramolecular polymer (SP) (Xiaoa et al., 2020). Xu et al. developed a cationic polymer CD-PGEA by the combination of β-cyclodextrin (β-CD) cores and poly(glycidyl methacrylate) (PGEA) arms, which are proposed to design and fabricate various organic/inorganic nanoscale hybrid and functional materials while using host/guest interaction. In the main process, the guest molecules of adamantine (Ad) were

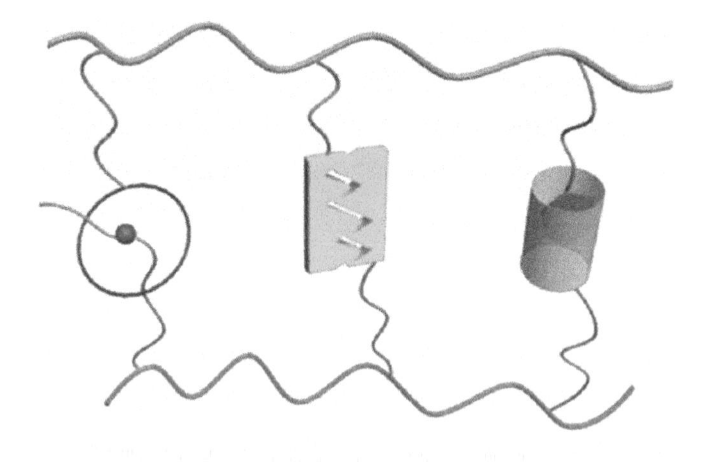

Fig. 16 Illustration of supramolecular polymers fabricated by orthogonal self-assembly based on multiple hydrogen bonding and macrocyclic host/guest interactions. *(Reprinted from Xiaoa, T., Zhoua, L., Suna, X.-Q., Huangb, F., Linc, C., Wang, L., 2020. Supramolecular polymers fabricated by orthogonal self-assembly based on multiple hydrogen bonding and macrocyclic host–guest interactions. Chin. Chem. Lett. 31, 1–9 with permission from Elsevier. Copyright © 2019 Published by Elsevier B.V. on behalf of Chinese Chemical Society and Institute of Materia Medica, Chinese Academy of Medical Sciences.)*

successfully immobilized on silica nanoparticles (Hu et al., 2013; Zhang et al., 2017; Zhao et al., 2015b). For example, in the initial stage 3-(glycidoxy propyl) tri ethoxy silane (GPTS) was attached effectively on star like hollow silica nanoparticles' surfaces to generate terminal epoxy functional groups; these epoxy groups then combine with cystamine through a chemical reaction and introduce disulfide bonds (Fig. 17).

After this, the adamantine (Ad) was combined by the reaction Ad-COOH with carboxylic groups and amino groups in the presence of both NHS and EDC, as shown in Fig. 16. At the end, CD-PGEA can be combined by host/guest connection with many disulfides bond-linked adamantine (Ad) guests on silica nanoparticles' surfaces. pH-sensitive poly(pyridyl disulfide ethyl methacrylate) (PDEAEMA) and poly(2-hydroxyethyl methacrylate) (PHPMA), which is hydrophilic in nature, can grow on the surface of silica nanoparticles using a bifunctional hetero initiator, as explained in Fig. 18. The RAFT agent is shown in green, the ATRP initiator in red, and the connected amino groups were assigned a blue color on the silica nanoparticle surface, which shows a very useful design of biofunctionalized silica nanoparticles (Huang et al., 2012e).

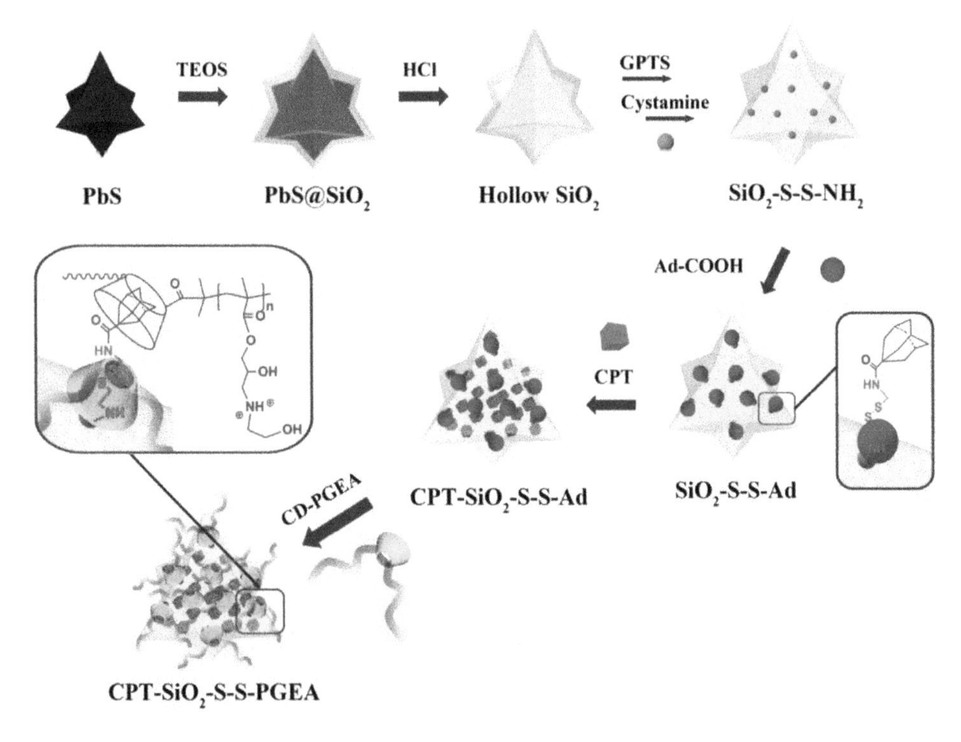

Fig. 17 Schematic presentation of the synthesis of star-like nanohybrids through host/guest interaction of cyclodextrin-adamantine (CD-Ad). *(Reprinted from Zhao, N., Lin, X., Zhang, Q., Ji, Z., Xu, F.J., 2015b. Redox-triggered gatekeeper-enveloped starlike hollow silica nanoparticles for intelligent delivery systems. Small 11, 6467–6479 with permission from John Wiley and Sons. Copyright © 2015 WILEY-VCH Verlag GmbH & Co. KGaA, Weinheim.)*

4.5 Wrapping

Wrapping is a common but versatile strategy to construct organic/inorganic nanoscale hybrid and functional materials via covalent interaction of inorganic and organic components. In most cases, the organic part encapsulates complete or some parts of inorganic nanoparticles through chemical reaction or precipitation processes. Self-assembled materials of the organic component can also encapsulate inorganic nanoparticles and generate functional materials. Wang et al. reported the synthesis of two kinds of nanoscale hybrid and functional materials, i.e., $NiCo_2S_4$@PNTs and $Ni_{1/3}Co_{2/3}(CO_3)_{0.5}OH \cdot H_2O$@PNTs, both of the functional materials exhibiting promising capacitive properties (Wang et al., 2019a,b).

Wang and his coworkers used a wrapping strategy for a nickel-cobalt metal-organic framework and polypyrrole nanotube for the preparation of (NiCo-MOF@PNTs) (Figs. 19 and 20). Wang's report provided inspiration for Yuexin Liu, who extended

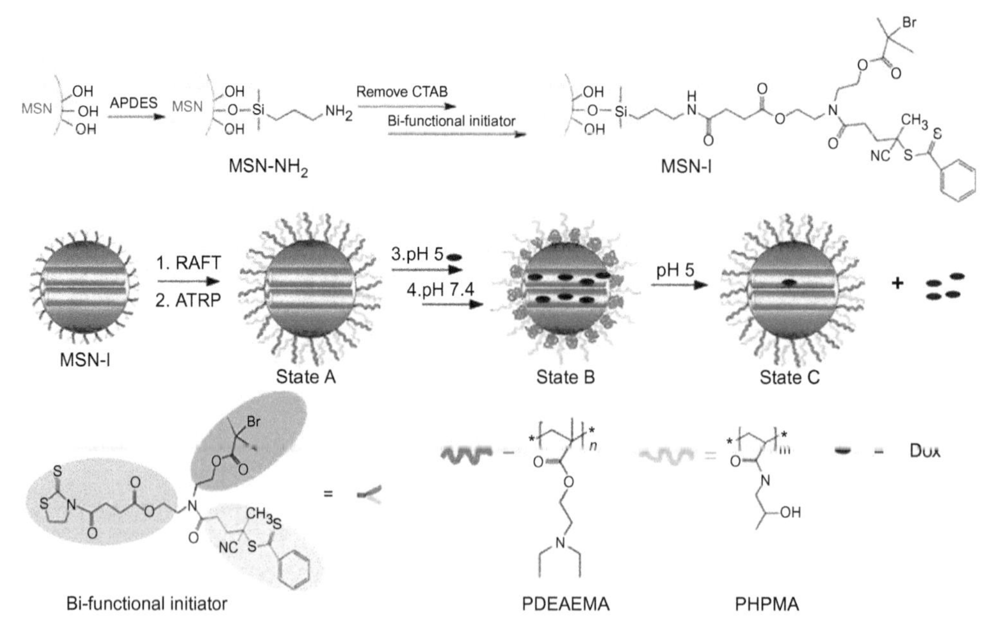

Fig. 18 Explanation of the synthetic route for the functionalization of mesoporous silica nanoparticles (MSNs) with poly(2-hydroxyethyl methacrylate) (PHPMA) and poly(pyridyldisulfide ethyl methacrylate) PDEAEMA. *(Reprinted from Huang, X., Hauptmann, N., Appelhans, D., Formanek, P., Frank, S., Kaskel, S., Temme, A., Voit, B., 2012c. Synthesis of hetero-polymer functionalized nanocarriers by combining surface-initiated ATRP and RAFT polymerization. Small 8, 3579–3583 with permission from John Wiley and Sons. Copyright © 2012 WILEY-VCH Verlag GmbH & Co. KGaA, Weinheim.)*

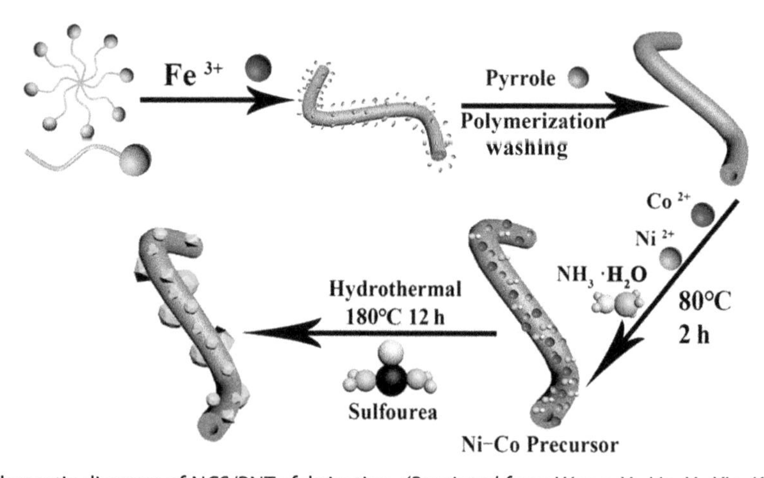

Fig. 19 Schematic diagram of NCS/PNTs fabrication. *(Reprinted from Wang, Y., Liu, Y., Xia, K., Zhang, Y., Yang, J., 2019a. NiCo2S4 nanoparticles anchoring on polypyrrole nanotubes for high-performance supercapacitor electrodes. J. Electroanal. Chem. 840, 242–248 with permission from Elsevier. Copyright © 2019 Elsevier B.V.)*

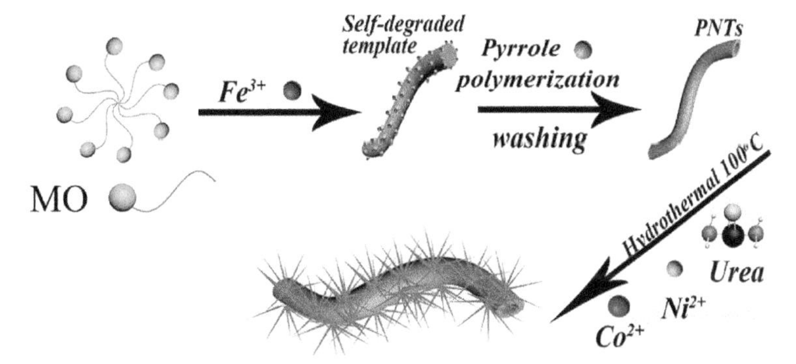

Fig. 20 Schematic diagram of NCC/PNTs fabrication. *(Reprinted from Wang, Y., Chen, Y., Liu, Y., Liu, W., Zhao, P., Li, Y., Dong, Y., Wang, H., Yang, J., 2019b. Urchin-like $Ni_{1/3}Co_{2/3}(CO_3)_{0.5}OH{\cdot}0.11H_2O$ anchoring on polypyrole nanotubes for super capacitor electrodes Electrochim. Acta, 295, 989–996 with permission from Elsevier. Copyright © 2018 Elsevier Ltd.)*

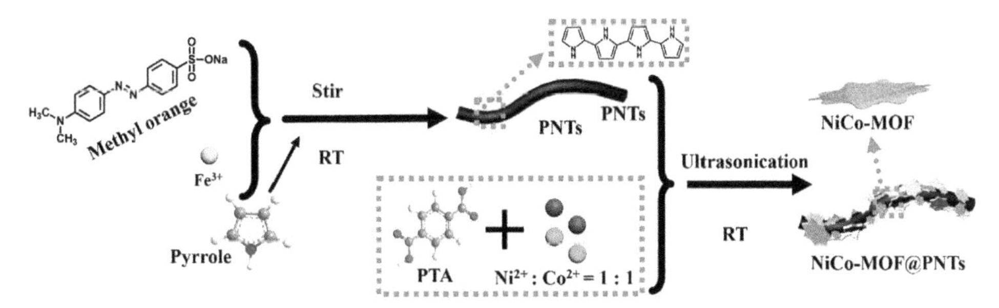

Fig. 21 Fabrication of NiCo-MOF@PNTs nanosheets. *(Reprinted from Liu, Y., Wang, Y., Chen, Y., Wang, C., Guo, L., 2020. NiCo-MOF nano sheets wrapping polypyrrole nanotubes for high-performance super capacitors. Appl. Surf. Sci. 507, 145089 with permission from Elsevier. Copyright © 2019 Elsevier B.V.)*

the experimental process of Wang et al. and prepared nanosheets of NiCo-MOF wrapping polypyrrole nanotube (PNTs) using the ultrasonic technique at room temperature (Fig. 21). Polypyrrole nanotube (PNTs) improve the capacitive ability of NiCo-MOF via multiple roles: (i) working as a substrate for the NiCo-MOF nanosheets in situ growth; (ii) producing more active sights and larger surface area and preventing the NiCo-MOF agglomeration; and (iii) improving the electrical conductivity of NiCo-MOF nanosheets (Liu et al., 2020).

The same wrapping strategy was used for the construction of polyacrylic acid (PAA) and a variety of metal oxide nanoparticles such as SnO_2, Fe_2O_3, and Fe_3O_4 and silver (Ag) nanoparticles while using isopropyl alcohol as a solvent medium (Li et al., 2013). This novel and general technique can also be used for the construction of UCNP@PAA nanoparticles with a core-shell and Janus morphology Au-PAA nanoparticles in which the gold (Au) spherical nanoparticles were further grown to branches of Au (Fig. 22)

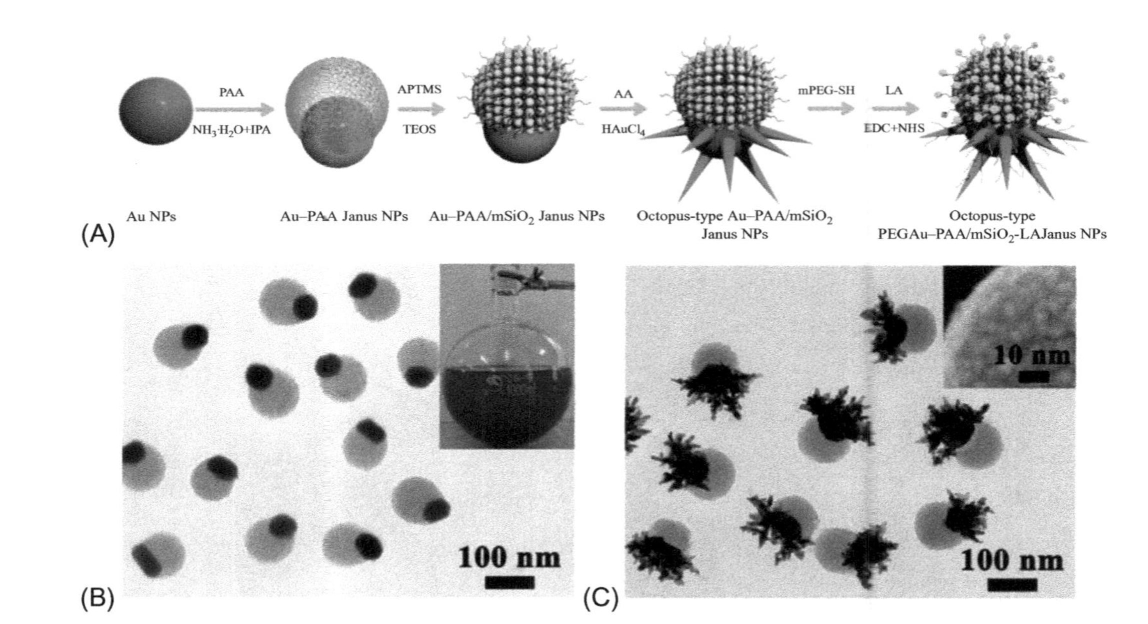

Fig. 22 Construction of PEGlyated Janus PAA-Au nano-scale hybrids (A). TEM image of PAA and Au spherical nanoparticles (B) Au branches (C). *(Reprinted from Zhang, L., Chen, Y., Li, Z., Li, L., Saint-Cricq, P., Li, C., Lin, J., Wang, C., Su, Z., Zink, J.I., 2016. Tailored synthesis of octopus-type Janus nanoparticles for synergistic actively-targeted and chemo-photo thermal therapy. Angew. Chem. Int. Ed. 55, 2118–2121 with permission from John Wiley and Sons. Copyright © 2016 WILEY-VCH Verlag GmbH & Co. KGaA, Weinheim.)*

(Zhang et al., 2016; Liu et al., 2015b). Next polyethylene glycol (PEG) with lactobionic acid can be conjugated to polyacrylic acid (PAA) and gold (Au) domains to improve biocompatibility and stability.

5. Concluding remarks and perspectives

Evidently, as discussed above with suitable examples, research has been conducted on the construction of hybrid and functional materials. With the speedy growth in various synthetic and fabricating strategies for both inorganic nanoparticles and organic parts of the material, their construction has also made considerable progress. It appears from the literature that preparation of inorganic nanoparticles is yet to mature, and functional materials prepared through surface functionalization are still prevalent. Recently, other construction techniques such as wrapping and one-pot synthesis have become appreciated due to their facile and simple approaches. With respect to the attributes and function of both inorganic and organic components, the importance of nanoscale functional and hybrid materials are well-founded. Typically, the biocompatibility, stability, dispersibility and distinguished magnetic and optical characteristics of functional and hybrid materials are irreplaceable. The role of inorganic nanoparticles is already recognized and the function of the organic component of the functional material is expected to be exploited further. For instance, the role of conducting and semiconducting polymers in hybrid and functional materials is currently restricted.

Aside from various formal and proven methods for the assembly of functional materials covered in this chapter, some less investigated techniques have great potential for the construction of next-generation materials. Development in the manufacturing techniques provides substantial opportunities to fabricate materials with a variety of new compositions for applications where multifunctional characteristics of those variations will be useful. Controllability and high accuracy of these fabrication techniques allow the engineered geometries and compositions that can be utilized in a variety of applications.

Acknowledgments

The authors are grateful to their institutes for providing research data, including literature access and other related resources.

Conflict of interest

The author(s) declare no potential conflict of interest.

References

Abareshi, M., Goharshadi, E.K., Mojtaba Zebarjad, S., Khandan Fadafan, H., Youssefi, A., 2010. Fabrication, characterization and measurement of thermal conductivity of Fe_3O_4 nanofluids. J. Magn. Magn. Mater. 322, 3895–3901.

Adhikari, R., Damm, C., Michler, G.H., Munstedt, H., Balta-Calleja, F.J., 2008. Processing and mechanical performance of SBS block copolymer/layered silicate nanocomposites. Compos. Interfaces 15, 453–463.

Albanese, A., Tang, P.S., Chan, W.C.W., 2012. The effect of nanoparticle size, shape, and surface chemistry on biological systems. Annu. Rev. Biomed. Eng. 14, 1–16.

Ali, N., Zhang, B., Zhang, H., Zaman, W., Li, W., Zhang, Q., 2014. Key synthesis of magnetic janus nanoparticles using a modified facile method. Particuology 17, 59–65.

Ali, N., Zhang, B., Zhang, H., Li, W., Zaman, W., Tian, L., Zhang, Q., 2015a. Novel Janus magnetic micro particle synthesis and its applications as a demulsifier for breaking heavy crude oil and water emulsion. Fuel 141, 258–267.

Ali, N., Zhang, B., Zhang, H., Zaman, W., Li, X., Li, W., Zhang, Q., 2015b. Interfacially active and magnetically responsive composite nanoparticles with raspberry like structure; synthesis and its applications for heavy crude oil/water separation. Colloids Surf. A Physicochem. Eng. Asp. 472, 38–49.

Ali, N., Baoliang, Z., Zhang, H., Zaman, W., Ali, S., Ali, Z., Li, W., Zhang, Q., 2015c. Iron oxide-based polymeric magnetic microspheres with a core shell structure: from controlled synthesis to demulsification applications. J. Polym. Res. 22, 219–230.

Ali, N., Zhang, B., Zhang, H., Zaman, W., Ali, S., Ali, Z., Li, W., Zhang, Q., 2015d. Monodispers and multifunctional magnetic composite core shell microspheres for demulsification applications. J. Chin. Chem. Soc. 62, 695–702.

An, Q., Zhang, P., Li, J.M., Ma, W.F., Guo, J., Hu, J., et al., 2012. Silver-coated magnetite–carbon core–shell microspheres as substrate-enhanced SERS probes for detection of trace persistent organic pollutants. Nanoscale 4, 5210–5216.

Arami, H., Khandhar, A., Liggitt, D., Krishnan, K.M., 2015. In vivo delivery, pharmacokinetics, biodistribution and toxicity of iron oxide nanoparticles. Chem. Soc. Rev. 44, 8576–8607.

Bagheri, A., Arandiyan, H., Adnan, N.N.M., Boyer, C., Lim, M., 2017. Controlled direct growth of polymer shell on upconversion nanoparticle surface via visible light regulated polymerization. Macromolecules 50, 7137–7147.

Balazs, A.C., 2007. Modeling self-assembly and phase behavior in complex mixtures. Annu. Rev. Phys. Chem. 58, 211–233.

Balazs, A.C., Emrick, T., Russell, T.P., 2006. Nanoparticle polymer composites: where two small worlds meet. Science 314, 1107–1110.

Basuki, J.S., Esser, L., Zetterlund, P.B., Whittaker, M.R., Boyer, C., Davis, T.P., 2013a. Grafting of P(OEGA) onto magnetic nanoparticles using Cu(0) mediated polymerization: comparing grafting "from" and "to" approaches in the search for the optimal material, design of nanoparticle MRI contrast agents. Macromolecules 46, 6038–6047.

Basuki, J., Duong, H.T.T., Macmillan, A., Erlich, R.B., Esser, L., Akerfeldt, M.C., Whan, R.M., Kavallaris, M., Boyer, C., Davis, T.P., 2013b. Using fluorescence lifetime imaging microscopy to monitor theranostic nanoparticle uptake and intracellular doxorubicin release. ACS Nano 7, 10175–10189.

Basuki, J.S., Esser, L., Duong, H.T.T., Zhang, Q., Wilson, P., Whittaker, M.R., Haddleton, D.M., Boyer, C., Davis, T.P., 2014a. Magnetic nanoparticles with diblock glycopolymer shells give lectin concentration-dependent MRI signals and selective cell uptake. Chem. Sci. 5, 715–726.

Basuki, J.S., Jacquemin, A., Esser, L., Li, Y., Boyer, C., Davis, T.P., 2014b. A block copolymer-stabilized co-precipitation approach to magnetic iron oxide nanoparticles for potential use as MRI contrast agents. Polym. Chem. 5, 2611–2620.

Behzadi, S., Serpooshan, V., Tao, W., Hamaly, M.A., Alkawareek, M.Y., Dreaden, E.C., Brown, D., Alkilany, A.M., Farokhzad, O.C., Mahmoudi, M., 2017. Cellular uptake of nanoparticles: journey inside the cell. Chem. Soc. Rev. 46, 4218–4244.

Bektas, N., Agim, B.A., Kara, S., 2004. Kinetic and equilibrium studies in removing lead ions from aqueous solutions by natural sepiolite. J. Hazard. Mater. 112 (1–2), 115–122.

Bever, M.B., Duwez, P.E., 1970. On gradient composites. In: ARPA Materials Summer Conferencepp. 117–140.
Bockstaller, M.R., Mickiewicz, R.A., Thomas, E.L., 2005. Block copolymer nanocomposites: perspectives for tailored functional materials. Adv. Mater. 17, 1331–1349.
Bombalski, L., Listak, J., Bockstaller, M.R., 2006. Structure, properties, and opportunities of block copolymer/particle nanocomposites. Annu. Rev. Nano Res. 1, 295–336.
Boyer, C., Corrigan, N.A., Jung, K., Nguyen, D., Nguyen, T.K., Adnan, N.N.M., Oliver, S., Shanmugam, S., Yeow, J., 2016. Copper-mediated living radical polymerization (atom transfer radical polymerization and copper(0) mediated polymerization): from fundamentals to bioapplications. Chem. Rev. 116, 1803–1949.
Burda, C., Chen, X.B., Narayanan, R., El-Sayed, M.A., 2005. Chemistry and properties of nanocrystals of different shapes. Chem. Rev. 105, 1025–1102.
Cannillo, V., et al., 2007. Prediction of the elastic properties profile in glass-alumina functionally graded materials. J. Eur. Ceram. Soc. 27 (6), 2393–2400.
Cao, Q., Ma, C., Bai, H., Li, X., Yan, H., Zhao, Y., Ying, W., Qian, X., 2014. Multivalent hydrazide-functionalized magnetic nanoparticles for glycopeptide enrichment and identification. Analyst 139, 603–609.
Chauhan, V.P., Popovic, Z., Chen, O., Cui, J., Fukumura, D., Bawendi, M.G., Jain, R.K., 2011. Fluorescent nanorods and nanospheres for real-time in vivo probing of nanoparticle shape-dependent tumor penetration. Angew. Chem. Int. Ed. 50, 11417–11420.
Chen, Q., de Leon, A., Advincula, R.C., 2015. Inorganic-organic thiol–ene coated mesh for oil/water separation. ACS Appl. Mater. Interfaces 7, 18566–18573.
Chen, G., Roy, I., Yang, C., Prasad, P.N., 2016. Nanochemistry and nanomedicine for nanoparticle-based diagnostics and therapy. Chem. Rev. 116, 2826–2885.
Chiu, H.Y., Gößl, D., Haddick, L., Engelke, H., Bein, T., 2018. Clickable multifunctional large-pore mesoporous silica nanoparticles as nano carriers. Chem. Mater. 30, 644–654.
Choi, K.H., Lee, S.H., Kim, Y.R., Malkinski, L., Vovk, A., Barnakov, Y., Park, J.H., Jung, Y.K., Jung, J.S., 2007. Magnetic behavior of Fe_3O_4 nanostructure fabricated by template method. J. Magn. Magn. Mater. 310, e861–e863.
Crosby, A.J., Lee, J.Y., 2007. Polymer nanocomposites: the "nano" effect on mechanical properties. Polym. Rev. 47, 217–229.
Cui, W., Li, J., Decher, G., 2016. Self-assembled smart nanocarriers for targeted drug delivery. Adv. Mater. 28, 1302–1311.
Dan, S., Jing, L., Xiao, Z., Yue, P., Zheng, W., Ming, Z., Chen, Q.X., Dong, W.F., Li, C., 2016. Janus "nano-bullets" for magnetic targeting liver cancer chemotherapy. Biomaterials 100, 118–133.
Darling, S.B., 2007. Directing the self-assembly of block copolymers. Prog. Polym. Sci. 32, 1152–1204.
Deng, R., Liang, F., Qu, X., Wang, Q., Zhu, J., Yang, Z., 2015. Diblock copolymer based janus nanoparticles. Macromolecules 48, 750–755.
Deng, L., Li, Q., Al-Rehili, S., Omar, H., Almalik, A., Alshamsan, A., Zhang, J., Khashab, N.M., 2016. Hybrid iron oxide-graphene oxide-polysaccharides microcapsule: a micro-matryoshka for on-demand drug release and antitumor therapy in vivo. ACS Appl. Mater. Interfaces 8, 6859–6868.
Dong, H., Huang, J., Koepsel, R., Ye, P., Russell, A., Matyjaszewski, K., 2011. Recyclable antibacterial magnetic nanoparticles grafted with quaternized poly(2-(dimethylamino)ethyl methacrylate) brushes. Biomacromolecules 12, 1305–1311.
Dong, X., Gao, S., Huang, J., 2019. A self-roughened and biodegradable superhydrophobic coating with UV shielding, solar-induced self-healing and versatile oil–water separation ability. J. Mater. Chem. A 7, 2122–2128.
Du, B.J., Liu, J.H., Ding, G.Y., Han, X., Li, D., Wang, E.K., Wang, J., 2017. Positively charged graphene/Fe_3O_4/polyethylenimine with enhanced drug loading and cellular uptake for magnetic resonance imaging and magnet-responsive cancer therapy. Nano Res. 10, 2280–2295.
Elacqua, E., Zheng, X., Shillingford, C., Liu, M., Weck, M., 2017. Molecular recognition in the colloidal world. Acc. Chem. Res. 50, 2756–2766.

Emadi, M., Shams, E., Amini, M.K., 2013. Removal of zinc from aqueous solutions by magnetite silica core-shell nanoparticles. J Chem. 2013,, 787682 10 pageshttps://doi.org/10.1155/2013/787682.

Fan, Q.L., Neoh, K.G., Kang, E.T., Shuter, B., Wang, S.C., 2007. Solvent-free atom transfer radical polymerization for the preparation of poly(poly(ethyleneglycol) monomethacrylate)-grafted Fe_3O_4 nanoparticles: synthesis, characterization and cellular uptake. Biomaterials 28, 5426–5436.

Forster, S., Plantenberg, T., 2002. From self-organizing polymers to nano hybrid and biomaterials. Angew. Chem. Int. Ed. 41, 689–714.

Ganesan, V., Ellison, C.J., Pryamitsyn, V., 2010. Mean-field models of structure and dispersion of polymer-nanoparticle mixtures. Soft Matter 6, 4010–4025.

Ge, J., Hu, Y., Biasini, M., Dong, C., Guo, J., Beyermann, W.P., Yin, Y., 2007a. One-step synthesis of highly water-soluble magnetite colloidal nano crystals. Chem. Eur. J. 13, 7153–7161.

Ge, J., Hu, Y., Zhang, T., Yin, Y., 2007b. Superparamagnetic composite colloids with anisotropic structures. J. Am. Chem. Soc. 129, 8974–8975.

Ge, S., Shi, X., Sun, K., Li, C., Uher, C., Baker, J.R., Holl, M.M.B., Orr, B.G., 2009. Facile hydrothermal synthesis of iron oxide nanoparticles with tunable magnetic properties. J. Phys. Chem. C 113, 13593–13599.

Geng, Y., Dalhaimer, P., Cai, S., Tsai, R., Tewari, M., Minko, T., Discher, D.E., 2007. Shape effects of filaments versus spherical particles in flow and drug delivery. Nat. Nanotechnol. 2, 249–255.

Gottron, J., Harries, K.A., Xu, Q., 2014. Creep behavior of bamboo. Constr. Build. Mater. 66, 79–88.

Hamley, I.W., 2003. Nanostructure fabrication using block copolymers. Nanotechnology 14, R39–R54.

Han, H.S., Choi, K.Y., Lee, H., Lee, M., An, J.Y., Shin, S., Kwon, S., Lee, D.S., Park, J.H., 2016. Gold-nanoclustered hyaluronan nano-assemblies for photothermally maneuvered photodynamic tumor ablation. ACS Nano 10, 10858–10868.

Haraguchi, K., 2007. Nanocomposite hydrogels. Curr. Opin. Solid State Mater. Sci. 11, 47–54.

Haryono, A., Binder, W.H., 2006. Controlled arrangement of nanoparticle arrays in block-copolymer domains. Small 2, 600–611.

Hinde, E., Thammasiraphop, K., Duong, H.T.T., Yeow, J., Karagoz, B., Boyer, C., Gooding, J.J., Gaus, K., 2017. Pair correlation microscopy reveals the role of nanoparticle shape in intracellular transport and site of drug release. Nat. Nanotechnol. 12, 81–89.

Hong, C.Y., Li, X., Pan, C.Y., 2009. Fabrication of smart anocontainers with a mesoporous core and a pH-responsive shell or controlled uptake and release. J. Mater. Chem. 19, 5155–5160.

Hu, H., Wang, X.B., Xu, S., Yang, W.T., Xu, F.J., Shen, J., Mao, C., 2012. Preparation and evaluation of well-defined hemocompatible layered double hydroxide-poly(sulfobetaine) nanohybrids. J. Mater. Chem. 22, 15362–15369.

Hu, Y., Chai, M.Y., Yang, W.T., Xu, F.J., 2013. Supramolecular host-guest pseudocomb conjugates composed of multiple star polycations tied tunably with a linear polycation backbone for gene transfection. Bioconjug. Chem. 24, 1049–1056.

Huang, X., Li, L., Liu, T., Hao, N., Liu, H., Chen, D., Tang, F., 2011a. The shape effect of mesoporous silica nanoparticles on bio distribution, clearance, and biocompatibility in vivo. ACS Nano 5, 5390–5399.

Huang, C., Neoh, K., Kang, E.T., Shuter, B., 2011b. Surface modified superparamagnetic iron oxide nanoparticles (Spions) for high efficiency folate-receptor targeting with low uptake by macrophages. J. Mater. Chem. 21, 16094–16102.

Huang, C., Neoh, K.G., Kang, E.T., 2012a. Combined ATRP and 'click' chemistry for designing stable tumor-targeting superparamagnetic iron oxide nanoparticles. Langmuir 28, 563–571.

Huang, X., Appelhans, D., Formanek, P., Simon, F., Voit, B., 2012b. Tailored synthesis of intelligent polymer nanocapsules: an investigation of controlled permeability and pH-dependent degradability. ACS Nano 6, 9718–9726.

Huang, X., Hauptmann, N., Appelhans, D., Formanek, P., Frank, S., Kaskel, S., Temme, A., Voit, B., 2012c. Synthesis of hetero-polymer functionalized nanocarriers by combining surface-initiated ATRP and RAFT polymerization. Small 8, 3579–3583.

Huang, B., Wang, D., Wang, G., Zhang, F., Zhou, L., 2017. Enhancing the colloidal stability and surface functionality of molybdenum disulfide (MoS_2) nanosheets with hyperbranched polyglycerol for photothermal therapy. J. Colloid Interface Sci. 508, 214–221.

Hyeon, T., 2003. Chemical synthesis of magnetic nanoparticles. Chem. Commun. 927–934.
Hyeon, T., Lee, S.S., Park, J., Chung, Y., Na, H.B., 2001. Synthesis of highly crystalline and monodisperse maghemite nanocrystallites without a size-selection process. J. Am. Chem. Soc. 123, 12798–12801.
Jha, D.K., Kant, T., Singh, R.K., 2013. A critical review of recent research on functionally graded plates. Compos. Struct. 96, 833–849.
Ji, Y., Marshall, J.E., Terentjev, E.M., 2012. Nanoparticle-liquid crystalline elastomer composites. Polymers 4, 316–340.
Jiang, B., Zhang, H.J., Zhang, L.H., Sun, Y.L., Xu, L.D., Sun, Z.N., Gu, W.H., Chen, Z.X., Yang, H.W., 2017. Novel one-step, in situ thermal polymerization fabrication of robust superhydrophobic mesh for efficient oil/water separation. Ind. Eng. Chem. Res. 56, 11817–11826.
Joseph, S.T.S., Ipe, B.I., Pramod, P., Thomas, K.G., 2006. Gold nanorods to nanochains: mechanistic investigations on their longitudinal assembly using α,ω-alkanedithiols and interplasmon coupling. J. Phys. Chem. B 110, 150–157.
Joshi, R.K., et al., 2015. Graphene oxide: the new membrane material. Appl. Mater. Today 1 (1), 1–12.
Kao, J., Thorkelsson, K., Bai, P., Rancatore, B.J., Xu, T., 2013. Toward functional nanocomposites: taking the best of nanoparticles, polymers, and small molecules. Chem. Soc. Rev. 42, 2654–2678.
Karaca, E., Şatır, M., Kazan, S., Açıkgöz, M., Öztürk, E., Gürdağ, G., Ulutaş, D., 2015. Synthesis, characterization and magnetic properties of Fe_3O_4 doped chitosan polymer. J. Magn. Magn. Mater. 373, 53–59.
Karagoz, B., Yeow, J., Esser, L., Prakash, S.M., Kuchel, R.P., Davis, T.P., Boyer, C., 2014. An efficient and highly versatile synthetic route to prepare iron oxide nanoparticles/nanocomposites with tunable morphologies. Langmuir 30, 10493–10502.
Kawasaki, A., Watanabe, R., 1997. Concept and P/M fabrication of functionally gradient materials. Ceram. Int. 23 (1), 73–83.
Kieback, B., Neubrand, A., Riedel, H., 2003. Processing techniques for functionally graded materials. Mater. Sci. Eng. A 362 (1–2), 81–106.
Kim, D., Lee, N., Park, M., Kim, B.H., An, K., Hyeon, T., 2009. Synthesis of uniform ferrimagnetic magnetite nanocubes. J. Am. Chem. Soc. 131, 454–455.
Kim, B., Han, G., Toley, B.J., Kim, C.K., Rotello, V.M., Forbes, N.S., 2010. Tuning payload delivery in tumour cylindroids using gold nanoparticles. Nat. Nanotechnol. 5, 465–472.
Klokkenburg, M., Vonk, C., Claesson, E.M., Meeldijk, J.D., Erné, B.H., Philipse, A.P., 2004. Direct imaging of zero-field dipolar structures in colloidal dispersions of synthetic magnetite. J. Am. Chem. Soc. 126, 16706–16707.
Koizumi, M., 1997. FGM activities in Japan. Compos. Part B 28 (1–2), 1–4.
Kokate, M., Garadkar, K., Gole, A., 2013. One pot synthesis of magnetite–silica nanocomposites: applications as tags, entrapment matrix and in water purification. J. Mater. Chem. A 1, 2022–2029.
Korth, B.D., Keng, P., Shim, I., Bowles, S.E., Tang, C.B., Kowalewski, T., Nebesny, K.W., Pyun, J., 2006. Polymer-coated ferromagnetic colloids from well-defined macromolecular surfactants and assembly into nanoparticle chains. J. Am. Chem. Soc. 128, 6562–6563.
Laurent, S., Forge, D., Port, M., Roch, A., Robic, C., Vander Elst, L., Muller, R.N., 2008. Magnetic iron oxide nanoparticles: synthesis, stabilization, vectorization, physicochemical characterizations, and biological applications. Chem. Rev. 108, 2064–2110.
Lazzari, M., Lopez-Quintela, M.A., 2003. Block copolymers as a tool for nano-material fabrication. Adv. Mater. 15, 1583–1594.
Lee, S.M., Jun, Y.W., Cho, S.N., Cheon, J., 2002. Single-crystalline star-shaped nanocrystals and their evolution: programming the geometry of nano-building blocks. J. Am. Chem. Soc. 124, 11244–11245.
Levchenko, I., Bazaka, K., Keidar, M., Xu, S., Fang, J., 2018. Hierarchical multicomponent inorganic metamaterials: intrinsically driven self-assembly at the nanoscale. Adv. Mater. 30, 1702226.
Li, Z., Barnes, J., Bosoy, A., Stoddart, J.F., Zink, J.I., 2012. Mesoporous silica nanoparticles in biomedical applications. Chem. Soc. Rev. 41, 2590–2605.
Li, L., Zhang, L., Xing, S., Wang, T., Luo, S., Zhang, X., Liu, C., Su, Z., Wang, C., 2013. Generalized approach to the synthesis of reversible concentric and eccentric polymer-coated nanostructures. Small 9, 825–830.

Li, J.J., Kawazoe, N., Chen, G., 2015a. Gold nanoparticles with different charge and moiety induce differential cell response on mesenchymal stem cell osteogenesis. Biomaterials 54, 226–236.

Li, Y., Sun, Y., Li, J., Su, Q., Yuan, W., Dai, Y., Han, C., Wang, Q., Feng, W., Li, F., 2015b. Ultrasensitive near-infrared fluorescence enhanced probe for in vivo nitroreductase imaging. J. Am. Chem. Soc. 137, 6407–6416.

Li, J., Yan, L., Li, H., Li, W., Zha, F., Lei, Z., 2015c. Underwater superoleophobic palygorskite coated meshes for efficient oil/water separation. J. Mater. Chem. A 3, 14696–14702.

Li, J., Cheng, H.M., Chan, C.Y., Ng, P.F., Chen, L., Fei, B., Xin, J.H., 2015d. Superhydrophilic and underwater superoleophobic mesh coating for efficient oil-water separation. RSC Adv. 5, 51537–51541.

Li, Z., Hu, Y., Howard, K.A., Jiang, T., Fan, X., Miao, Z., Sun, Y., Besenbacher, F., Yu, M., 2016. Multifunctional bismuth selenide nanocomposites for antitumor thermo-chemotherapy and imaging. ACS Nano 10, 984–997.

Li, J., Tenjimbayashi, M., Zacharia, N.S., Shiratori, S., 2018. One-step dipping fabrication of Fe_3O_4/PVDF-HFP composite 3D porous sponge for magnetically controllable oil – water separation. ACS Sustain. Chem. Eng. 6, 10706–10713.

Lin, X., Lu, F., Chen, Y., Liu, N., Cao, Y., Xu, L., Wei, Y., Feng, L., 2015a. One-step breaking and separating emulsion by tungsten oxide coated mesh. ACS Appl. Mater. Interfaces 7, 8108–8113.

Lin, Y., Wang, S., Zhang, Y., Gao, J., Hong, L., Wang, X., Wu, W., Jiang, X., 2015b. Ultra-high relaxivity iron oxide nanoparticles confined in polymer nanospheres for tumor MR imaging. J. Mater. Chem. B 3, 5702–5710.

Lin, X., Zhao, N., Yan, P., Hu, H., Xu, F.J., 2015c. The shape and size effects of polycation functionalized silica nanoparticles on gene transfection. Acta Biomater. 11, 381–392.

Ling, D., Hyeon, T., 2013. Chemical design of biocompatible iron oxide nanoparticles for medical applications. Small 9, 1450–1466.

Liu, J., Zhang, L., Shi, S., Chen, S., Zhou, N., Zhang, Z., Cheng, Z., Zhu, X., 2010. A novel and universal route to SiO_2-supported organic/inorganic hybrid noble metal nanomaterials via surface RAFT polymerization. Langmuir 26, 14806–14813.

Liu, X.L., Yang, Y., Ng, C.T., Zhao, L.Y., Zhang, Y., Bay, B.H., Fan, H.M., Ding, J., 2015a. Magnetic vortex nanorings: a new class of hyperthermia agent for highly efficient in vivo regression of tumors. Adv. Mater. 27, 1939–1944.

Liu, B., Chen, Y.Y., Li, C.X., He, F., Hou, Z.Y., Huang, S.S., Zhu, H.M., Chen, X.Y., Lin, J., 2015b. Poly (acrylic acid) modification of Nd^{3+}-sensitized upconversion nanophosphors for highly efficient UCL imaging and pH-responsive drug delivery. Adv. Funct. Mater. 25, 4717–4729.

Liu, Y., Hu, J., Yu, X., Xu, X., Gao, Y., Li, H., Liang, F., 2017. Preparation of Janus-type catalysts and their catalytic performance at emulsion interface. J. Colloid Interface Sci. 490, 357–364.

Liu, Y., Liu, N., Jing, Y., Jiang, X., Yu, L., Yan, X., 2019. Surface design of durable and recyclable superhydrophobic materials for oil/water separation. Colloids Surf. A Physicochem. Eng. Asp. 567, 128–138.

Liu, Y., Wang, Y., Chen, Y., Wang, C., Guo, L., 2020. NiCo-MOF nano sheets wrapping polypyrrole nanotubes for high-performance super capacitors. Appl. Surf. Sci. 507, 145089.

Lu, X., Jiang, R., Fan, Q., Zhang, L., Zhang, H., Yang, M., Ma, Y., Wang, L., Huang, W., 2012. Fluorescent-magnetic poly(poly-(ethyleneglycol)monomethacrylate)-grafted Fe_3O_4 nanoparticles from post-atom-transfer-radical-polymerization modification: synthesis, characterization, cellular uptake and imaging. J. Mater. Chem. 22, 6965–6973.

Lu, X., Yu, M., Wang, G., Zhai, T., Xie, S., Ling, Y., et al., 2013. H-TiO_2@MnO_2//H-TiO_2@C core–shell nanowires for high performance and flexible asymmetric supercapacitors. Adv. Mater. 25, 267–272.

Lutz, J.F., Stiller, S., Hoth, A., Kaufner, L., Pison, U., Cartier, R., 2006. One-pot synthesis of PEGylated ultrasmall iron-oxide nanoparticles and their in vivo evaluation as magnetic resonance imaging contrast agents. Biomacromolecules 7, 3132–3138.

Lynch, I., Dawson, K.A., 2008. Protein-nanoparticle interactions. Nano Today 3, 40–47.

Ma, X., Zhao, Y., Liang, X.J., 2011. Theranostic nanoparticles engineered for clinic and pharmaceutics. Acc. Chem. Res. 44, 1114–1122.

Ma, X., Qu, Q., Zhao, Y., 2015. Targeted delivery of 5-aminolevulinic acid by multifunctional hollow mesoporous silica nanoparticles for photodynamic skin cancer therapy. ACS Appl. Mater. Interfaces 7, 10671–10676.

Matsushita, Y., 2007. Creation of hierarchically ordered nano phase structures in block polymers having various competing interactions. Macromolecules 40, 771–776.
Matyjaszewski, K., Tsarevsky, N.V., 2009. Nanostructured functional materials prepared by atom transfer radical polymerization. Nat. Chem. 1, 276–288.
Mi, P., Kokuryo, D., Cabral, H., Wu, H., Terada, Y., Saga, T., Aoki, I., Nishiyama, N., Kataoka, K., 2016. A pH-activatable nanoparticle with signal-amplification capabilities for non-invasive imaging of tumour malignancy. Nat. Nanotechnol. 11, 724–730.
Min, K.H., Min, H.S., Lee, H.J., Park, D.J., Yhee, J.Y., Kim, K., Kwon, I.C., Jeong, S.Y., Silvestre, O.F., Chen, X., Hwang, Y.S., Kim, E.C., Lee, S.C., 2015. pH-controlled gas-generating mineralized nanoparticles: a theranostic agent for ultrasound imaging and therapy of cancers. ACS Nano 9, 134–145.
Na, H.B., Palui, G., Rosenberg, J.T., Ji, X., Grant, S.C., Mattoussi, H., 2012. Multidentate catechol-based polyethylene glycol oligomers provide enhanced stability and biocompatibility to iron oxide nanoparticles. ACS Nano 6, 389–399.
Nagao, D., Goto, K., Ishii, H., Konno, M., 2011. Preparation of asymmetrically nanoparticle-supported, monodisperse composite dumbbells by protruding a smooth polymer bulge from rugged spheres. Langmuir 27 (21), 13302–13307.
Naseem, R., Tahir, S.S., 2001. Removal of Pb(II) from aqueous/acidic solutions by using bentonite as an adsorbent. Water Res. 35 (16), 3982–3986.
Nguyen, K.T., Zhao, Y., 2015. Engineered hybrid nanoparticles for on-demand diagnostics and therapeutics. Acc. Chem. Res. 48, 3016–3025.
Ohno, K., Mori, C., Akashi, T., Yoshida, S., Tago, Y., Tsujii, Y., Tabata, Y., 2013. Fabrication of contrast agents for magnetic resonance imaging from polymer-brush-afforded iron oxide magnetic nanoparticles prepared by surface-initiated living radical polymerization. Biomacromolecules 14, 3453–3462.
Oz, Y., Arslan, M., Gevrek, T.N., Sanyal, R., Sanyal, A., 2016. Modular fabrication of polymer brush coated magnetic nanoparticles: engineering the interface for targeted cellular imaging. ACS Appl. Mater. Interfaces 8 (30), 19813–19826.
Palchoudhury, S., Lead, J.R., 2014. A facile and cost-effective method for separation of oil – water mixtures using polymer-coated iron oxide nanoparticles. Environ. Sci. Technol. 48, 14558–14563.
Park, J.H., Von Maltzahn, G., Zhang, L., Derfus, A.M., Simberg, D., Harris, T.J., Ruoslahti, E., Bhatia, S.N., Sailor, M.J., 2009. Systematic surface engineering of magnetic nanoworms for in vivo tumor targeting. Small 5, 694–700.
Peiris, P.M., Schmidt, E., Calabrese, M., Karathanasis, E., 2011. Assembly of linear nano-chains from iron oxide nanospheres with asymmetric surface chemistry. PLoS ONE. 6, e15927.
Peng, J., Liu, Q., Xu, Z., Masliyah, J., 2012. Novel magnetic demulsifier for water removal from diluted bitumen emulsion. Energy Fuel 26, 2705–2710.
Pyun, J., 2007. Nanocomposite materials from functional polymers and magnetic colloids. Polym. Rev. 47, 231–263.
Qin, S., Wang, L., Zhang, X., Su, G., 2010. Grafting poly(ethylene glycol) monomethacrylate onto Fe_3O_4 nanoparticles to resist nonspecific protein adsorption. Appl. Surf. Sci. 257, 731–735.
Qiu, J., Yang, R., Li, M., Jiang, N., 2005. Preparation and characterization of porous ultrafine Fe_2O_3 particles. Mater. Res. Bull. 40, 1968–1975.
Reddy, L.H., Arias, J.L., Nicolas, J., Couvreur, P., 2012. Magnetic nanoparticles: design and characterization, toxicity and biocompatibility, pharmaceutical and biomedical applications. Chem. Rev. 112, 5818–5878.
Roghanizad, A., Abdolmaleki, M.K., Ghoreishi, S.M., Dinaric, M., 2020. One-pot synthesis of functionalized mesoporous fibrous silica nanospheres for dye adsorption: isotherm, kinetic, and thermodynamic studies. J. Mol. Liq. 300, 112367.
Rosaria, C., Sciortino, M., Alonzo, G., Schrijver, A.D., Pagliaro, M., 2011. From molecules to systems: sol-gel microencapsulation in silica-based materials. Chem. Rev. 111, 765–789.
Ruiz, A., Hernández, Y., Cabal, C., González, E., Veintemillas Verdaguer, S., Martínez, E., Morales, M.P., 2013. Biodistribution and pharmacokinetics of uniform magnetite nanoparticles chemically modified with polyethylene glycol. Nanoscale 5, 11400–11408.
Ruiz-Hitzky, E., Darder, M., Aranda, P., Ariga, K., 2010. Advances in biomimetic and nanostructured biohybrid materials. Adv. Mater. 22, 323–336.

Sanchez, C., Julián, B., Belleville, P., Popall, M., 2005. Applications of hybrid organic – inorganic nanocomposites. J. Mater. Chem. 15, 3559–3592.

Sandhu, K.K., McIntosh, C.M., Simard, J.M., Smith, S.W., Rotello, V.M., 2002. Gold nanoparticle-mediated transfection of mammalian cells. Bioconjug. Chem. 13, 3–6.

Schadler, L.S., Kumar, S.K., Benicewicz, B.C., Lewis, S.L., Harton, S.E., 2007. Designed interfaces in polymer nanocomposites: a fundamental view point. MRS Bull. 32, 335–340.

Shen, M., Bever, M.B., 1972. Gradients in polymeric materials. J. Mater. Sci. 7 (7), 741–746.

Shenhar, R., Norsten, T.B., Rotello, V.M., 2005. Polymer-mediated nanoparticle assembly: structural control and applications. Adv. Mater. 17, 657–669.

Shim, M.S., Kwon, Y.J., 2012. Stimuli-responsive polymers and nanomaterials for gene delivery and imaging applications. Adv. Drug Deliv. Rev. 64, 1046–1059.

Smart, T., Lomas, H., Massignani, M., Flores-Merino, M.V., Perez, L.R., Battaglia, G., 2008. Block copolymer nanostructures. Nano Today 3, 38–46.

Song, G., Wang, Q., Wang, Y., Lv, G., Li, C., Zou, R., Chen, Z., Qin, Z., Huo, K., Hu, R., et al., 2013. A low-toxic multifunctional nano platform based on $Cu_9S_5@mSiO_2$ core-shell nanocomposites: combining photothermal- and chemotherapies with infrared thermal imaging for cancer treatment. Adv. Funct. Mater. 23, 4281–4292.

Srivastava, S.K., Tyagi, R., Pant, N., Pal, N., 1989. Studies on the removal of some toxic metal ions. Part II. Removal of lead and cadmium by montmorillonite and kaolinite. Environ. Technol. Lett. 10 (3), 275–282.

Stober, W., Fink, A., Bohn, E., 1968. Controlled growth of monodisperse silica spheres in the micron size range. J. Colloid Interface Sci. 26, 62–69.

Sun, Z.K., Yang, J.P., Wang, J.X., Li, W., Kaliaguine, S., Hou, X.F., et al., 2014. A versatile designed synthesis of magnetically separable nano-catalysts with well-defined core-shell nanostructures. J. Mater. Chem. A 2, 6071–6074.

Sun, Q., You, Q., Pang, X., Tan, X., Wang, J., Liu, L., Guo, F., Tan, F., Li, N., 2017. A photo responsive and rod-shape nano carrier: single wavelength of light triggered photo thermal and photodynamic therapy based on AuNRs-capped & Ce6-doped mesoporous silica nanorods. Biomaterials 122, 188–200.

Tang, F., Li, L., Chen, D., 2012. Mesoporous silica nanoparticles: synthesis, biocompatibility and drug delivery. Adv. Mater. 24, 1504–1534.

Tenzer, S., Docter, D., Kuharev, J., Musyanovych, A., Fetz, V., Hecht, R., Schlenk, F., Fisher, D., Kiouptsi, K., Reinhardt, C., et al., 2013. Rapid formation of plasma protein corona critically affects nanoparticle pathophysiology. Nat. Nanotechnol. 8, 772–781.

Tian, J., Zheng, F., Zhao, H., 2011. Nanoparticles with Fe_3O_4-nanoparticle cores and gold-nanoparticle coronae prepared by self-assembly approach. J. Phys. Chem. C 115, 3304–3312.

Tjong, S.C., 2006. Structural and mechanical properties of polymer nanocomposites. Mater. Sci. Eng. R. Rep. 53, 73–197.

Topham, P.D., Parnell, A.J., Hiorns, R.C., 2011. Block copolymer strategies for solar cell technology. J. Polym. Sci. B Polym. Phys. 49, 1131–1156.

Vaia, R.A., Maguire, J.F., 2007. Polymer nanocomposites with prescribed morphology: going beyond nanoparticle-filled polymers. Chem. Mater. 19, 2736–2751.

Verma, A., Uzun, O., Hu, Y., Hu, Y., Han, H.S., Watson, N., Chen, S., Irvine, D.J., Stellacci, F., 2008. Surface-structure-regulated cell-membrane penetration by monolayer-protected nanoparticles. Nat. Mater. 7, 588–595.

Walkey, C.D., Olsen, J.B., Guo, H., Emili, A., Chan, W.C.W., 2012. Nanoparticle size and surface chemistry determine serum protein adsorption and macrophage uptake. J. Am. Chem. Soc. 134, 2139–2147.

Wang, Y., Xu, H., Ma, Y., Guo, F., Wang, F., Shi, D., 2011. Facile one-pot synthesis and morphological control of asymmetric superparamagnetic composite nanoparticles. Langmuir 27, 7207–7212.

Wang, H., Yang, J., Li, Y., Sun, L., Liu, W., 2013a. Combining magnetic field/temperature dual stimuli to significantly enhance gene transfection of nonviral vectors. J. Mater. Chem. B 1, 43–51.

Wang, F., Pauletti, G.M., Wang, J., Zhang, J., Ewing, R.C., Wang, Y., Shi, D., 2013c. Dual surface-functionalized janus nanocomposites of polystyrene/$Fe_3O_4@SiO_2$ for simultaneous tumor cell targeting and stimulus-induced drug release. Adv. Mater. 25, 3485–3489.

Wang, Y., Wang, K., Zhang, R., Liu, X., Yan, X., Wang, J., Wagner, E., Huang, R., 2014. Synthesis of core-shell graphitic carbon@silica nanospheres with dual-ordered mesopores for cancer-targeted photothermochemotherapy. ACS Nano 8, 7870–7879.

Wang, S., Li, K., Chen, Y., Chen, H., Ma, M., Feng, J., Zhao, Q., Shi, J., 2015a. Biocompatible PEGylated MoS_2 nanosheets: controllable bottom-up synthesis and highly efficient photothermal regression of tumor. Biomaterials 39, 206–217.

Wang, C., Chen, J.C., Zhou, X., Li, W., Liu, Y., Yue, Q., et al., 2015b. Magnetic yolk-shell structured anatase-based microspheres loaded with Au nanoparticles for heterogeneous catalysis. Nano Res. 8, 238–245.

Wang, H., Han, R.L., Yang, L.M., Shi, J.H., Liu, Z.J., Hu, Y., Wang, Y., Liu, S.J., Gan, Y., 2016a. Design and synthesis of core-shell-shell upconversion nanoparticles for NIR-induced drug release, photodynamic therapy, and cell imaging. ACS Appl. Mater. Interfaces 8, 4416–4423.

Wang, G., Zhang, F., Tian, R., Zhang, L., Fu, G., Yang, L., Zhu, L., 2016b. Nanotubes-embedded indocyanine green-hyaluronic acid nanoparticles for photoacoustic-imaging-guided phototherapy. ACS Appl. Mater. Interfaces 8, 5608–5617.

Wang, H., Han, R.L., Yang, L.M., Shi, J.H., Liu, Z.J., Hu, Y., Wang, Y., Liu, S.J., Gan, Y., 2016c. Design and synthesis of core-shell-shell upconversion nanoparticles for NIR-induced drug release, photo dynamic therapy, and cell imaging. ACS Appl. Mater. Interfaces 8, 4416–4423.

Wang, Z., Chang, Z., Lu, M., Shao, D., Yue, J., Yang, D., Li, M., Dong, W.F., 2017. Janus silver/silica nanoplatforms for light-activated liver cancer chemo/photothermal therapy. ACS Appl. Mater. Interfaces 9, 30306–30317.

Wang, Y., Liu, Y., Xia, K., Zhang, Y., Yang, J., 2019a. $NiCo_2S_4$ nanoparticles anchoring on polypyrrole nanotubes for high-performance supercapacitor electrodes. J. Electroanal. Chem. 840, 242–248.

Wang, Y., Chen, Y., Liu, Y., Liu, W., Zhao, P., Li, Y., Dong, Y., Wang, H., Yang, J., 2019b. Urchin-like $Ni_{1/3}Co_{2/3}(CO_3)_{0.5}OH{\cdot}0.11H_2O$ anchoring on polypyrole nanotubes for super capacitor electrodes. Electrochim. Acta 295, 989–996.

Wen, Q., Di, J., Jiang, L., Yu, J., Xu, R., 2013. Zeolite-coated mesh film for efficient oil-water separation. Chem. Sci. 4, 591–595.

Wu, J., Chen, J., Qasim, K., Xia, J., Lei, W., Wang, B.P., 2012. A hierarchical mesh film with superhydrophobic and superoleophilic properties for oil and water separation. J. Chem. Technol. Biotechnol. 87, 427–430.

Wu, S.H., Mou, C.Y., Lin, H.P., 2013. Synthesis of mesoporous silica nanoparticles. Chem. Soc. Rev. 42, 3862–3875.

Wu, Z.C., Li, W.P., Luo, C.H., Su, C.H., Yeh, C.S., 2015. Rattle-type Fe_3O_4@CuS developed to conduct magnetically guided photoinduced hyperthermia at first and second NIR biological windows. Adv. Funct. Mater. 25, 6527–6537.

Wu, M., Meng, Q., Chen, Y., Zhang, L., Li, M., Cai, X., Li, Y., Yu, P., Zhang, L., Shi, J., 2016. Large pore-sized hollow mesoporous organosilica for redox-responsive gene delivery and synergistic cancer chemotherapy. Adv. Mater. 28, 1963–1969.

Xia, Y., Yang, P., Sun, Y., Wu, Y., Mayers, B., Gates, B., Yin, Y., Kim, F., Yan, H., 2003. One-dimensional nanostructures: synthesis, characterization, and applications. Adv. Mater. 15, 353–389.

Xiaoa, T., Zhoua, L., Suna, X.-Q., Huangb, F., Linc, C., Wang, L., 2020. Supramolecular polymers fabricated by orthogonal self-assembly based on multiple hydrogen bonding and macrocyclic host–guest interactions. Chin. Chem. Lett. 31, 1–9.

Xin, F., Rui-Rui, G., Zhang, Q., Rao, S.-J., Zheng, X.-L., Qu, D.-H., Tiana, H., 2016. Phototriggered supramolecular polymerization of a [c2]daisy chain rotaxane. Polym. Chem. 7, 2166–2170.

Xu, Z.Z., Wang, C.C., Yang, W.L., Deng, Y.H., Fu, S.K., 2004. Encapsulation of nanosized magnetic iron oxide by polyacrylamide via inverse miniemulsion polymerization. J. Magn. Magn. Mater. 277, 136–143.

Xu, Z.P., Zeng, Q.H., Lu, G.Q., Yu, A.B., 2006. Inorganic nanoparticles as carriers for efficient cellular delivery. Chem. Eng. Sci. 61, 1027–1040.

Yan, P., Zhao, N., Hu, H., Lin, X., Liu, F., Xu, F.J., 2014. A facile strategy to functionalize gold nanorods with polycation brushes for biomedical applications. Acta Biomater. 10, 3786–3794.

Yang, X., Grailer, J.J., Rowland, I.J., Javadi, A., Hurley, S.A., Steeber, D.A., Gong, S., 2010. Multifunctional SPIO/DOX-loaded wormlike polymer vesicles for cancer therapy and MR imaging. Biomaterials 31, 9065–9073.
Yang, P., Gai, S., Lin, J., 2012. Functionalized mesoporous silica materials for controlled drug delivery. Chem. Soc. Rev. 41, 3679–3698.
Yang, W., Liu, S., Bai, T., Keefe, A.J., Zhang, L., Ella-Menye, J.R., Li, Y., Jiang, S., 2014a. Poly(-carboxybetaine) nanomaterials enable long circulation and prevent polymer-specific antibody production. Nano Today 9, 10–16.
Yang, H.C., Pi, J.K., Liao, K.J., Huang, H., Wu, Q.Y., Huang, X.J., Xu, Z.K., 2014b. Silica-decorated polypropylene microfiltration membranes with a mussel-inspired intermediate layer for oil-in-water emulsion separation. ACS Appl. Mater. Interfaces 6, 12566–12572.
Yang, P., Ding, Y., Lin, Z., Chen, Z., Li, Y., Qiang, P., et al., 2014c. Low-cost high-performance solid-state asymmetric supercapacitors based on MnO_2 nanowires and Fe_2O_3 nanotubes. Nano Lett. 14, 731–736.
Yang, Y., Liu, X., Lv, Y., Herng, T.S., Xu, X., Xia, W., Zhang, T., Fang, J., Xiao, W., Ding, J., 2015a. Orientation mediated enhancement on magnetic hyperthermia of Fe_3O_4 nanodisc. Adv. Funct. Mater. 25, 812–820.
Yang, S., Cao, C., Sun, Y., Huang, P., Wei, F., Song, W., 2015b. Nanoscale magnetic stirring bars for heterogeneous catalysis in microscopic systems. Angew. Chem. Int. Ed. 54, 2661–2664.
Yang, H.Y., Jang, M.S., Gao, G.H., Lee, J.H., Lee, D.S., 2016. pH responsive biodegradable polymeric micelles with anchors to interface magnetic nanoparticles for MR imaging in detection of cerebral ischemic area. Nanoscale 8, 12588–12598.
Ye, M., Zorba, S., He, L., Hu, Y., Maxwell, R.T., Farah, C., Zhang, Q., Yin, Y., 2010. Self-assembly of superparamagnetic magnetite particles into peapod-like structures and their application in optical modulation. J. Mater. Chem. 20, 7965–7969.
Ye, F., Barrefelt, A., Asem, H., Abedi-Valugerdi, M., El-Serafi, I., Saghafian, M., Abu-Salah, K., Alrokayan, S., Muhammed, M., Hassan, M., 2014. Biodegradable polymeric vesicles containing magnetic nanoparticles, quantum dots and anticancer drugs for drug delivery and imaging. Biomaterials 35, 3885–3894.
Yue, Q., Zhang, Y., Wang, C., Wang, X.Q., Sun, Z.K., Hou, X.F., et al., 2015. Magnetic yolk–shell mesoporous silica microspheres with supported Au nanoparticles as recyclable high-performance nanocatalysts. J. Mater. Chem. A 3, 4586–4594.
Zhang, T., Ge, J., Hu, Y., Zhang, Q., Aloni, S., Yin, Y., 2008. Formation of hollow silica colloids through a spontaneous dissolution-regrowth process. Angew. Chem. 120, 5890–5895.
Zhang, L., Xue, H., Cao, Z., Keefe, A., Wang, J., Jiang, S., 2011. Multifunctional and degradable zwitterionic nanogels for targeted delivery, enhanced MR imaging, reduction-sensitive drug release, and renal clearance. Biomaterials 32, 4604–4608.
Zhang, F., Zhang, W.B., Shi, Z., Wang, D., Jin, J., Jiang, L., 2013a. Nanowire-haired inorganic membranes with superhydrophilicity and underwater ultralow adhesive superoleophobicity for high-efficiency oil/water separation. Adv. Mater. 25, 4192–4198.
Zhang, L., Zhong, Y., Cha, D., Wang, P., 2013b. A self-cleaning underwater superoleophobic mesh for oil-water separation. Sci. Rep. 3, 2326.
Zhang, E., Cheng, Z., Lv, T., Qian, Y., Liu, Y., 2015a. Anti-corrosive hierarchical structured copper mesh film with superhydrophilicity and underwater low adhesive superoleophobicity for highly efficient oil-–water separation. J. Mater. Chem. A 3, 13411–13417.
Zhang, Y., Ang, C.Y., Li, M., Tan, S.Y., Qu, Q., Luo, Z., Zhao, Y., 2015b. Polymer-coated hollow mesoporous silica nanoparticles for triple-responsive drug delivery. ACS Appl. Mater. Interfaces 7, 18179–18187.
Zhang, L., Chen, Y., Li, Z., Li, L., Saint-Cricq, P., Li, C., Lin, J., Wang, C., Su, Z., Zink, J.I., 2016. Tailored synthesis of octopus-type Janus nanoparticles for synergistic actively-targeted and chemo-photo thermal therapy. Angew. Chem. Int. Ed. 55, 2118–2121.
Zhang, Q., Shen, C.N., Zhao, N., Xu, F.J., 2017. Redox-responsive and drug-embedded silica nanoparticles with unique self-destruction features for efficient gene/drug codelivery. Adv. Funct. Mater. 27, 1606229.

Zhang, Y., Ren, K., Zhang, X., Chao, Z., Yang, Y., Ye, D., Dai, Z., Liu, Y., Ju, H., 2018a. Photo-tearable tape close-wrapped up conversion nano capsules for near-infrared modulated efficient siRNA delivery and therapy. Biomaterials 163, 55–66.

Zhang, J.-G., Zhang, X.-Y., Yu, H., Luo, Y.-L., Xu, F., Chen, Y.-S., 2018b. Preparation, self-assembly and performance modulation of gold nanoparticles decorated ferrocene-containing hybrid block copolymer multifunctional materials. J. Ind. Eng. Chem. 65, 224–235.

Zhao, Z., Liang, F., Zhang, G., Ji, X., Wang, Q., Qu, X., Song, X., Yang, Z., 2015a. Dually responsive janus composite nanosheets. Macromolecules 48, 3598–3603.

Zhao, N., Lin, X., Zhang, Q., Ji, Z., Xu, F.J., 2015b. Redox-triggered gatekeeper-enveloped starlike hollow silica nanoparticles for intelligent delivery systems. Small 11, 6467–6479.

Zheng, X., Yan, B., Wu, F., Zhang, J., Qu, S., Zhou, S., Weng, J., 2016. Supercooling self-assembly of magnetic shelled core/shell supraparticles. ACS Appl. Mater. Interfaces 8, 23969–23977.

Zhou, W., He, W., Zhong, S., Wang, Y., Zhao, H., Li, Z., Yan, S., 2009. Biosynthesis and magnetic properties of mesoporous Fe_3O_4 composites. J. Magn. Magn. Mater. 321, 1025–1028.

Zhou, J., Meng, L., Feng, X., Zhang, X., Lu, Q., 2010. One-pot synthesis of highly magnetically sensitive nanochains coated with a highly cross-linked and biocompatible polymer. Angew. Chem. Int. Ed. 49, 8476–8479.

Zhou, J., Zhang, W., Hong, C., Pan, C., 2015. Silica nanotubes decorated by pH-responsive diblock copolymers for controlled drug release. ACS Appl. Mater. Interfaces 7, 3618–3625.

Zhu, Y.K., Chen, D.J., 2017. Novel clay-based nanofibrous membranes for effective oil/water emulsion separation. Ceram. Int. 43, 9465–9471.

Zou, H., Wu, S., Ran, Q., Shen, J., 2008. A simple and low-cost method for the preparation of monodisperse hollow silica spheres. J. Phys. Chem. C 112, 11623–11629.

CHAPTER 5

Characterization and physicochemical properties of nanomaterials

Zia Ur Rahman Farooqi[a], Abdul Qadeer[a], Muhammad Mahroz Hussain[a], Nukshab Zeeshan[a], and Predrag Ilic[b]
[a]Institute of Soil and Environmental Sciences, Faculty of Agriculture, University of Agriculture Faisalabad, Faisalabad, Pakistan
[b]PSRI Institute for Protection and Ecology of the Republic of Srpska, Banja Luka, Banja Luka, Republic of Srpska, Bosnia and Herzegovina

1. Introduction

Nanomaterials are a set of substances having at least one dimension less than 100 nm (nanometer) in size, where 1 nm is one millionth part of 1 mm, which is approximately 100,000 times smaller than the diameter of a human hair. Nanomaterials have unique qualities and attract scientists due to their unique optical, magnetic, and electrical properties, which can be used in electronics, medicine, and environmental clean-up. These materials are the cause of the emergence of a new subject area, called nanotechnology. This subject area has the potential to revolutionize the existing electronics, medicinal, agricultural, and environmental technologies. These materials also have significant commercial importance and can increase these benefits if properly managed and applied (Saleh, 2017; Kalarikkal et al., 2018). Nanotechnology is an emerging field of science that includes synthesis and development of various nanomaterials. "Nano" originates from a Latin word, which means dwarf. Presently, nanoparticles are being produced from variety of feed-stock like different heavy metals (copper, zinc, titanium, magnesium, gold, alginate, and silver, etc.) (Dong et al., 2015), and different plant tissues and extracts (Singh et al., 2016b).

Different technologies and processes are used for the synthesis of nanoparticles; for example, hydrothermal synthesis in nanoparticles is generated at a wide temperature range from room temperature to very high temperatures (Singh et al., 2016b), the solvothermal method, in which solvent is treated under moderate to high pressure (between 1 and 10,000 atm) and temperature between 100°C and 1000°C (Nunes et al., 2019), combustion, in which thermally induced redox reaction takes place between fuel and oxidant to convert feedstock to nanoparticles (Houshiar et al., 2014), sol gel method, which uses a wet-chemical technique for nanoparticles production (Aziz et al., 2013), and biosynthesis, which is the conversion of living organism's products into nanoparticles using microbes, etc. (Malik et al., 2017). Nanoparticles are being used for diverse

Nanomaterials: Synthesis, Characterization, Hazards and Safety
https://doi.org/10.1016/B978-0-12-823823-3.00005-7

purposes, from medical treatments, use in various branches of industry production such as solar and oxide fuel batteries for energy storage, to wide incorporation into diverse materials for everyday use such as cosmetics or clothes (Dubchak et al., 2010). Nanotechnology is quite a hotspot because of its applications in food sciences technology (Ranjan et al., 2014), biotechnology, and its other allied fields, e.g., biomedical engineering, delivery of drugs in the body, and tissue engineering (Dasgupta et al., 2017). Metal-based nanoparticles (NPs) are being used as a part of our regular life activities in different forms like cosmetics, electronic devices (Luo et al., 2006), textile fabrics (Patra and Gouda, 2013), and antimicrobial paints (Kumar et al., 2008). Metal oxide nanoparticles have also been used in biomedicine and are being applied in the pharmaceutical industry, for example, protein detection, gene therapy, drug delivery, tissues engineering, and cancer therapy (Brigger et al., 2012). Metallic oxide nanoparticles are also used in environmental clean-up acting as remedial tools for heavy metals from soils and water and as a carbon sequestration source (cerium oxide nanoparticles and carbon nanotubes and nanofibers), act as an eco-friendly approach for pollution remediation (Yoshikawa et al., 2014; Recillas et al., 2010; Lau et al., 2016), and as a source of soil fertility via supply of nano-fertilizers to nutrient-deficient soils (Chhipa, 2017).

If we talk about the advantages of nanoparticles, they are one of the best choices as an eco-friendly approach for environmental pollutants remediation as they act effectively to remove pollutants in contaminated soils and supply nutrients to soils, providing farmers with monetary as well as yield benefits. However, despite the benefits, different types of nanomaterials can cause adverse effects to the environment, humans, and other organisms by causing toxicity. Many studies done on test animals have indicated that prolonged exposure to nanoparticles can cause multiple organ toxicities like microbial and plant death (Tang et al., 2018; Cox et al., 2016), and humans and animals can face hepatotoxicity and carcinogenicity through food, contaminated water, and direct ingestion (Tortella et al., 2019).

2. Nano-materials: History, use and types

2.1 History

2.1.1 Natural occurrence

Naturally, nanomaterials are produced through phenomena of cosmological (Pais, 2005), geological (Simakov, 2018), meteorological, and biological origins. A significant fraction of these is nano-sized interplanetary dust, which is continuously falling on the earth at the rate of thousands of tons per year (Simakov et al., 2015) the same phenomenon produces dust particles in the air and soil. In addition, many viruses have diameters in the nanomaterials range.

2.1.2 Preindustrial technology

Historically, nanomaterials were unknowingly used without comprehension of their nature, i.e., in by ancient glassmakers and potters (John and Plane, 2012). Later, nanomaterials used for this purpose were identified as silver and copper nanoparticles dispersed in the glassy glaze.

2.2 Use of nanomaterials

Because of the unique and discrete properties, which are transitional to those of bulk matter and individual molecules, nanomaterials have gained increased attention. Distinguished on the basis of these properties and core material nanoparticles can be either engineered or natural (Rajput et al., 2017). The use of nanomaterials has increased in the medical field, the construction industry, agriculture, analytical chemistry, and electronic fabrication, etc. because of their peculiar properties, which had increased their production up to 2000 tons by 2004 and was estimated to reach up to 60,000 tons by 2012–14 (Table 1). Nanowaste management and application as fertilizers are considered to present a risk of introducing nanoparticles into the soil, in addition to exposure through intended use (Ebbs et al., 2016).

Nanometals are usually used in soil ecosystems because of their strong antimicrobial and antiviral characteristics, which cause adversative impacts on the persistence, mobility functions, and reproduction of soil microorganisms (Cornelis et al., 2014), dependent on the concentrations of NPs used, their size, their shape, and their reactivity within the soil setting (Cumberland and Lead, 2009). The property of constancy in nanomaterials control their toxicity and there are many factors that influence their stability in any media, such as environmental pH, capping agent, configuration of background electrolyte, and ionic strength (Pachapur et al., 2016). Nanomaterials' behavior across the soil ecosystem also fluctuates with various forms of nanomaterials, which cause chemical changes, possible alterations, and bio-availability in the soil setting (Nowack et al., 2015).

The concentrations of most nanomaterials are unknown but according to exposure modeling, it is proposed that the amounts for these nanomaterials are greater in soil in contrast to water and air because of their ultimate use as fertilizers (Milani et al., 2012). Study of

Table 1 Categories of nanomaterials.

Sr. no.	Nanomaterial category	Examples	References
1	Nanoenabled	Food products	Flari et al. (2011)
2	Nanofiber	Nanodrugs, nanosensors, nanofilters	De Jong and Geus (2000)
3	Nanoparticle	Metal nanoparticles, e.g., silver nanoparticles	Khan et al. (2019)
4	Nanoplate	Nickel phosphides nanoplates	Yu et al. (2016)
5	Nanostructure	Titanium dioxide nanostructures	Shen et al. (2018)
6	Nanotube	Carbon nanotubes	O'connell (2018)
7	Nanowire	Metallic nanowires, e.g., copper nanowires	Li et al. (2017)

the fate of nanomaterials and their toxicological hazards into the soil is an imperative goal. Nanoparticles like zinc oxide, titanium oxide, and copper oxide have shown influential impacts on organic carbon of soil and the number of microorganism species present in the soil. The influence of nanomaterials upon activity, composition, and biomass of soil microbial communities should be known as they are sensitive indicators of environmental stresses like antimicrobial agents and heavy metals' presence and their response in the soil. Soil microorganisms are very sensitive to change in soil environment and they respond negatively toward environmental stresses when they are exposed to, for example, iron oxide nanoparticles (NPs). Iron oxide NPs can stimulate soil enzyme activities. The change in the state of oxidation of Fe, i.e., the so-called redox wheel, may play a key part in P, N, and S accessibility and uptake in plants as well as playing a vital role in nutrient recycling (Tourinho et al., 2012).

2.3 Types of nanomaterials, uses, and their safety concerns

2.3.1 Carbon nanotubes and fibers

Carbon nanotubes (CNTs) are used in electric equipment such as transistors, sensors, capacitors, electrodes, organic light emitting diodes, and in solar cells (Schnorr and Swager, 2011) and for wastewater treatment (Chen et al., 2016). Derivatives like carbon nanotubes-nanofluids are also used as solar energy collectors (Sidik et al., 2017).

There are some drawbacks to using CNTs. These side effects occur when CNTs are not used with proper safety measures and in permissible concentrations. Some drawbacks and health effects associated with CNTs are breathing problems after their accidental inhalation, asthma, lung disorders, renal and hepatic malfunctioning, as well as neural damage (Hassan et al., 2020; Koiranen et al., 2017).

2.3.2 Ceramic nanoparticles

Ceramic nanoparticles are used as treatment aids in radiation therapies and bone repair (McKinnon, 2018) and in energy supply and storage (Kaur et al., 2017). Despite all the benefits, they are not fully commercialized due to safety concerns. Some toxicity concerns related to ceramic nanoparticles are induction of oxidative stress and inflammation, although this varies from cell to cell. The associative effects of oxidative and inflammation effects include cytotoxic activity in the lungs, liver, heart, and brain. The effects can also include prethrombotic effects in heart function, and could also induce genotoxicity, carcinogenicity, and teratogenicity in exposed organisms. It has also been reported that NPs could cross the blood-brain barrier and cause toxicity in the brain (Singh et al., 2016a; Bessa et al., 2018).

2.3.3 Metal oxide nanoparticles

Among all types of nanomaterials, NPs are the most discussed nanomaterials having wide applicability and benefits ranging from agriculture to environmental clean-up. They have various types, as described below.

2.3.4 Silver NPs

Silver NPs are among the most effective NPs due to their scientifically proved antimicrobial activities (Gong et al., 2007) and beneficial uses in the textile industry, wastewater treatment, and sunscreen lotions, etc. They can be synthesized through the process of biosynthesis by plants such as *Azadirachta indica*, *Capsicum annuum*, and *Carica papaya* (Jha and Prasad, 2010).

2.3.5 Gold NPs

Gold NPs are widely used in biochemical studies to identify the interactions between proteins and act as tracers in DNA fingerprinting for possible detection of DNA presence in samples. These are also used as identification materials in the presence of antibodies like streptomycin, gentamycin, and neomycin. Another form of gold NP, gold nanorods, are used in detection of cancerous cells, their diagnosis, and treatment as well as identification of bacterial classes (Avnika and Garg, 2013).

2.3.6 Alloy NPs

Alloy NPs contain a mixture of one or more metals or their NPs. So, they can exhibit structural properties different from the individual metals (Ceylan et al., 2006). As silver has the highest electrical conductivity among metals, its NPs with associated metals are widely used. Alloy NPs properties are in between the properties of their constituent metals, hence showing advantages over ordinary metallic NPs (Mohl et al., 2011).

2.3.7 Magnetic NPs

Magnet-based NPs like Fe_3O_4 (magnetite) and Fe_2O_3 (hematite) are known to be biocompatible with each other. These have been extensively studied for their potential use for treatment of cancer patients, stem cell sorting and manipulation, drug delivery, gene therapy, DNA analysis, and magnetic resonance imaging (MRI) (Fan et al., 2009).

2.3.8 Paramagnetic NPs

Paramagnetic NPs are nanomaterials that can be manipulated with a magnetic field. These NPs are manufactured using metals having magnetic properties, i.e., cobalt, nickel, iron, and their respective oxides, such as magnetite, cobalt ferrite, and chromium dioxide. Para-magnetic NPs have greater magnetic ability than other conventional agents and are used in the medical field for diagnostic and therapeutic purposes (Cuenca et al., 2006).

2.3.9 Carbon-based NPs

The most studied carbon-based NPs are carbon nanotubes (CNTs). These are graphene sheets that are wrapped or rolled into a tube. Due to their strong structural stability (100 times stronger than steel) and lighter weight, they are used as alternatives to steel and metals that are mainly used for the structural reinforcement. Additionally, CNTs are unique in that they are conductors along the length and nonconductors crossways. Due to electrical conductivity, structure, high strength, and electron affinity, CNTs have wide commercial applications (Khan et al., 2017).

2.3.10 Polymeric NPs

Polymeric NPs are organic in nature. They can be nanocapsular or nanospherical according to their method of preparation. Nanospherical NPs have structure resembling a matrix, and capsular NPs have a core-shell like structure. They are advantageous due to their control release property, protection of drug molecules, ability to combine in therapy and imaging, having specific targets, high biodegradability, and biocompatibility (Khan et al., 2017).

2.3.11 Semiconductor NPs

Semiconductor NPs are the NPs with properties between the metals and metalloids/ nonmetals. They are widely used in photo-catalysis, electrical equipment and devices, etc. Some examples are GaN, GaP, InP, ZnO, ZnS, CdS, CdSe, and CdTe (Klaine et al., 2008).

Numerous studies have shown that the transportation of NPs happens in natural environments through biotic and the abiotic components present in an ecosystem. The fate, transportation kinetics, and harmfulness of NPs may be influenced by physical processes (disaggregation, aggregation, and adsorption), biological processes, and chemical dissolution (complication, oxidation, and reduction reactions). Since the environment has a great influence on transportation and accumulation of contaminants in soil, we can say that NPs are also affected by their respective environment. The transformed physicochemical features of NPs may affect the transport and removal of contaminants bound with the NPs (Joo and Zhao, 2017).

3. Synthesis of nanomaterials

Scientists are developing and synthesizing novel nanomaterials with better properties, high applicability, and efficiencies for environmental clean-up and sustaining life on the Earth (Shibata et al., 1998). In general, there are only two approaches for synthesis of nanomaterials: top-down and bottom-up. Top-down uses crushing, milling, or grinding, while bottom-up synthesizes the nanomaterial atom-by-atom, molecule-by-molecule, or cluster-by-cluster.

3.1 Hydrothermal synthesis

Hydrothermal synthesis is the production of nano-materials using autoclave by reacting the feedstock with aqueous solution. Autoclaving raises the temperature to boiling and the pressure to the point of saturation of vapors. This method is specially used for the synthesis of titanium oxide NPs (Yang et al., 2017). In this method, grain size, particle morphology, crystalline phase, and surface chemistry can be controlled through manipulation of the solution composition, reaction temperature, pressure, solvent properties, additives, and aging time of the aqueous phase in the autoclave (Carp et al., 2004).

3.2 Solvothermal method

The solvothermal method is similar to the hydrothermal method with one exception. In this method, numerous types of solvents can be used, in contradiction to hydrothermal synthesis, in which only one solvent can be used. This method is considered to be the most versatile method for nanomaterial synthesis as it produces nanomaterials with a variety of size and shape distributions. Titanium oxide NPs are produced using this method (Nagaveni et al., 2002).

3.3 Chemical vapor deposition (CVD)

This method is used in semiconductor and related industries for production of highly pure and high performance NPs and films. In this process, the feedstock is subjected to volatile substances and their reaction decomposes the feedstock and NPs are produced (Kim et al., 2004). Cao et al. (2005) prepared Sn^{4+} and TiO_2 NPs following this method. The advantage of using this method is that it produces uniform coating of NPs. However, it has some limitations, as this process requires a high temperature, and is thus difficult to scale-up (Gracia et al., 2004).

3.4 Combustion

The process of combustion produces substances with crystalline structure and greater surface area. This phenomenon, which is used for NPs production, involves rapid heating of a solution containing redox groups up to 650°C for 1–2 min, which makes the material crystalline. Since the time is so short, the transition from anatase to rutile is inhibited (Nagaveni et al., 2002).

3.5 Conventional sol-gel method

Sol-gel is a synthetic method for nanomaterials synthesis. In this process, colloidal suspension, called sol, is generated from hydrolysis and polymerization of the feedstocks (Pierre, 1998). This method has the advantage of controlling the texture, chemical, and morphological properties of the nanomaterials. It is also a convenient method as it allows impregnation or co-precipitation and includes molecular scale mixing, high purity of the precursors, and homogeneity of the sol-gel products with a high purity of physical, morphological, and chemical properties (Kolen'ko et al., 2005). Gel is a factor that affects the properties of the product only (Chen et al., 2009).

3.6 Biosynthesis

Biosynthesis of nanomaterials through microorganisms is an eco-friendly technology. In this method, a variety of microbes are used to synthesize metallic NPs. The synthesis of NPs may be intracellular or extracellular (Hulakoti and Taranath, 2014). This method produces NPs by uptake of metal ions into microbe cells and NPs production by the action of certain enzymes (Narayanan and Sakthivel, 2010).

4. Use of nanomaterials in environmental clean-up

Rapid advancements in the field of science and technology, especially nanotechnology, has led to increase in the scope of its applications in environmental clean-up, especially heavy metals contaminated soil and water (Cai et al., 2013). The growing interest in nanomaterials applications for pollutants remediation is applied on a wide scale. Additionally, this is an eco-friendly and cost-effective tool (Fatisson et al., 2010) with advantages over the high costs and secondary pollution associated with old approaches (Ben-Moshe et al., 2010).

4.1 Soil remediation and water treatment

Synthetically produced NPs are widely used in remediation of environmental pollutants and have greater applicability for this purpose owing to the cost-effectiveness and rapid action as compared to the standard methods (Fig. 1). Numerous types of NPs are used for remediation of pollutants. Some examples of the NPs are as follows:

(i) Nano-sized calcium peroxide is used for remediation of petroleum hydrocarbons.
(ii) Nano-sized zerovalent iron is used for halogenated compounds remediation.
(iii) Nano-sized metal oxides are used for heavy metals remediation.
(iv) CNTs, bio-nanoparticles, polymeric nanoparticles, etc. are used for remediation of organic and inorganic contaminants (Sarkar et al., 2012).

4.1.1 Nano-sized calcium peroxide

Chemical oxidation is used as a remediation method to get rid of contaminants in soils. There is an increasing tendency to move from conventional to advanced methods, which are easy to use and leave no contaminant when applied. Nano-sized calcium peroxide releases the peroxide at a slower pace, which results in a longer time working as remediation of contaminants (Khodaveisi et al., 2011). Calcium peroxide NPs act speedily by increasing surface to volume ratio and reducing agglomeration of the individual moieties (Khodaveisi et al., 2011). Calcium peroxide NPs are used to remediate oil-contaminated soils. Recently two American companies have used these for the successful removal of gasoline, heating oil, and methyl tertiary butyl ether from soil (Mueller and Nowack, 2010). The oxygen produced as a result of this degrading reaction between the NPs

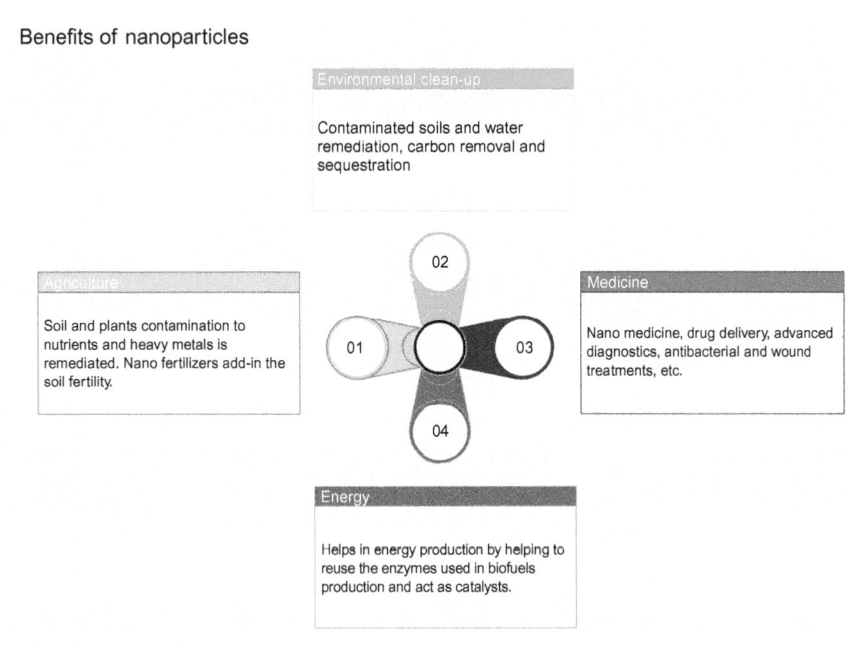

Fig. 1 Benefits of nanoparticles.

and the contaminant facilitates in creating an aerobic environment suitable for bioremediation (Mueller and Nowack, 2010), so these NPs can be used for hydrocarbon degradation over a wide pH spectrum in less time (Arienzo, 2000).

4.1.2 Nanoscale zerovalent iron

Nanoscale zerovalent iron (nZVI) is widely used in soil and water remediation. It has a range of advantages in that it can remediate a wide range of soil and water contaminants (Zhang and Elliott, 2006) and can degrade polycyclic aromatic hydrocarbons (PAHs) at room temperature. According to Zhang (2003), these NPs can effectively remediate contaminants due to the fact that they can modify the pH according to requirements. In a study, Zhang (2003) achieved treatment efficiency up to 99% of TCE (trichloroethylene).

4.1.3 Metal oxide NPs

Metal oxide NPs are known to be highly effective remediants of pharmaceuticals from contaminated water. Han et al. (2015) used manganese oxide NPs for metal contaminated soil remediation effectively. He observed that at pH around 6–7, manganese oxide NPs acted fast and effectively degraded pollutants. The reactive action becomes even more pronounced when used for treating soil-sorbed estradiol.

Cerium oxide NPs are the most recently synthesized NPs. They have wide applicability and have proved useful in stopping bioaccumulation contamination in soils and plants (Pradhan and Parida, 2010). A mixture of iron and cerium oxide NPs is reported to have the potential for tannery and other industrial pollutants like heavy metals (Vivekananthan et al., 2019). Cerium oxide nanoparticles, being semiconducting in nature, display photo-catalysis. This phenomenon is also important for the degradation of dyes that contaminate soil (Korsvik et al., 2007).

4.1.4 Other nanomaterials

There are numerous other nanomaterials that are not mentioned or less discussed in this chapter, including CNTs and associated types like single and double walled CNTs, polymeric nanoparticles (nano-wires), nanocomposites (nanocomposite of polyethylene oxide and polyethyleneimine), and bio-nanoparticles (virus, plasmids, proteins, etc.) (Zaman et al., 2014). Their properties have enabled us to remediate polluted sites and offer a broad range of environmental applications like sorbents, high-flux membranes, depth filters, antimicrobial agents, environmental sensors, renewable energy technologies, and pollution prevention strategies (Mauter and Elimelech, 2008).

4.1.5 Sorption of contaminants

Nanomaterials are widely used in remediation of organic and inorganic contaminants through sorption mechanisms. Magnetite NPs are commonly used for this purpose due to high reactivity and surface area. Sorption is a process commonly used for the remediation of pollutants that are in dissolved forms, like metals, organic, and biological materials in soil or water mediums. Thus, the magnetic behavior of the magnetite is used to extract the metals and other pollutants from the soil or water medium in a remediation processes. This method has several advantages, i.e., it can remediate pollution in situ as well as generate valuable metals through recovery (Saleh, 2016).

4.1.6 Inorganic compounds

With the recent interest in using NPs for removal of toxic metals from soil and water, this method has demonstrated advantages and cost-effectiveness over conventional remediation strategies (Zhao et al., 2016). NPs can effectively remove metals like arsenic and chromium from contaminated water through adsorption, as the capacity of the NPs is increased when size is decreased (Liu et al., 2015) as compared to conventional sorbent/adsorbents (Mayo et al., 2007; Mohan and Pittman Jr., 2007). Variations in pH, temperature, and ionic strength can also affect adsorption capacity of NPs, as arsenic sorption is decreased when ionic strength is increased and decreased when there is increase in particle aggregation and decrease in surface area. High temperature increases adsorption of arsenic due to increase in the kinetic activity and increase in the mass transfer rate from solution to surface (Mayo et al., 2007).

4.1.7 Organic contaminants

Generally, organic compounds are attached to each other through hydrogen bonding, van der Waals, surface complexation reactions, i.e., green rust (Kone et al., 2011), as well as hydrophobic interactions (Li-Zhi and Hans Christian, 2016). Contaminants adsorption using NPs has been studied using a wide range of pH (Illés and Tombácz, 2006) and they have been shown to remove tetracyclines efficiently from contaminated water. A total of 439,440 kinetics and extents of sorption of nalidixic acid (quinolone antibiotic) have been recently investigated by Usman et al. While the effect of ionic strength was of less significance, the adsorption was found to be strongly affected by particle size and surface properties of nanoparticles (Usman et al., 2014).

4.2 Soil fertility management (nanofertilizers, etc.)

The leaf and crop residues litter decomposition in soil is a critical step in the mineralization of organic nutrients, the production of soil organic matter (SOM), and the balance of carbon in terrestrial ecosystems, which has a vital role in the promotion of the balance of nutrients in soil and normal material cycle (Rashid et al., 2017). In soil ecosystems, maintaining the soil fertility and ensuring the availability of nutrients is crucial. Soil is an excellent carrier of litter, so by improving the rates of litter decomposition can stimulate the nutrient cycle of soil and soil quality can also be improved (Fang et al., 2012). While the quality and quantity of litter and its quantity used plays an imperative role in the nutrient cycle and soil carbon storage (Kaye et al., 2005), there is considerable proof that the litter substrate quality and decomposition rates of litter in soil have prominent effects on soil nutrients in the soil (Salamanca et al., 2003).

4.3 Climate change mitigation

Global warming has observable impacts on climate change, which is causing adverse impacts not only on the environment but also on life on Earth (Peñuelas et al., 2013). Global warming is causing undeniable effects on the environment like sea level rise, melting of glaciers, shifting of global cropping patterns, habitat loss, and extinction of animals and plants. Loss of diversity is becoming more and more common because of global warming and climate change (Willis et al., 2004). Some metal-based NPs tend to persist in the environment. As they have minor dissolution, they get accumulated in soil and water media (Pradhan et al., 2011). Indeed, NPs get partly dissolved and form agglomerates and aggregates in soil media. Kinetic experiments have shown that nanometal oxides of TiO_2 are very stable after forming agglomerates and aggregates in soil solutions. Nanoparticles with such long-term stability can be freely taken up, if they are present in soil's water pores and cause bioaccumulation (Zhao et al., 2013). Bio-magnification may also occur if these NPs interact with organic components inside the living organisms, e.g., carbohydrates and lipids (Tourinho et al., 2012). Many studies have reported the role of magnetite modifications in improving soil microbial biomass carbon and extracellular enzymes activities: a slight change in concentration of any compound in the soil causes an impression on soil health and also greatly influences the magnetism of soil. Magnetite nanoparticles not only increase the rate of degradation of organic material in soil but also enhance the microbial biomass carbon in soil (Fang et al., 2012).

Soil ecosystems are very susceptible to changes because microbial communities respond to exotic stresses sensitively. Furthermore, there are several instances proving changes like alteration in microbial communities within soils because of minor differences in soil functions, physical properties, and chemical characteristics (Lin et al., 2018). Therefore, the impact of exotic pressures on native microbial population conformation and the function aspects of these communities are often recounted. For example, the stress of heavy metals and their oxides in soil significantly decreases the soil microbial populations. Pollution of heavy metals in the soil is one of the major factors disturbing the microbial communities present in soil (Chen et al., 2014).

Limited studies have demonstrated the opposite effects related to nanoparticles and their impact on the environment, both generally and on soil characteristics in specific. Elliott and Zhang (2001) observed the impacts of bimetallic (Fe/Pd) nanoparticles when these NPs were injected into the soil to test the area to determine the reduction of chlorinated aliphatic hydrocarbons and trichloroethene. Nanoparticles do not affect the porosity of soil, and clogging, has also been shown to be minimal. Fullerene NPs (carbon nanotubes) have been observed to have a slight influence on the soil's populations of microbes and on soil microbial progressions (Nafees et al., 2020). There is a notable reduction in the soil microbial biomass carbon and a decrease in nitrogen as well as extracellular enzyme activity. Metal-based NPs have also been investigated to find the effect of ZnO and TiO_2 on soil microbial populations and it is observed that both the nanoparticles reduce the microbial mass in soil and their diversity in soil. Pradhan et al. (2011) examined the impacts of silver nanoparticles and copper oxide on decomposition of leaf litter showing that exposure of the microbes to these nanoparticles may lead to a reduction in rate of leaf decay. Nanoparticles cause changes in microbial species diversity as well as causing changes in their ecological functions. Nanoparticles not only disturb the native microbial diversity but also hinder the activities of these communities by inhibiting enzyme activity in the soil. It is shown that silver nanoparticles hinder the actions of dehydrogenase in the soil (Burke et al., 2015), while iron oxide nanoparticles can stimulate the enzymatic activities when added to the soil; however different behaviors are observed on different concentration. Growing use of NPs in the environment has raised concerns over how the release of NPs in the surroundings would affect the health of ecosystems and safety of humans. Soil enzymatic activities are the main biological indicators of soil health. These enzymes are usually acid phosphatases, arylsulfatase, β-glucosaminidase, β-glucosidase, and peroxidases. Nanoparticles are found to affect these enzymes activities by causing an increase or decrease in activity depending upon the concentration used. Zinc oxide NPs are used to enhance nutrient availability—this regulates nutrient availability but also causes adverse impacts on soil microbial biomass (You et al., 2018).

Greenhouse gases emission is responsible for global warming, causing worldwide climate change. The major greenhouse gases are carbon dioxide, nitrous oxide, methane, water vapors, chlorofluorocarbons, and ozone, which are responsible for this global climate change. The largest contributors of these gases to the environment are agricultural practices, industrial processes, and the burning of fossil fuels (Ravindranath and Sathaye, 2002). Carbon dioxide contributes 81%, methane contributes 10%, nitrous oxide contributes to climate change by 6%, and fluorinated carbons contribute 3% toward global warming which causes global climate change. There is a direct link between technology and climate change. Advancements in technology cause the changes in the environment, which in turn place an adverse impact on the environment (Smita et al., 2012). Carbon dioxide is the product of everyday economic activities (Bin and Dowlatabadi, 2005). Global concentrations of CO_2 are continuously increasing and out of all causes, nanotechnology plays an indirect part in the release of CO_2 and affects the organic matter present in the soil matrix. Addition of crop leaves litter in soil plays a substantial part in these emissions, which is influenced by various factors including litter quality, litter concentrations and soil microbial biomass carbon, which is the one factor of utmost importance (Kaye et al., 2005).

5. Advantages of nanomaterials use

Nanoparticles have been proposed as a cost-effective and environmentally friendly method for pollutants remediation from soil and water, being an effective tool against these pollutants and providing an easy and reliable approach toward sustainability. These are effective and suitable as they are eco-friendly as well as cost-effective (Aziz et al., 2013; DiStasio et al., 2018).

Nanotechnology deals with the remediation of environmental pollutants. Their nanoscale dimensions provide them larger surface area compared to volume, so they have very specific properties. For example, zinc oxide nanoparticles has great potential for applications due to antibacterial, antifungal, antidiabetic, antiinflammatory, wound healing, and antioxidant properties. Toxic chemicals are remediated by these nanoparticles by adsorption, immobilization, conversion to other forms, or binding to a specific animal or plant tissues (Agarwal et al., 2017; Wang et al., 2019; Chauhan, 2020). Magnetic nanoparticles are used in production of energy, i.e., biogas, bioethanol, and biodiesel, by increasing the production volume of biogas alongside the enhanced methane content, immobilizing biofuels-producing enzymes at the end of the process so that these can be reused again, acting as a catalyst, increasing the thermal and storage stability of the enzymes (Muhammad and Badshah, 2019). These nanoparticles also act as antibacterial agents to kill any unwanted bacteria by production of reactive oxygen species, cation release, biomolecule damages, ATP depletion, and membrane interaction (Slavin et al., 2017).

6. Safety concerns of nanomaterials

Behind the use of every useful product, there are disadvantages when it is not used properly or used in excess. Nanoparticles work best when these are correctly applied for the desired operation. When their concentration exceeds safe limits, they can pose numerous health and environmental risks. Different organisms in the environment, including microbes, plants, animals, and humans, exposed to nanomaterials can suffer adverse health outcomes as it is seen that, sooner or later, toxicity establishes (Fig. 2).

6.1 Sources of exposure and routes of entry

Nanomaterials enter into the human body through the dermal route, ingestion, or inhalation. Humans are exposed to nanomaterials during their production, use, or disposal stages (Fig. 3).

6.2 Impacts on humans and animals

Nanoparticles cause toxicity to humans and animals, affecting cells and organs and causing cytotoxicity and even organ failure. They cause allergy, cytotoxicity, genotoxicity, carcinogenicity, immunotoxicity, teratogenicity, neurotoxicity, and hepatotoxicity.

6.2.1 Allergic effects

Inhalation of MONPs leads to pulmonary toxicity, i.e., ZnO nanoparticles induce pulmonary oxidative stress as well as inflammation in mice. It has been shown that release of Zn^{2+} from ZnO nanoparticles causes allergies (Horie et al., 2015; Vennemann et al., 2017; Huang et al., 2015). Bronchocentric interstitial inflammation was observed after inhalation of Ag nanoparticles. Ag nanoparticles are accumulated in the lungs and induce inflammatory responses in the peritoneum (Chuang et al., 2013).

6.2.2 Cytotoxic effects

Cytotoxicity is the toxicity in the cells of organisms. Cytotoxicity of nanoparticles was first discovered in microbes. Nanoparticles of Fe_3O_4, TiO_2, CuO, and ZnO proved to be toxic for different microbial species. These nanoparticles have antibacterial effect by inducing toxicity to the bacterial cells and causing their mortality (Stankic et al., 2016; Jeong et al., 2018).

To study the effects of nanoparticles on human health, quantitative structure-activity relationship and other models were used to predict the cytotoxicity potential. It was observed that nanoparticles can affect human health by introducing cytotoxicity in cells (Pan et al., 2016; Simeone and Costa, 2019). Results in another study indicated that nanoparticles like ZnO, CuO, and MgO produce cytotoxicity in the human vascular endothelial cells, if the exposure to these is increased, by producing intracellular reactive oxygen species and affecting permeability of plasma membrane (Sun et al., 2011).

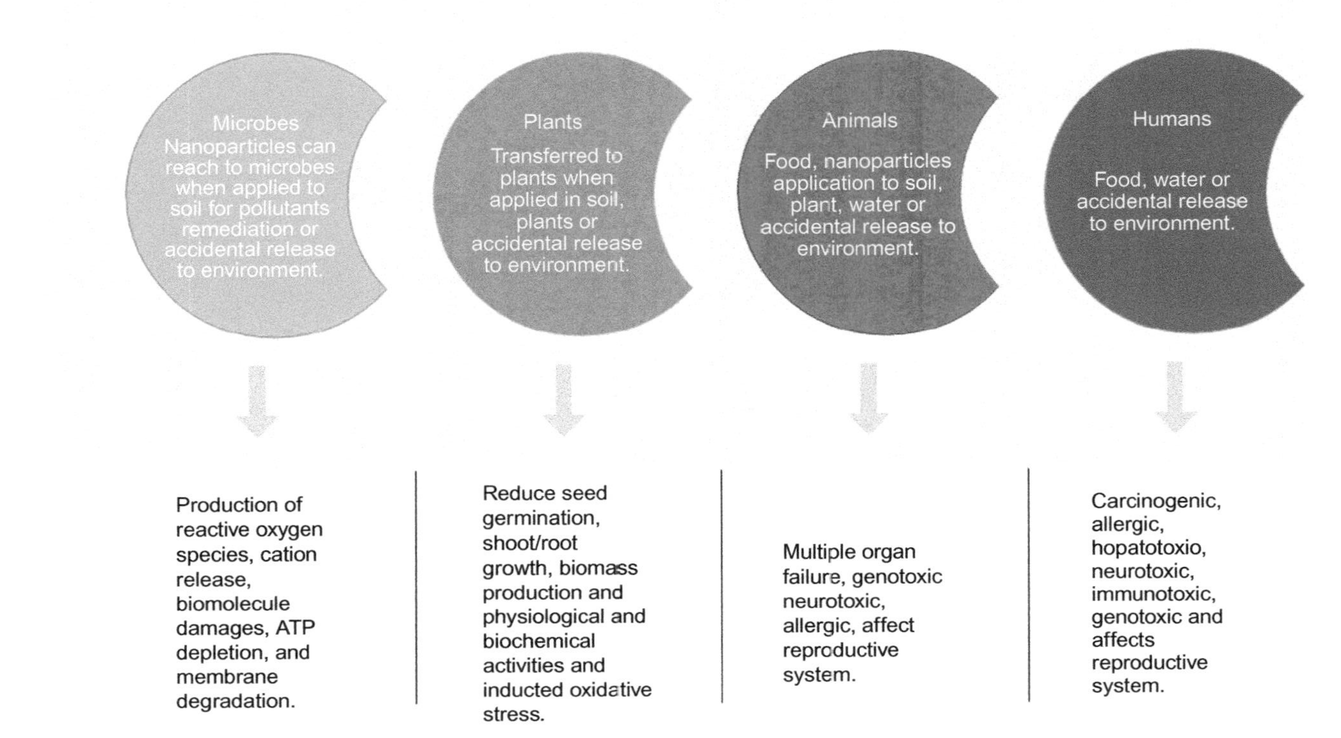

Fig. 2 Toxicities of nanoparticles to organisms.

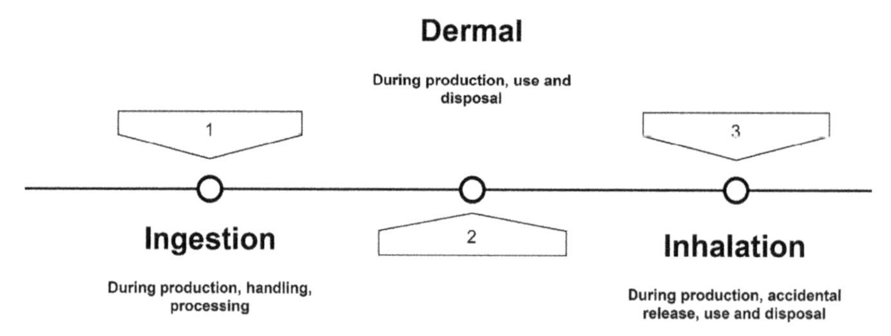

Fig. 3 Sources and routes of human exposure to nanomaterials.

Metal oxide nanoparticles were used to check the cytotoxicity of these on human epithelial cells. It was revealed that nanoparticles were able to generate ROS in these cells and induce oxidative stress (Fahmy and Cormier, 2009). The same type of experiment was done on bronchial epithelial cells and resulted in cytotoxicity (Liu et al., 2011). In another study, human and catfish cells were exposed to nanoparticles (ZnO, TiO_2, CuO and Co_3O_4) and resulted in cytotoxicity, which caused ROS production and destruction of cell and mitochondrial membranes (Wang et al., 2011). Among the nanoparticles, ZnO nanoparticles are most dangerous to human cells and have proved lethal (Lozano et al., 2011). To benefit from the cytotoxic potential of NPs, scientists are trying to destroy cancerous cells using NPs, but these studies have unconfirmed results (Rasmussen et al., 2010).

6.2.3 *Carcinogenic effects*

Due to the small size of MONPs, they are easily transported into the biological systems. In a study, genotoxic and carcinogenicity potential of cobalt, nickel, and copper nanoparticles was observed in vivo mammalian cells. It was observed that these NPs can be a potential genotoxic and carcinogenic source in animals via abnormal apoptosis, oxidative stress, and pro-inflammatory effects (Magaye et al., 2012). Nickel-based nanoparticles are also known to be carcinogenic in nature, initiating inflammation, DNA damage, micronucleus formation, and causing sarcoma and other activates that accelerate cancer (Magaye and Zhao, 2012). The same effects are summarized in another study, i.e., NPs induce genotoxicity through damaging DNA, chromosomal aberrations, DNA strand breaks, oxidative DNA damage, and mutations (Kwon et al., 2014).

6.2.4 *Neurotoxicity*

In vitro neurotoxicity studies of CeO_2-nanoparticles showed dose-dependent neurotox icity effects (Darroudi et al., 2013). Significant evidence indicates that NPs elicit toxicity in humans exposed to them by introducing blood-brain barrier oxidative stress and inflammatory mechanisms (Karmakar et al., 2014). Results of another study showed that

ZnO-nanoparticles increase the production of pro-inflammatory cytokines in the serum and the brains of mice, increased oxidative stress level, impaired learning and memory abilities, and hippocampal pathological changes (Tian et al., 2015). Some observations from laboratory experiments showed that Ag, Cu, and Al nanoparticles induce neurotoxicity by altering sensory, motor, and cognitive functions at the time of development of brain pathologies (Shanker Sharma and Sharma, 2012). In another study, NPs-induced brain damage was studied in mice indicating that NPs can even cause death (Sharma et al., 2013).

6.2.5 Immunotoxic effects

Immunotoxicity refers to reduced immune system efficiency against microbial diseases. Different NPs were found to be immunotoxic, especially CeO, ZnO, and FeO-nanoparticles. Effects on lymphocytes (Easo and Mohanan, 2015), lysosomal destabilization, phagocytosis, changes in the functions of the phagocytic cells, which ultimately decreases the ability of animals to defend themselves against pathogens and infectious diseases (Jovanović and Palić, 2012), induce mortality in fish via bioaccumulation (Gagnon et al., 2018), and can react with biological systems to reduce immune system response when surrounded by a protein corona (Corbo et al., 2016).

6.2.6 Hepatoxicity

NPs are extensively used in industry, agriculture, medicine, and cosmetic products, with promising results. But, they can pose high potential risks for human health via inducing hepatotoxicity. ZnO-nanoparticles exhibited inflammation in cells, necrosis, hydropic degeneration, hepatocytes apoptosis, anisokaryosis, karyolysis, nuclear membrane irregularity, glycogen content depletion, and hemosidrosis, which indicates that ZnO-nanoparticles can produce oxidative stress in cells, which may affect the functions of many organs and systems (Almansour et al., 2017) including disturbances to antioxidant enzymes (Volkovova et al., 2015) and damage to liver cells, as shown by increased oxidative stress and inflammation in mice (Wei et al., 2016; Gao et al., 2017).

6.2.7 Teratogenicity and reproductive disorders

Nanoparticle's extensive usage results in environmental toxicity and causes different threats. CuO nanoparticles cause decreases in antioxidant enzymes, and induce oxidative stress, apoptosis, developmental abnormalities (Ganesan et al., 2016), and even transgenerational toxicity in rats (Kadammattil et al., 2018). In zebra fish, reduced embryo and larvae survival, delay in hatching rate, tissue ulceration, and malformation and mortality are observed (Zhu et al., 2008, 2012).

7. Summary

Being an emerging technology, nanotechnology can support environmental sustainability through its wide-application. It can use nanoparticles for various beneficial uses that support sustaining life in the biosphere. It helps in pollution removal like soil and water pollutants, removal of heavy metals, and bacteria; it helps in improving climate change-affected soil fertility through nanofertilizers, acts as a climate mitigation tool by helping carbon capture and storage using cerium oxide nanoparticles, carbon nanotubes and nanofibers. NPs are also effectively used in and as medicines, and used as drug delivery agents, rapid diagnostic tools, and fast wound-healing agents. These nanoparticles can be hazardous when not properly administered or used. They reduce seed germination and plant growth, kill beneficial bacteria in the soils, and reduce its fertility. They can be released into the environment through accidental release, excessive application, or during disposal. They enter into the human body through inhalation, ingestion, or dermal contact. Their release in the environment can affect a wide array of organisms, especially animals and humans, by causing antibacterial action cell and organ disorders through cytotoxicity, immunotoxicity, neurotoxicity, and even death. So, future research directions should be directed toward their safe limits of use and proper administration so that there should be minimum hazards while using these beneficial nanoparticles.

References

Agarwal, H., Venkat Kumar, S., Rajeshkumar, S., 2017. A review on green synthesis of zinc oxide nanoparticles—an eco-friendly approach. Resour. Efficient Technol. 3, 406–413.

Almansour, M.I., Alferah, M.A., Shraideh, Z.A., Jarrar, B.M., 2017. Zinc oxide nanoparticles hepatotoxicity: histological and histochemical study. Environ. Toxicol. Pharmacol. 51, 124–130.

Arienzo, M., 2000. Degradation of 2,4,6-trinitrotoluene in water and soil slurry utilizing a calcium peroxide compound. Chemosphere 40, 331–337.

Avnika, T., Garg, G., 2013. Short review on application of gold nanoparticles. Glob. J. Pharmacol. 7, 34–38.

Aziz, M., Abbas, S.S., Baharom, W.R.W., 2013. Size-controlled synthesis of SnO_2 nanoparticles by sol–gel method. Mater. Lett. 91, 31–34.

Ben-Moshe, T., Dror, I., Berkowitz, B., 2010. Transport of metal oxide nanoparticles in saturated porous media. Chemosphere 81, 387.

Bessa, M.J., Fraga, S., Brandão, F., Fokkens, P., Boere, J., Leseman, D., Salmatonidis, A., Viana, M., Cassee, F., Teixiera, J.P., 2018. Pairwise toxicity evaluation of ceramic nanoparticles in submerged and air-liquid interface cultures of human alveolar epithelial A549 cells. In: 20th International Congress on In Vitro Toxicology (ESTIV2018), 15–18 October 2018.

Bin, S., Dowlatabadi, H., 2005. Consumer lifestyle approach to US energy use and the related CO_2 emissions. Energy Policy 33, 197–208.

Brigger, I., Dubernet, C., Couvreur, P., 2012. Nanoparticles in cancer therapy and diagnosis. Adv. Drug Deliv. Rev. 64, 24–36.

Burke, D.J., Pietrasiak, N., Situ, S.F., Abenojar, E.C., Porche, M., Kraj, P., Lakliang, Y., Samia, A.C.S., 2015. Iron oxide and titanium dioxide nanoparticle effects on plant performance and root associated microbes. Int. J. Mol. Sci. 16, 23630–23650.

Cai, L.M.T., Ma, H., Kim, H., 2013. Cotransport of titanium dioxide and fullerene nanoparticles in saturated porous media. Environ. Sci. Technol. 47, 5703.

Cao, J., Lee, S., Zhang, X., Chow, J.C., An, Z., Ho, K., Watson, J.G., Fung, K., Wang, Y., Shen, Z., 2005. Characterization of airborne carbonate over a site near Asian dust source regions during spring 2002 and its climatic and environmental significance. J. Geophys. Res. Atmos. 110, 1–8.

Carp, O., Huisman, C.L., Reller, A., 2004. Photoinduced reactivity of titanium dioxide. Prog. Solid State Chem. 32, 33.

Ceylan, A., Jastrzembski, K., Shah, S.I., 2006. Enhanced solubility Ag-Cu nanoparticles and their thermal transport properties. Metall. Mater. Trans. A 37, 2033.

Chauhan, P.S., 2020. Lignin nanoparticles: eco-friendly and versatile tool for new era. Bioresour. Technol. Rep. 9, 100374.

Chen, F.C., Wu, J.L., Lee, C.L., Hong, Y., Kuo, C.H., Huang, M.H., 2009. Plasmonic-enhanced polymer photovoltaic devices incorporating solution-processable metal nanoparticles. Appl. Phys. Lett. 95, 182.

Chen, J., He, F., Zhang, X., Sun, X., Zheng, J., Zheng, J., 2014. Heavy metal pollution decreases microbial abundance, diversity and activity within particle-size fractions of a paddy soil. FEMS Microbiol. Ecol. 87, 164–181.

Chen, J., Zhang, L., Huang, T., Li, W., Wang, Y., Wang, Z., 2016. Decolorization of azo dye by peroxymonosulfate activated by carbon nanotube: radical versus non-radical mechanism. J. Hazard. Mater. 320, 571–580.

Chhipa, H., 2017. Nanofertilizers and nanopesticides for agriculture. Environ. Chem. Lett. 15, 15–22.

Chuang, H.-C., Hsiao, T.-C., Wu, C.-K., Chang, H.-H., Lee, C.-H., Chang, C.-C., Cheng, T.-J., 2013. Allergenicity and toxicology of inhaled silver nanoparticles in allergen-provocation mice models. Int. J. Nanomedicine 8, 4495.

Corbo, C., Molinaro, R., Parodi, A., Furman, N.E.T., Salvatore, F., Tasciotti, E., 2016. The impact of nanoparticle protein corona on cytotoxicity, immunotoxicity and target drug delivery. Nanomedicine 11, 81–100.

Cornelis, G., Hund-Rinke, K., Kuhlbusch, T., van den Brink, N., Nickel, C., 2014. Fate and bioavailability of engineered nanoparticles in soils: a review. Crit. Rev. Environ. Sci. Technol. 44, 2720–2764.

Cox, A., Venkatachalam, P., Sahi, S., Sharma, N., 2016. Silver and titanium dioxide nanoparticle toxicity in plants: a review of current research. Plant Physiol. Biochem. 107, 147–163.

Cuenca, A.G., Jiang, H., Hochwald, S.N., Delano, M., Cance, W.G., Grobmyer, S.R., 2006. Emerging implications of nanotechnology on cancer diagnostics and therapeutics. Cancer 107, 459–466.

Cumberland, S.A., Lead, J.R., 2009. Particle size distributions of silver nanoparticles at environmentally relevant conditions. J. Chromatogr. A 1216, 9099–9105.

Darroudi, M., Hakimi, M., Sarani, M., Kazemi Oskuee, R., Khorsand Zak, A., Gholami, L., 2013. Facile synthesis, characterization, and evaluation of neurotoxicity effect of cerium oxide nanoparticles. Ceram. Int. 39, 6917–6921.

Dasgupta, N., Ranjan, S., Ramalingam, C., 2017. Applications of nanotechnology in agriculture and water quality management. Environ. Chem. Lett. 15, 591–605.

De Jong, K.P., Geus, J.W., 2000. Carbon nanofibers: catalytic synthesis and applications. Catal. Rev. 42, 481–510.

DiStasio, N., Lehoux, S., Khademhosseini, A., Tabrizian, M., 2018. The multifaceted uses and therapeutic advantages of nanoparticles for atherosclerosis research. Materials 11, 754.

Dong, H., Chen, Y.-C., Feldmann, C., 2015. Polyol synthesis of nanoparticles: status and options regarding metals, oxides, chalcogenides, and non-metal elements. Green Chem. 17, 4107–4132.

Dubchak, S., Ogar, A., Mietelski, J.W., Turnau, K., 2010. Influence of silver and titanium nanoparticles on arbuscular mycorrhiza colonization and accumulation of radiocaesium in *Helianthus annuus*, Spain. J. Agric. Res. 8, 103–108.

Easo, S.L., Mohanan, P.V., 2015. In vitro hematological and in vivo immunotoxicity assessment of dextran stabilized iron oxide nanoparticles. Colloids Surf. B: Biointerfaces 134, 122–130.

Ebbs, S.D., Bradfield, S.J., Kumar, P., White, J.C., Musante, C., Ma, X., 2016. Accumulation of zinc, copper, or cerium in carrot (*Daucus carota*) exposed to metal oxide nanoparticles and metal ions. Environ. Sci. Nano 3, 114–126.

Elliott, D.W., Zhang, W.-X., 2001. Field assessment of nanoscale bimetallic particles for groundwater treatment. Environ. Sci. Technol. 35, 4922–4926.

Fahmy, B., Cormier, S.A., 2009. Copper oxide nanoparticles induce oxidative stress and cytotoxicity in airway epithelial cells. Toxicol. in Vitro 23, 1365–1371.

Fan, T.X., Chow, S.K., Zhang, D., 2009. Biomorphic mineralization. MATL. Sci. 54, 542–659.

Fang, G., Si, Y., Tian, C., Zhang, G., Zhou, D., 2012. Degradation of 2, 4-D in soils by Fe_3O_4 nanoparticles combined with stimulating indigenous microbes. Environ. Sci. Pollut. Res. 19, 784–793.

Fatisson, J., Ghoshal, S., Tufenkji, N., 2010. Deposition of carboxymethylcellulose-coated zero-valent iron nanoparticles onto silica: roles of solution chemistry and organic molecules. Langmuir 26, 12832.

Flari, V., Chaudhry, Q., Neslo, R., Cooke, R., 2011. Expert judgment based multi-criteria decision model to address uncertainties in risk assessment of nanotechnology-enabled food products. J. Nanopart. Res. 13, 1813–1831.

Gagnon, C., Bruneau, A., Turcotte, P., Pilote, M., Gagné, F., 2018. Fate of cerium oxide nanoparticles in natural waters and immunotoxicity in exposed rainbow trout. J. Nanomed. Nanotechnol. 9, 489.

Ganesan, S., Anaimalai Thirumurthi, N., Raghunath, A., Vijayakumar, S., Perumal, E., 2016. Acute and sub-lethal exposure to copper oxide nanoparticles causes oxidative stress and teratogenicity in zebrafish embryos. J. Appl. Toxicol. 36, 554–567.

Gao, S., Wang, X., Wang, S., Zhu, S., Rong, R., Xu, X., 2017. Complex effect of zinc oxide nanoparticles on cadmium chloride-induced hepatotoxicity in mice: protective role of metallothionein. Metallomics 9, 706–714.

Gong, P., Li, H., He, X., Wang, K., Hu, J., Tan, W., Tan, S., Zhang, X.Y., 2007. Preparation and antibacterial activity of Fe_3O_4@Ag nanoparticles. Nanotechnology 18, 604–611.

Gracia, F., Holgado, J.P., Caballero, A., 2004. Gonzalez-Elipe, structural, optical, and photoelectrochemical properties of Mn+-TiO2 model thin film photocatalysts. J. Phys. Chem. B 108, 17466–17476.

Han, B., Zhang, M., Zhao, D., Feng, Y., 2015. Degradation of aqueous and soil-sorbed estradiol using a new class of stabilized manganese oxide nanoparticles. Water Res. 70, 288–299.

Hassan, A., Saeed, A., Afzal, S., Shahid, M., Amin, I., Idrees, M., 2020. Applications and hazards associated with carbon nanotubes in biomedical sciences. Inorg. Nano-Metal Chem. 1–12.

Horie, M., Stowe, M., Tabei, M., Kuroda, E., 2015. Pharyngeal aspiration of metal oxide nanoparticles showed potential of allergy aggravation effect to inhaled ovalbumin. Inhal. Toxicol. 27, 181–190.

Houshiar, M., Zebhi, F., Razi, Z.J., Alidoust, A., Askari, Z., 2014. Synthesis of cobalt ferrite ($CoFe_2O_4$) nanoparticles using combustion, coprecipitation, and precipitation methods: a comparison study of size, structural, and magnetic properties. J. Magn. Magn. Mater. 371, 43–48.

Huang, K.-L., Lee, Y.-H., Chen, H.-I., Liao, H.-S., Chiang, B.-L., Cheng, T.-J., 2015. Zinc oxide nanoparticles induce eosinophilic airway inflammation in mice. J. Hazard. Mater. 297, 304–312.

Hulakoti, N.I., Taranath, T.C., 2014. Biosynthesis of nanoparticles using microbes: a review. Colloids Surf. B: Biointerfaces 121, 474–483.

Illés, E., Tombácz, E., 2006. The effect of humic acid adsorption on pH-dependent surface charging and aggregation of magnetite nanoparticles. J. Colloid Interface Sci. 295, 115–123.

Jeong, J., Kim, S.-H., Lee, S., Lee, D.-K., Han, Y., Jeon, S., Cho, W.-S., 2018. Differential contribution of constituent metal ions to the cytotoxic effects of fast-dissolving metal-oxide nanoparticles. Front. Pharmacol. 9.

Jha, A.K., Prasad, K., 2010. Physical chemistry. Int. J. Green Nanotechnol. 1, 110–117.

John, M., Plane, C., 2012. Cosmic dust in the earth's atmosphere. Chem. Soc. Rev. 41, 6507–6565.

Joo, S.H., Zhao, D., 2017. Environmental dynamics of metal oxide nanoparticles in heterogeneous systems: a review. J. Hazard. Mater. 322, 29–47.

Jovanović, B., Palić, D., 2012. Immunotoxicology of non-functionalized engineered nanoparticles in aquatic organisms with special emphasis on fish—review of current knowledge, gap identification, and call for further research. Aquat. Toxicol. 118–119, 141–151.

Kadammattil, A.V., Sajankila, S.P., Prabhu, S., Rao, B.N., Rao, B.S.S., 2018. Systemic toxicity and teratogenicity of copper oxide nanoparticles and copper sulfate. J. Nanosci. Nanotechnol. 18, 2394–2404.

Kalarikkal, N., Thomas, S., Koshy, O., 2018. Nanomaterials: Physical, Chemical, and Biological Applications. .

Karmakar, A., Zhang, Q., Zhang, Y., 2014. Neurotoxicity of nanoscale materials. J. Food Drug Anal. 22, 147–160.

Kaur, M., Ishii, S., Shinde, S.L., Nagao, T., 2017. All-ceramic microfibrous solar steam generator: TiN plasmonic nanoparticle-loaded transparent microfibers. ACS Sustain. Chem. Eng. 5, 8523–8528.

Kaye, J.P., Mcculley, R., Burke, I., 2005. Carbon fluxes, nitrogen cycling, and soil microbial communities in adjacent urban, native and agricultural ecosystems. Glob. Chang. Biol. 11, 575–587.

Khan, I., Saeed, K., Khan, I., 2017. Nanoparticles: properties, applications and toxicities. Arab. J. Chem. 20.

Khan, I., Saeed, K., Khan, I., 2019. Nanoparticles: properties, applications and toxicities. Arab. J. Chem. 12, 908–931.

Khodaveisi, J., Banejad, H., Afkhami, A., Olyaie, E., Lashgari, S., Dashti, R., 2011. Synthesis of calcium peroxide nanoparticles as an innovative reagent for in situ chemical oxidation. J. Hazard. Mater. 192, 1437–1440.

Kim, C.S., Okuyama, K., Nakaso, K., Shimada, M., 2004. Direct measurement of nucleation and growth modes in titania nanoparticles generation by a CVD method. J. Chem. Eng. Jpn 37, 1379.

Klaine, S.J., Alvarez, P.J., Batley, G.E., Fernandes, T.F., Handy, R.D., Lyon, D.Y., Mahendra, S., Mclaughlin, M.J., Lead, J.R., 2008. Nanomaterials in the environment: behavior, fate, bioavailability, and effects. Environ. Toxicol. Chem. 27, 1825–1851.

Koiranen, T., Nevalainen, T., Virkki-Hatakka, T., Aalto, H., Murashko, K., Backfolk, K., Kraslawski, A., Pyrhönen, J., 2017. The risk assessment of potentially hazardous carbon nanomaterials for small scale operations. Appl. Mater. Today 7, 104–111.

Kolen'ko, Y.V., Kovnir, K.A., Gavrilov, A.I., Garshev, A.V., Meskin, P.E., Churagulov, B.R., Bouchard, M., Colbeaujustin, C., Lebedev, O.I., Vantendeloo, G., Yoshimura, M., 2005. Structural, textural, and electronic properties of a nanosized mesoporous $Zn_xTi_{1-x}O_{2-x}$ solid solution prepared by a supercritical drying route. J. Phys. Chem. B 109, 20303–20309.

Kone, T., Hanna, K., Usman, M., 2011. Interactions of synthetic Fe (II)-Fe (III) green rusts with pentachlorophenol under various experimental conditions. Colloids Surf. A Physicochem. Eng. Asp. 385, 152–158.

Korsvik, C., Patil, S., Seal, S., Self, W.T., 2007. Superoxide dismutase mimetic properties exhibited by vacancy engineered ceria nanoparticles. Chem. Commun. 10, 1056–1058.

Kumar, A., Vemula, P.K., Ajayan, P.M., John, G., 2008. Silver-nanoparticle-embedded antimicrobial paints based on vegetable oil. Nat. Mater. 7, 236.

Kwon, J.Y., Koedrith, P., Seo, Y.R., 2014. Current investigations into the genotoxicity of zinc oxide and silica nanoparticles in mammalian models in vitro and in vivo: carcinogenic/genotoxic potential, relevant mechanisms and biomarkers, artifacts, and limitations. Int. J. Nanomedicine 9, 271.

Lau, J., Dey, G., Licht, S., 2016. Thermodynamic assessment of CO_2 to carbon nanofiber transformation for carbon sequestration in a combined cycle gas or a coal power plant. Energy Convers. Manag. 122, 400–410.

Li, Y., Cui, F., Ross, M.B., Kim, D., Sun, Y., Yang, P., 2017. Structure-sensitive CO_2 electroreduction to hydrocarbons on ultrathin 5-fold twinned copper nanowires. Nano Lett. 17, 1312–1317.

Lin, C., Ma, R., Xiong, J., 2018. Can the watershed non-point phosphorus pollution be interpreted by critical soil properties? A new insight of different soil P states. Sci. Total Environ. 628, 870–881.

Liu, R., Rallo, R., George, S., Ji, Z., Nair, S., Nel, A.E., Cohen, Y., 2011. Classification nanoSAR development for cytotoxicity of metal oxide nanoparticles. Small 7, 1118–1126.

Liu, C.-H., Chuang, Y.-H., Chen, T.-Y., Tian, Y., Li, H., Wang, M.-K., Zhang, W., 2015. Mechanism of arsenic adsorption on magnetite nanoparticles from water: thermodynamic and spectroscopic studies. Environ. Sci. Technol. 49, 7726–7734.

Li-Zhi, H., Hans Christian, B.H., 2016. Synthesis and reactivity of surfactant intercalated layered iron(II)-iron(III) hydroxides. Curr. Inorg. Chem. 6, 68–82.

Lozano, T., Rey, M., Rojas, E., Moya, S., Fleddermann, J., Estrela-Lopis, I., Donath, E., Wang, B., Mao, Z., Gao, C., González-Fernández, Á., 2011. Cytotoxicity effects of metal oxide nanoparticles in human tumor cell lines. J. Phys. Conf. Ser. 304, 012046.

Luo, X., Morrin, A., Killard, A.J., Smyth, M.R., 2006. Application of nanoparticles in electrochemical sensors and biosensors. Electroanalysis 18, 319–326.

Magaye, R., Zhao, J., 2012. Recent progress in studies of metallic nickel and nickel-based nanoparticles' genotoxicity and carcinogenicity. Environ. Toxicol. Pharmacol. 34, 644–650.

Magaye, R., Zhao, J., Bowman, L., Ding, M., 2012. Genotoxicity and carcinogenicity of cobalt-, nickel-and copper-based nanoparticles. Exp. Ther. Med. 4, 551–561.
Malik, B., Pirzadah, T.B., Kumar, M., Rehman, R.U., 2017. Biosynthesis of nanoparticles and their application in pharmaceutical industry. In: Prasad, R., Kumar, V., Kumar, M. (Eds.), Nanotechnology: Food and Environmental Paradigm. Springer Singapore, Singapore.
Mauter, M.S., Elimelech, M., 2008. Environmental applications of carbon-based nanomaterials. Environ. Sci. Technol. 42, 5843–5859.
Mayo, J.T., Yavuz, C., Yean, S., Cong, L., Shipley, H., Yu, W., Falkner, J., Kan, A., Tomson, M., Colvin, V.L., 2007. The effect of nanocrystalline magnetite size on arsenic removal. Sci. Technol. Adv. Mater. 8, 71–75.
McKinnon, S.R., 2018. Investigation of the Use of Novel Ceramic Oxide Nanoparticles to Improve the Clinical Outcome of Radiation Therapies. .
Milani, N., Mclaughlin, M.J., Stacey, S.P., Kirby, J.K., Hettiarachchi, G.M., Beak, D.G., Cornelis, G., 2012. Dissolution kinetics of macronutrient fertilizers coated with manufactured zinc oxide nanoparticles. J. Agric. Food Chem. 60, 3991–3998.
Mohan, D., Pittman Jr., C.U., 2007. Arsenic removal from water/wastewater using adsorbents-a critical review. J. Hazard. Mater. 142, 1–53.
Mohl, M., Dobo, D., Kukovecz, A., Konya, Z., Kordas, K., Wei, J., Vajtai, R., Ajayan, P.M., 2011. J. Phys. Chem. 115, 9403.
Mueller, N.C., Nowack, B., 2010. Nanoparticles for remediation: solving big problem with little particles. Elements. 6, 395–400.
Muhammad, A., Badshah, M., 2019. Magnetic nanoparticles: eco-friendly application in biofuel production. In: Abd-Elsalam, K.A., Mohamed, M.A., Prasad, R. (Eds.), Magnetic Nanostructures: Environmental and Agricultural Applications. Springer International Publishing, Cham.
Nafees, M., Ali, S., Rizwan, M., Aziz, A., Adrees, M., Hussain, S.M., Ali, Q., Junaid, M., 2020. Effect of nanoparticles on plant growth and physiology and on soil microbes. In: Nanomaterials and Environmental Biotechnology. Springer, Cham, pp. 65–85.
Nagaveni, K., Gayen, A., Subbanna, G.N., Hegde, M.S., 2002. Pd-coated Ni nanoparticles by the polyol method: an efficient hydrogenation catalyst. J. Mater. Chem. 12, 3147–3151.
Narayanan, K.B., Sakthivel, A., 2010. Biological synthesis of metal nanoparticles by microbes. Adv. Colloid Interface Sci. 156, 1–13.
Nowack, B., Baalousha, M., Bornhöft, N., Chaudhry, Q., Cornelis, G., Cotterill, J., Gondikas, A., Hassellöv, M., Lead, J., Mitrano, D.M., 2015. Progress towards the validation of modeled environmental concentrations of engineered nanomaterials by analytical measurements. Environ. Sci. Nano 2, 421–428.
Nunes, D., Pimentel, A., Santos, L., Barquinha, P., Pereira, L., Fortunato, E., Martins, R., 2019. Synthesis, design, and morphology of metal oxide nanostructures. In: Nunes, D., Pimentel, A., Santos, L., Barquinha, P., Pereira, L., Fortunato, E., Martins, R. (Eds.), Metal Oxide Nanostructures. Elsevier (Chapter 2).
O'connell, M.J., 2018. Carbon Nanotubes: Properties and Applications. CRC press.
Pachapur, V.L., Larios, A.D., Cledon, M., Brar, S.K., Verma, M., Surampalli, R.Y., 2016. Behavior and characterization of titanium dioxide and silver nanoparticles in soils. Sci. Total Environ. 563, 933–943.
Pais, A., 2005. Subtle Is the Lord: The Science and the Life of Albert Einstein. Oxford University Press. ISBN 978-0-19-280672-7 Retrieved 6 December 2016.
Pan, Y., Li, T., Cheng, J., Telesca, D., Zink, J.I., Jiang, J., 2016. Nano-QSAR modeling for predicting the cytotoxicity of metal oxide nanoparticles using novel descriptors. RSC Adv. 6, 25766–25775.
Patra, J.K., Gouda, S., 2013. Application of nanotechnology in textile engineering: an overview. J. Eng. Technol. Res. 5, 104–111.
Peñuelas, J., Sardans, J., Estiarte, M., Ogaya, R., Carnicer, J., Coll, M., Barbeta, A., Rivas-Ubach, A., Llusià, J., Garbulsky, M., 2013. Evidence of current impact of climate change on life: a walk from genes to the biosphere. Glob. Chang. Biol. 19, 2303–2338.
Pierre, A.C., 1998. Introduction to Sol-Gel Processing. Kluwer Academic Publishers, Boston, p. 394.
Pradhan, G.K., Parida, K.M., 2010. Fabrication of iron-cerium mixed oxide: an efficient photo catalyst for dye degradation. Int. J. Eng. Sci. Technol. 2, 9.

Pradhan, A., Seena, S., Pascoal, C., Cássio, F., 2011. Can metal nanoparticles be a threat to microbial decomposers of plant litter in streams? Microb. Ecol. 62, 58–68.
Rajput, V.D., Minkina, T., Sushkova, S., Tsitsuashvili, V., Mandzhieva, S., Gorovtsov, A., Nevidomskyaya, D., Gromakova, N., 2017. Effect of nanoparticles on crops and soil microbial communities. J. Soils Sediments 1–9.
Ranjan, S., Dasgupta, N., Chakraborty, A.R., Samuel, S.M., Ramalingam, C., Shanker, R., Kumar, A., 2014. Nanoscience and nanotechnologies in food industries: opportunities and research trends. J. Nanopart. Res. 16, 2464.
Rashid, M.I., Shahzad, T., Shahid, M., Ismail, I.M., Shah, G.M., Almeelbi, T., 2017. Zinc oxide nanoparticles affect carbon and nitrogen mineralization of *Phoenix dactylifera* leaf litter in a sandy soil. J. Hazard. Mater. 324, 298–305.
Rasmussen, J.W., Martinez, E., Louka, P., Wingett, D.G., 2010. Zinc oxide nanoparticles for selective destruction of tumor cells and potential for drug delivery applications. Expert Opin. Drug Deliv. 7, 1063–1077.
Ravindranath, N.H., Sathaye, J.A., 2002. Climate Change and Developing Countries. Springer.
Recillas, S., Colón, J., Casals, E., González, E., Puntes, V., Sánchez, A., Font, X., 2010. Chromium VI adsorption on cerium oxide nanoparticles and morphology changes during the process. J. Hazard. Mater. 184, 425–431.
Salamanca, E.F., Kaneko, N., Katagiri, S., 2003. Rainfall manipulation effects on litter decomposition and the microbial biomass of the forest floor. Appl. Soil Ecol. 22, 271–281.
Saleh, T.A., 2016. Nanocomposite of carbon nanotubes/silica nanoparticles and their use for adsorption of Pb (II): from surface properties to sorption mechanism. Desalin. Water Treat. 57, 10730–10744.
Saleh, T.A., 2017. Advanced Nanomaterials for Water Engineering, Treatment, and Hydraulics. IGI Global.
Sarkar, S., Guibal, E., Quignard, F., SenGupta, A.K., 2012. Polymer-supported metals and metal oxide nanoparticles: synthesis, characterization, and applications. J. Nanopart. Res. 14, 715.
Schnorr, J.M., Swager, T.M., 2011. Emerging applications of carbon nanotubes. Chem. Mater. 23, 646–657.
Shanker Sharma, H., Sharma, A., 2012. Neurotoxicity of engineered nanoparticles from metals. CNS Neurol. Disord. Drug Targets 11, 65–80.
Sharma, A., Muresanu, D.F., Patnaik, R., Sharma, H.S., 2013. Size- and age-dependent neurotoxicity of engineered metal nanoparticles in rats. Mol. Neurobiol. 48, 386–396.
Shen, S., Chen, J., Wang, M., Sheng, X., Chen, X., Feng, X., Mao, S.S., 2018. Titanium dioxide nanostructures for photoelectrochemical applications. Prog. Mater. Sci. 98, 299–385.
Shibata, S., Aoki, K., Yano, T., Yamane, M., 1998. Preparation of silica microspheres containing Ag nanoparticles. J. Sol-Gel Sci. Technol. 11, 279–287.
Sidik, N.A.C., Yazid, M.N.A.W.M., Samion, S., 2017. A review on the use of carbon nanotubes nanofluid for energy harvesting system. Int. J. Heat Mass Transf. 111, 782–794.
Simakov, S.K., 2018. Nano- and micron-sized diamond genesis in nature: an overview. Geosci. Front. 9, 1849–1858.
Simakov, S.K., Kouchi, A., Mel Nik, N.N., Scribano, V., Kimura, Y., Hama, T., Suzuki, N., Saito, H., Yoshizawa, T., 2015. Nanodiamond finding in the hyblean shallow mantle xenoliths. Nat. Sci. Rep. 5, 1–8.
Simeone, F.C., Costa, A.L., 2019. Assessment of cytotoxicity of metal oxide nanoparticles on the basis of fundamental physical–chemical parameters: a robust approach to grouping. Environ. Sci. Nano 6, 3102–3112.
Singh, D., Singh, S., Sahu, J., Srivastava, S., Singh, M.R., 2016a. Ceramic nanoparticles: recompense, cellular uptake and toxicity concerns. Artif Cells Nanomed. Biotechnol. 44, 401–409.
Singh, P., Kim, Y.-J., Zhang, D., Yang, D.-C., 2016b. Biological synthesis of nanoparticles from plants and microorganisms. Trends Biotechnol. 34, 588–599.
Slavin, Y.N., Asnis, J., Häfeli, U.O., Bach, H., 2017. Metal nanoparticles: understanding the mechanisms behind antibacterial activity. J. Nanobiotechnol. 15, 65.
Smita, S., Gupta, S.K., Bartonova, A., Dusinska, M., Gutleb, A.C., Rahman, Q., 2012. Nanoparticles in the environment: assessment using the causal diagram approach. Environ. Health 11, S13.

Stankic, S., Suman, S., Haque, F., Vidic, J., 2016. Pure and multi metal oxide nanoparticles: synthesis, antibacterial and cytotoxic properties. J. Nanobiotechnol. 14, 73.

Sun, J., Wang, S., Zhao, D., Hun, F.H., Weng, L., Liu, H., 2011. Cytotoxicity, permeability, and inflammation of metal oxide nanoparticles in human cardiac microvascular endothelial cells. Cell Biol. Toxicol. 27, 333–342.

Tang, J., Wu, Y., Esquivel-Elizondo, S., Sørensen, S.J., Rittmann, B.E., 2018. How microbial aggregates protect against nanoparticle toxicity. Trends Biotechnol. 36, 1171–1182.

Tian, L., Lin, B., Wu, L., Li, K., Liu, H., Yan, J., Liu, X., Xi, Z., 2015. Neurotoxicity induced by zinc oxide nanoparticles: age-related differences and interaction. Sci. Rep. 5, 16117.

Tortella, G., Rubilar, O., Durán, N., Diez, M., Martínez, M., Parada, J., Seabra, A., 2019. Silver nanoparticles: toxicity in model organisms as an overview of its hazard for human health and the environment. J. Hazard. Mater. 390, 121974.

Tourinho, P.S., van Gestel, C.A., Lofts, S., Svendsen, C., Soares, A.M., Loureiro, S., S., 2012. Metal-based nanoparticles in soil: fate, behavior, and effects on soil invertebrates. Environ. Toxicol. Chem. 31, 1679–1692.

Usman, M., Martin, S., Cimetière, N., Giraudet, S., Chatain, V., Hanna, K., 2014. Sorption of nalidixic acid onto micrometric and nanometric magnetites: experimental study and modeling. Appl. Surf. Sci. 299, 136–145.

Vennemann, A., Alessandrini, F., Wiemann, M., 2017. Differential effects of surface-functionalized zirconium oxide nanoparticles on alveolar macrophages, rat lung, and a mouse allergy model. Nano 7, 280.

Vivekananthan, V., Chandrasekhar, A., Alluri, N.R., Purusothaman, Y., Kim, W.J., Kang, C.N., Kim, S.J., 2019. A flexible piezoelectric composite nanogenerator based on doping enhanced lead-free nanoparticles. Mater. Lett. 249, 73–76.

Volkovova, K., Handy, R.D., Staruchova, M., Tulinska, J., Kebis, A., Pribojova, J., Ulicna, O., Kucharská, J., Dusinska, M., 2015. Health effects of selected nanoparticles in vivo: liver function and hepatotoxicity following intravenous injection of titanium dioxide and Na-oleate-coated iron oxide nanoparticles in rodents. Nanotoxicology 9, 95–105.

Wang, Y., Aker, W.G., Hwang, H.-M., Yedjou, C.G., Yu, H., Tchounwou, P.B., 2011. A study of the mechanism of in vitro cytotoxicity of metal oxide nanoparticles using catfish primary hepatocytes and human HepG2 cells. Sci. Total Environ. 409, 4753–4762.

Wang, X., Huang, C., Li, X., Xie, C., Yu, S., 2019. PVA-encapsulated palladium nanoparticles: eco-friendly and highly selective catalyst for hydrogenation of nitrobenzene in aqueous medium. Chem. Asian J. 14, 2266–2272.

Wei, Y., Li, Y., Jia, J., Jiang, Y., Zhao, B., Zhang, Q., Yan, B., 2016. Aggravated hepatotoxicity occurs in aged mice but not in young mice after oral exposure to zinc oxide nanoparticles. NanoImpact 3–4, 1–11.

Willis, K., Bennett, K., Walker, D., 2004. The evolutionary legacy of the ice ages-papers of a discussion meeting held at the Royal Society on 21 and 22 May 2003-introduction. Philos. Trans. R. Soc. Lond. B Biol. Sci. 359, 157–158.

Yang, Q., Xu, Q., Jiang, H.L., 2017. Metal–organic frameworks meet metal nanoparticles: synergistic effect for enhanced catalysis. Chem. Soc. Rev. 46, 4774–4808.

Yoshikawa, K., Sato, H., Kaneeda, M., Kondo, J., 2014. Synthesis and analysis of CO_2 adsorbents based on cerium oxide. J. CO2 Util. 8, 34–38.

You, T., Liu, D., Chen, J., Yang, Z., Dou, R., Gao, X., Wang, L., 2018. Effects of metal oxide nanoparticles on soil enzyme activities and bacterial communities in two different soil types. J. Soils Sediments 18, 211–221.

Yu, X.-Y., Feng, Y., Guan, B., Lou, X.W.D., Paik, U., 2016. Carbon coated porous nickel phosphides nanoplates for highly efficient oxygen evolution reaction. Energy Environ. Sci. 9, 1246–1250.

Zaman, M., Ahmad, E., Qadeer, A., Rabbani, G., Khan, R.H., 2014. Nanoparticles in relation to peptide and protein aggregation. Int. J. Nanomedicine 9, 899.

Zhang, W.X., 2003. Nanoscale iron particles for environmental remediation: an overview. J. Nanopart. Res. 5, 323–332.

Zhang, W.X., Elliott, D.W., 2006. Applications of iron nanoparticles for groundwater remediation. Remediation 16, 7–21.

Zhao, L., Sun, Y., Hernandez-Viezcas, J.A., Servin, A.D., Hong, J., Niu, G., Peralta-Videa, J.R., Duarte-Gardea, M., Gardea-Torresdey, J.L., 2013. Influence of CeO_2 and ZnO nanoparticles on cucumber physiological markers and bioaccumulation of Ce and Zn: a life cycle study. J. Agric. Food Chem. 61, 11945–11951.
Zhao, X., Liu, W., Cai, Z., Han, B., Qian, T., Zhao, D., 2016. An overview of preparation and applications of stabilized zero-valent iron nanoparticles for soil and groundwater remediation. Water Res. 100, 245–266.
Zhu, X., Zhu, L., Duan, Z., Qi, R., Li, Y., Lang, Y., 2008. Comparative toxicity of several metal oxide nanoparticle aqueous suspensions to Zebrafish (*Danio rerio*) early developmental stage. J. Environ. Sci. Health A 43, 278–284.
Zhu, X., Tian, S., Cai, Z., 2012. Toxicity assessment of iron oxide nanoparticles in zebrafish (*Danio rerio*) early life stages. PLoS ONE. 7, e46286.

CHAPTER 6

Nanotoxicology-toxicology of nanomaterials and incidental nanomaterials

Aisha A. Waris[a], Tabinda Athar[a], Hina Fatima[b], and Madiha Nisar[a]
[a]Institute of Soil and Environmental Sciences, Faculty of Agriculture, University of Agriculture Faisalabad, Faisalabad, Pakistan
[b]School of Applied Biosciences, Kyungpook National University, Daegu, South Korea

1. Introduction

The use of nanostructured elements by human beings from prehistoric times has been confirmed. In the last few decades, this technology has grown exponentially as nanoparticles and nanostructures have been introduced into multiple industrial processes and products, but the concern about their deleterious impacts is also growing as these nanoscale pieces react chemically with surfaces (Preethi et al., 2019). Currently, scientists and researchers are mainly focusing on the beneficial aspects of this technology, and very limited research has been carried out with regard to their interaction with and impact on biological systems (Sohail et al., 2019). Nanoparticles (NPs) can enter the body through inhalation, ingestion through food, and penetration into the skin cells and tissues, while the level of toxicity that different nanoparticles can cause depends on their physiochemical properties. There has been exponential growth in terms of production and use, but its toxicity risks on humans, animals, plants, and the overall environment are still not well enough researched (Elsaesser and Howard, 2012). We all know that nanomaterials are used in food packaging, in agricultural food production, as nanosensors for identification of pesticides, in the cosmetics industry, information technology, electronics, fertilizer production, for detection of impurities in water, and much more (Sohail et al., 2019). Therefore, it is essential that the toxicological side of these particles is explored, and potential risks are seriously considered. In recent times, many studies have been conducted in order to estimate the risks of nanoparticles, and they have led to the conclusion that their use may provide benefits but at the same time safety standards need to be established and maintained (Preethi et al., 2019). This assessment will not only help in understating the bio-nano interaction, but also provide a direction for safe and better designing of these particles to avoid toxicity to biological entities during and after use (Marmiroli et al., 2019).

Nanomaterials: Synthesis, Characterization, Hazards and Safety
https://doi.org/10.1016/B978-0-12-823823-3.00003-3

2. Natural, engineered, incidental nanoparticles

Natural nanoparticles are naturally abundant and are not harmful in any aspect. However, incidental nanoparticles have become unintentionally abundant regionally and globally due to human activities (Bursten et al., 2016), while engineered nanoparticles have been introduced very recently with improved properties so that maximum benefits could be derived from them (Hochella et al., 2019). Compared to natural and incidental nanoparticles, engineered nanoparticles pose advantages and disadvantages to our ecosystem. Since the introduction of engineered nanoparticles, they have changed in terms of type, reactivity, and complexity (Stöber et al., 1968). The nanoparticles, natural, incidental and engineered, all change physically and chemically with time, they pass into the ecosystem, alter their shape and size, and go through dissolution and evaporation processes, therefore, their exploration in biological systems needs an engineered and economic approach (McNutt, 2017).

3. Properties that lead to nanomaterial toxicity

The unprecedented use of nanoparticles in the food industry has raised severe concerns about their consequences on human health. Plentiful work has been done on risk assessment and safety issues of NPs addition to food products. However, owing to the lack of standard protocols for risk assessment and safety evaluation of NPs in food products, it is still a challenge and an unclear subject to identify the consequences of NPs incorporation in foods (Preethi et al., 2019). Cytotoxicity of NPs depends mostly on their physiochemical properties, i.e., toxicity of NPs varies with respect to their size, shape, doses, core/shell type, and surface characteristics, i.e., surface charge and catalytic properties (Anandharamakrishnan and Parthasarathi, 2019).

3.1 Size and shape

Morphological characteristics (size and shape) have always received great attention owing to their imperative influence on the properties of NPs. Both the size and shape of NPs have vital roles in controlling the reactivity, types, and distribution of reactive sites, toxicity, tendency for aggregation, and transportation physics. The slightest change in the surface characteristics of NPs plays a decisive role in the fate and reactivity of NPs (Hochella et al., 2019). For instance, smaller-sized NPs show greater toxicity compared to larger sized ones, as smaller-sized NPs have greater translocation tendency to the epithelial cells with protracted circulation and retention time followed by deep penetration in the tissues and prolonged retention in organs (Gnach et al., 2015). Mishra et al. (2016) conducted a study using three primary particle sizes—10, 50, and 100 nm—to explore the size-dependent cytotoxicity of Ag NPs in human liver-derived hepatoma (HepG2) cells. It was observed that Ag NPs with 10 nm size exhibited higher cellular toxicity by perturbations in the autophagy lysosomal system and inflammation activation compared to 50 nm-Ag NPs and 100 nm Ag NPs. Similarly, another study used Ag NPs of different sizes, i.e., 10, 20, 75, and 100 nm, at a dose of 20 mg/mL; among those, 10 nm Ag NPs displayed substantial alteration in the permeability of human gut epithelium (T84 Cells) (Williams et al., 2016) however, owing to their small size they can be utilized as antitumor drugs. Ag NPs with a particle size of 13.2 ± 4.72 nm and concentration range 0.5–2.0 μg Ag/mL showed high toxicity for two types of tumor cells: HeLa (adhesive cells) and U937 (suspension cells), suggesting the usefulness of Ag NPs for cancer therapy (Kaba and Egorova, 2015).

Additionally, the characteristics of NPs are not only dependent on size but are also shape-dependent, especially coercivity of nanocrystals, their catalytic activity, and melting points. In addition, the selectivity, electrical, and optical properties of NPs are highly shape-dependent (Ghosh Chaudhuri and Paria, 2012). NPs are not all of spherical shape but are also available in cubic, prism, hexagon, octahedron, disk, wire, rod, and tube shapes (Sukhanova et al., 2018). Nanoparticles' shape often varies with respect to the method of synthesis. For instance, iron oxide NPs fabricated by aerosol/vapor (pyrolysis) method displayed spherical shape with size 5–60 nm, whereas iron oxide NPs synthesized by microemulsion method had cubic shape with 4–15 nm size (Ali et al., 2016). Toxicity of NPs also varies with respect to the shape of NPs. Generally, spherical nanoparticles are less toxic compared to rod- and tube-shaped NPs. For instance, a study conducted to identify the impact of CeO_2 NPs shape on the in vitro toxicity revealed that CeO_2 NPs with rod shape showed significant LDH release and TNF-α production compared to octahedron/cube CeO_2 (Forest et al., 2017). Likewise, another study evaluated the toxicological effects of flat and spherical-shaped Ag NPs on zebrafish embryos. Both types of nanoparticles displayed considerable toxic effects on zebrafish embryos, whereas Ag NPs with flat shape were more hazardous compared to nanospheres (Abramenko et al., 2018).

3.2 Composition

NPs can largely be divided into two basic categories: simple or composite. Simple NPs are generally composed of one type of material and composite NPs (co-NPs) are made up of two different types of materials. Both the outer shell (outer material) and core (inner material) of NPs can be composed of inorganic or organic compounds. However, often the selection of material for the outer shell of the NPs depends on their end use. For instance, there are four possible combinations of core/shell NPs, i.e., inorganic/inorganic, organic/organic, organic/inorganic, and inorganic/organic (Ghosh Chaudhuri and Paria, 2012). Simple NPs could be further subdivided into organic, inorganic, and carbon-based NPs. Among organic NPs, the most common ones are dendrimers, liposomes, and micelles. Organic NPs like micelles and liposomes have a hollow core and are nontoxic, biodegradable, and sensitive to heat and light. Owing to these distinctive characteristics they are effectively being used in drug delivery systems with special use in targeted drug delivery (Ealias and Saravanakumar, 2017). Recently, Zhang et al. (2016) reported the use of SeBDP NPs (diselenide-containing fluorescent molecules) and SePTX NPs (containing antitumor drug paclitaxel). Assemblage of both SeBDP and SePTX in the form of co-NPs showed selectivity of toxicities between tumor and normal cells with higher proliferation inhibition in cancerous cells compared to normal cells.

Contrary to this, inorganic NPs are usually composed of metals and metal oxides. Metals like Cd, Cu, Co, Pb, Al, Au, Zn, and Fe can be converted to nanometer sizes by either destructive or constructive methods, which in turn impart unique characteristics like high surface area, surface charge, pore size, and surface charge density to the NPs. Metal NPs possess distinctive optical properties and owing to this, they find applications in numerous areas. For instance, use of Au NPs as a coating material is very common in scanning electron microscope sampling (Khan et al., 2019). Besides this, Ag NPs have found applications in the medicinal, environmental, and industrial sectors. Ag NPs have a tendency to destroy strains of both gram positive and negative bacteria. However, their biocidal property largely depends on their size and stability (Akter et al., 2018). Further, metal oxide NPs are synthesized to increase the reactivity and efficiency of the NPs compared to their respective metals. For example, Fe_2O_3 NPs were found to be more efficient compared to Fe NPs owing to increased reactivity. Apart from this, inorganic nanoparticles of metal oxides like Al_2O_3, TiO_2, MnO_2, ZrO_2, ZnO, MgO, and CeO_2 have been synthesized and are being used extensively for adsorption of noxious pollutants (Tyagi et al., 2017).

A third type of NPs are carbon-based NPs, which are composed solely of carbon. They can be categorized mainly as carbon nanotubes (CNTs), graphene, and fullerenes. CNTs are tube like structures composed of a single sheet of graphene (layer of carbon atoms) with a diameter of <1 nm. CNTs can be single layered, double layered, and multilayered (composed of multiple layers of interlinked nanotubes) with a length ranging from micrometers to millimeters (Yang et al., 2019). Moreover, fullerene is also an allotrope of carbon, which is generally composed of carbon atoms interlinked with each other by sp2 hybridization forming a closed or partially closed mesh. Fullerene molecules can be of various shapes, i.e., hollow spheres, tubes, and ellipsoids. Single layered fullerene is composed of 28–1500 carbon atoms with a diameter of 8.2 nm and multilayered fullerenes have a diameter in the range of 4–36 nm (Clancy et al., 2018; Ealias and Saravanakumar, 2017).

3.3 Solubility

The solubility of NPs is one of the vital factors for NPs risk assessments, since the toxicity of the NPs depends on solubility and dissolution rate. To date there are no standard protocols available to evaluate the dissolution rate and bio-durability (half-life) of the NPs (Arts et al., 2016). However, there are a couple of parameters like pH and temperature that have been used to assess the solubility of NPs. The solubility of the media under assessment can be influenced by the constituents of the media under investigation. For instance, organic and inorganic contents in the media under test can increase or decrease the dissolution rate. Hence, it is imperative to consider the effect of the media components on dissolution of NPs (Mu et al., 2014). Additionally, solubility and reactivity of NPs greatly depend on the physical and chemical properties of the environment surrounding. Hence, while measuring the solubility of NPs, the media used for measurement should simulate the target environment as closely as possible, since some specific interactions present in complex biological systems could be absent in simple solutions. Thus, deriving results from simple solutions while trying to predict the activity in complex biological liquids could potentially add errors to the results (Van der Merwe and Pickrell, 2018).

Moreover, a robust review of the literature shows that dissolution of NPs is strongly affected by the pH, particle size, and the crystal form of the NPs. A study recently reported the impact of particle size, pH, and crystal forms on the dissolution rate of Zn, ZnO, and TiO_2 NPs and their bulk analogues at body temperature. It was noticed that half-lives of bulk analogues were either greater of equal to their counter NPs. Also, dissolution behavior of NPs was significantly affected by their particle size and crystal forms (Avramescu et al., 2017). Leitner et al. (2019) conducted a study to identify the impact of particle size on the dissolution behavior of CuO NPs. Smaller and spherical NPs of 2 nm size revealed higher solubility compared to their counter bulk analogues. On the other hand, a decrease in the toxicity of nZVI was observed with increase in particle size. Crystal phase also plays an important role in the related toxicity and dissolution of NPs. For instance, among two crystal phases analyzed, α-Fe_2O_3 and γ-Fe_2O_3, α-Fe_2O_3 showed more toxicity toward the algal cells compared to γ-Fe_2O_3 NPs. NPs synthesized in oxygen-rich conditions displayed better solubility compared to their less oxidized counterparts owing to existence of oxygen induced defects in the highly oxidized NPs, having friable structure. An increase in the dissolution rate of NPs might be due to increase in the surface area. Higher dissolution rate in turn lead to higher cytotoxicity toward bacterial cells, suggesting their role as potential antibacterial agents (Helmlinger et al., 2016). The abovementioned discussion emphasizes the point that higher solubility of NPs is important for their proper reactivity. Hence, few studies solely highlight ways to improve the solubility of the NPs in order to improve their efficiency, bioavailability, and toxicity. Packaging of the drug in mesoporous silica NPs is increasingly popular because mesoporous silica NPs provide stability and solubility to the drugs (Meka et al., 2018). NPs have also been used to improve curcumin solubility and bioavailability by encapsulation in saponin-coated curcumin NPs. Bioavailability of curcumin NPs was recorded as 8.9-fold higher compared to free curcumin (Peng et al., 2018).

3.4 Surface characteristics

Surface and surface characteristics of NPs play a vital role in their activity, toxicity, and overall biological application. For instance, surface to volume ratio of NPs determines the numbers of active molecules available on the surface of nanoparticles. NPs with smaller sizes offer greater numbers of active surface molecules (Sukhanova et al., 2018). When nanoparticle size falls below 100 nm, the number of surface expressed molecules increases exponentially. Hence, there is an inverse relationship between the surface-active molecules and the size of the NPs. Decrease in size of NPs in turn increase the surface area of NPs with greater numbers of external surface expressed molecules compared to internal. Greater surface area means greater reactivity of NPs, which plays an essential role in defining chemical and biological properties of NPs (Krug and Wick, 2011).

Moreover, reduction in the size of the NPs could create irregular crystal plans, which can in turn cause disruption in the electronic configuration and eventually give rise to modified electronic properties. Following the change in the electronic configuration there could be generation of active, passive, hydrophobic, and hydrophilic functional groups on the surface of NPs depending on the composition of the NPs (Navya and Daima, 2016). Apart from this, surface ionic distribution, surface charge, stability of NPs, aggregation, and surface corona have specific roles in determining the surface reactivity of NPs and their use in toxicological and biomedical applications. Among those mentioned above, surface ionic distribution, charge, and surface area will be discussed in detail.

3.4.1 Surface charge

Surface charge of NPs is one of the most crucial physiochemical properties, having an imperative role in controlling the stability and aggregation of NPs. In addition, surface charge of NPs has an imperative role in determining their toxicological and biomedical applications also providing insight about their nano-bio interaction in different environmental systems (Navya and Daima, 2016). Apart from this, solution pH, ionic strength, and surface charge have vital roles in the aggregation of NPs especially under dispersed situations. In addition, electrostatic interactions on the bio-nano interface are greatly influenced by the type of surface charge. Charge on NPs' surfaces can be generated in numerous ways, i.e., by existence of crystal defects and presence of ionizable functional groups (Clavier et al., 2016). There can be three types of charges on NPs' surfaces, i.e., positive, negative, and neutral. Generally, positively charged NPs have shown greater toxicity compared to negative and neutral NPs. This could be due to the electrostatic attraction between the negatively charged cell glycoproteins and positively charged NPs. Owing to this, positively charged NPs have a greater tendency to enter the cells and cause toxicity. On the other hand, if the NPs have negative charge, then repulsion could be expected between the negative charge-bearing NPs and cell proteins. Hence, the process of nanoparticle entry could be slower compared to their counterparts (Sukhanova et al., 2018) It was confirmed by a study conducted to analyze the effect of positive and negative charge-bearing polystyrene NPs on HeLa and NIH/3T3 cells. It was observed that positively charged NPs were more toxic compared to negative charge NPs owing to their greater ability to penetrate and strong attachment with the negatively charged DNA (Liu et al., 2011). For instance, recently a study was conducted to assess the role of surface charge in the cellular uptake of NPs. For this purpose, seven different types of polystyrene NPs with identical cores were employed. A positive correlation was observed between the amount of internalized polystyrene NPs and the zeta potential. Hence, it was concluded that surface charge has a major role in the cellular uptake, and a minor role in other properties like aggregation, surface corona development, and effect of compositional elements (Jeon et al., 2018). To date, various compounds like polyethylene glycol, NH_2, or SH groups, methotrexate, polyethyleneimine, and dextran and folic acid have been used to alter the surface and surface charge of various NPs.

Surface ion distribution is another important characteristic of NPs. However, little importance has been given to this property compared to others. Keeping in mind the Monte Carlo Simulations, there are three different types of ion distribution of NPs, i.e., heterogeneous surface distribution, homogenous surface distribution, and one charge in the center (Landau and Binder, 2014). Moreover, each NP can have different numbers of sites ranging from 10 to 400 and can be represented by these configurations. A perfect crystal usually represents the homogeneous ionic, however, if the crystal has some impurity, defects, and passive reactive surface, this type of crystal usually exhibits heterogeneous ionic distribution. In general, heterogeneous distribution displays 100 reactive sites on the surface with charge $1.5/nm^2$. However, for homogenous it is imperative to employ heterogeneous distribution first, followed by changing all sites with a similar charge. Afterwards, the charge positions can be toned by Metropolis criterion to give the final configuration (Clavier et al., 2016).

3.4.2 Surface chemistry/exterior protein corona

When NPs come across with biological fluids, they often end up getting a surface coating, which is referred as a "protein corona." Proteins have certain compositions and contain a net surface charge depending on the pH of the medium in which they are dispersed (Navya and Daima, 2016; Kallay and Žalac, 2002). Further, the adsorption reaction of protein on an NP's surface is not only dependent on the nanoparticle's properties but also on the characteristics of proteins and the dispersion medium. Different types of interaction forces, like hydrogen bonding and van der Waals force, play vital roles in nano-bio interactions. However, the association of the proteins and NPs vary with respect to the longevity of the interaction. For instance, long term or irreversible attachment of proteins on a nanoparticle's surface is referred to as a "hard corona"; on the other hand, short term or quick, reversible attachment of the proteins is called a "soft corona" (Saptarshi et al., 2013). This coating has specific impacts on the interactions of NPs in biological systems. NPs, after entering the biological system, are highly unstable and prone to aggregate, since NPs have high free surface energy owing to their size in nano ranges (Akter et al., 2018). Hence, in order to provide stability to the NPs they need surface modification in the form of a shield. This shield provides protection to the NPs from the fluctuations in pH and electrolyte concentration in the dispersion media. In order to preserve the potential of NPs and prevent them aggregating, it is imperative to provide some surface coating to develop the right surface chemistry. In this regard, several researches have been published on functionalizing the surface chemistry of NPs for their stability in biological fluids (Dama et al., 2014). For colloidal stabilization, broadly, two major approaches are being followed. The first and most followed approach is achieving colloidal stabilization by electrostatic repulsion (i.e., citrate stabilized NPs). However, this approach is only feasible at low ionic strength, whereas at higher ionic strength, it fails to provide stability to the NPs. The second approach works by adding a layer of physical barrier on the nanoparticle's surface (steric stabilization). This could be possible by employing polymers like poly ethyl glycol (PEG) or poly vinyl pyrrolidone (PVP) in dispersions (Guerrini et al., 2018).

Inorganic NPs when combined with organic proteins produce certain changes in the structure and functioning of the adsorbed proteins and thus, in turn, have an effect on the overall nano-bio reactivity. The interaction of NPs and proteins also depends on the shape of the NPs. Curved NPs provide more surface area and cause certain conformational modifications in the secondary structure of the NPs compared to NPs with flat surfaces. Further, this type of nano-bio interaction is often irreversible. Adsorbed proteins involved in nano-bio interactions have been observed to play an imperative role in the surface bio-nano interactions (Ahsan et al., 2018). Moreover, a variation in the protein corona formation behavior was observed with respect to different sizes and shapes of NPs. For example, Piella et al. (2016) conducted a study on citrate-stabilized gold nanoparticles with size ranging from 3.5 to 150 nm in the presence of bovine serum. It was observed that the thickness and the density of protein corona were strongly dependent on the particle size of gold nanoparticles. Similarly, another study employing gold NPs also revealed that protein corona composition was strongly influenced by size and shape of Au NPs (García-Álvarez et al., 2018).

3.4.3 Surface area

The augmented use of nanoparticles is due to their small sizes in nano-scale, which in turn provide NPs with high surface area. The high surface area of NPs means high reactivity, which is imperative for their nano-biological interactions. Owing to this, greater importance has been given to the external surface of NPs compared to their core. Surface characteristic of the NPs can be tuned by functionalization with different compounds in order to improve their reactivity in the biological system (Elsaesser and Howard, 2012). Moreover, the toxicity of NPs varies with respect to their size. Smaller size NPs have been found to be more toxic compared to their larger counterparts. For instance, a study employed iron oxide-silica nanocomposites (MNCs) with four different sizes 20, 40, 100, 200 nm. It was observed that nanocomposites with size 20 and 40 nm were only involved in IL-6 production and inflammatory responses (Injumpa et al., 2017). TiO_2 NPs of two different sizes, 25 nm (nanotube morphology) and 60 nm (anatase morphology), were employed to identify their cytotoxicity in human respiratory cells with treatment concentrations 0.1–100 μg/mL. Greater reduction in the cell viability was observed for NPs with 25 nm size and nanotube morphology. Further, greater toxicity was triggered by smaller-sized NPs (25 nm) in both A549 and 16HBE cell lines compared to larger sized NPs (60 nm) (Ma et al., 2017). Moreover, since silver NPs are well-known for their role as antibacterial agents Zhang et al. (2018) conducted a study to identify the effect of different sizes (10 and 50 nm) of NPs on soil nitrogen fixation ability of *Azotobacter vinelandii* bacteria at a concentration of 10 mg/L for the duration of 12 h. Ag NPs with smaller size (10 nm) were found to be more toxic compared to larger size NPs, which was observed by a reduced number of cell lines. Further, Ag NPs were not only responsible for inhibiting nitrogen fixation of *Azotobacter vinelandii* but also induced the production of reactive oxygen species and hydroxyl radicals leading to enzyme inactivation and death by apoptosis.

Besides this, particle surface area can influence the pH and other properties of dispersion media, i.e., a study conducted to identify the effects of TiO_2 NPs of the same size and different concentration (15, 25, 50, 150, and 500 μg/mL) on the pH of the dispersion media. When TiO_2 NPs with different concentrations were dispersed in DI water, the solution pH decreased with the increase of nanoparticle surface area, since after the dispersion of TiO_2 NPs, hydroxyl ions covered the surface of the NPs, which eventually caused the generation of more H^+ ions. Thus, with the increase of mass concentration from 15 to 500 μg/mL, a downfall in the solution from 5.7 to 5.1 with an augmentation in the zeta potential from 29 to 38 mV was observed. Further, a decrease in average diameter from 756 to 412 nm was noticed (Suttiponparnit et al., 2011). Surface area of NPs can be enhanced following various approaches, i.e., recently, surfactant-free synthesis of silica NPs using rice husk was carried out following the Taguchi approach. By employing this approach, silica NPs with surface area as high as 740 m^2/g and particle size 43.04 nm were synthesized. These NPs with high surface area could be a favorable material for biomedical and catalytic applications (Song et al., 2018). In another study, Fe-Ti nanocomposites were synthesized by following AACVD technique. The synthesized NPs exhibited high surface area 107–204 m^2/g. This nanocomposite could be employed as a promising adsorbent (Monárrez-Cordero et al., 2018).

4. Parameters that decide fate and toxicity of nanomaterials

At present, nanomaterials are ubiquitous in life (Donia and Carbone, 2019). However, the use of nanomaterials in daily life is causing toxicities and this toxicity and the fate of nanomaterials is dependent on their physical and chemical properties. Nanomaterials have peculiar physiochemical properties such as charge, surface area, shape, size, and aggregation. Due to their smaller size, nanomaterials are much more reactive than the bulk material and their entry to the cell and potential toxicity is also greater. The toxicity associated with nanomaterials is size-dependent and they can easily penetrate and translocate in the organs of the body. The retention of nanomaterials in the organs leads to production of reactive oxygen species (Gnach et al., 2015).

The characterized shapes of nanomaterials are rods, cubes, ellipsoids, spheres, and cylinders, and these shapes cause different kinds of toxicities (Sukhanova et al., 2018). Roughness, charge, and hydrophobicity are important surface chemistry parameters of nanomaterials that can significantly control the toxicological effects of nanomaterials. These surface properties of nanomaterials can cause great alterations in the blood-brain barrier, immune system, cellular uptake, colloidal behavior, phagocytosis, and cellular uptake. The absorption rate of positively charged nanomaterials is significantly higher than of negatively charged nanomaterials. Positive charged nanomaterials, after crossing the membranes, bind strongly with DNA as it is negatively charged, which causes prolongation of G0/G1 phases of cells' life cycles. In addition to this, the positively charged nanomaterials have much greater affinity toward proteins and can cause significant alternation in the protein structure, which in turn may lead to the inhibition of enzymatic activities and then disruption of biological processes (Guerrini et al., 2018).

Nanomaterials that have cationic surface charge can easily interact with the genetic materials and biological membranes and can produce more toxicity than the neutral or anionic nanomaterials (Navya and Daima, 2016). Surface charge, size, and shape of nanomaterials can also cause alterations in the integrity of the blood-brain barrier through the agglomeration and aggregation of nanomaterials (Gatoo et al., 2014). The nanomaterials may be composed of either inorganic or organic substances, and in the production of these nanomaterials, many reagents are involved. So sometimes, due to the involvement of impurities or other unwanted materials, unexpected toxicity and side effects may result. Due to internal body changes in the pH and oxidation reduction reactions, the composition of nanomaterials may be changed in the body (Ali et al., 2016).

Solubility of nanomaterials is an important concern relating to nanotoxicity. It is greatly affected by temperature and pH changes, and these factors have important roles in the dissolution of nanomaterials. Soluble nanomaterials can be very toxic as compared to the insoluble ones. Irrespective of different physiochemical properties of nanomaterials, agglomeration can cause significant toxicity and in certain conditions, higher levels of exposure to nanomaterials can cause chronic diseases such as cancer and fibrosis

(Donaldson et al., 2006). Morphology of nanomaterials has a direct role in nanotoxicology. Compared to other inhalable fibers, nanofibers and nanomaterials can pose a serious threat of lung inflammation, and prolonged exposure can also cause cancer (Firme III and Bandaru, 2010). The results of various studies have shown that carbon nanotubes are much more toxic than silica dust or ultra-fine carbon black (Chalupa et al., 2004).

Solvent or medium conditions have a direct effect on dispersion and agglomeration of nanomaterials that in turn affects their toxicity. Additionally, the toxic effects of nanomaterials are also dependent on the composition of the medium in which they are suspended. The same nanomaterials in different mediums exhibit different kinds of toxic manifestations (Hou et al., 2013). Although some of the dispersing agents can improve the physicochemical properties of nanomaterials in the medium, some can exert adverse effects leading to toxicity.

The surface properties of nanomaterials play a direct role in their toxicity. Surface coating is helpful to reduce the toxicities of nanomaterials by modification of their physicochemical properties such as chemical reactivity and optical, magnetic, and electrical properties (Yin et al., 2005).

5. Exposure of nanomaterials on living things

5.1 Humans

Although nanotechnology has been revolutionary, due to the nanometers dimensions of NPs, the safety concerns for consumers, workers, and the human environment have been greatly increased (Sajid et al., 2015). The exposure of human beings to nanomaterials can be described through different mechanisms. Exposure to nanomaterials in the workplace can occur during the supply chain process, such as during production, processing, packaging, transportation, quality control, and these nanomaterials can originate from screening, cutting, engine exhaust, vacuum cleaners, heating units of forklift trucks, grinding, blending, mixing, and polishing (Jeevanandam et al., 2018). The worst scenario can happen in the production of nanomaterials in the powder stage and any accidental spillage or leaking during the manufacturing process presents serious risks (O'Shaughnessy, 2013).

The main route of exposure of human beings to nanomaterials is through the skin, inhalation through respiration, and ingestion along with food. The exposure of consumers to nanomaterials is becoming more common due to the wide-ranging use of nanomaterials in food and food packaging (Mackevica, 2016). After consumption, NPs become part of the blood circulation and penetrate into other body organs (Pattan and Kaul, 2014) and cross the blood-brain barrier (Mc Carthy et al., 2015). This exposure of human beings to nanomaterials can cause significant danger due to their potential toxicity (Oberbek et al., 2019). The penetration of nanomaterials is in the transcutaneous manner and in all cases, they can penetrate to the cell by endocytosis and can also penetrate to many cells by transcytosis. So, once inhaled, ingested, or absorbed they can reach to the olfactive epithelium, axons, and then to the olfactive bulbs where they directly affect the neurons. They also penetrate to the lungs, heart, spleen, lymph nodes, and bone marrow. Studies have shown that nanomaterials in the body can provoke anti-oxidant activities, pro-oxidant activities, oxidative stress, and causes inflammation. The severity of these events is directly related to the concentration of nanomaterial and their specific type.

The incorporation of nanocarriers or nanostructured food materials in the human body can cause significant complications in the digestion process and can also alter the body metabolism and absorption profile. Nanomaterials in the food undergo various physiochemical transformations and the bioavailability of the nutrients is altered (Luo et al., 2015). Nanomaterials, after transportation to the gastrointestinal tract, are exposed to different conditions such as changes in ionic composition, pH changes, surface activated compounds (proteins, fatty acids, phospholipids, bile salts), digestive enzymes (lipases, proteases, amylases), microflora, and biological surfaces such as the intestine, stomach, esophagus, and tongue. Soon after the nanomaterials are passed through the gastrointestinal tract, their characteristics are changed and then these modified nanomaterials are transported throughout the body and can cause toxicity based on these surface changes (McClements et al., 2016).

This toxicity of nanomaterials can be exhibited at different cellular and organ levels. The cytotoxicity can result in faulty cell signaling, abnormal growth, necrosis, or apoptosis. Nanotoxicology may also cause misfolding of proteins, inactivation of enzymes, protein aggregation, irregular electron transport in mitochondria, mutations, improper gene expression, mRNA degradation, and DNA damage. Metal nanomaterials have a significant role in carcinogenicity and mutagenesis as metal nanomaterials cause the production of reactive oxygen species, which cause genomic instability (Preethi et al., 2019).

5.2 Plants

According to exposure modeling, it is indicated that the concentration of nanomaterials in the soil is much more than in the air and environment, so plants are more prone to nanomaterial toxicity and soil is a major source of release of nanomaterials into the environment (Gottschalk et al., 2009). Due to the toxicity caused by nanomaterials, oxidative stress becomes evident (Hong et al., 2014). Nanomaterials exert direct influences on plant hormones as the production and regulation of plant hormones is greatly affected by the plant physiological conditions toward toxicity (Santner et al., 2009). The main physiological indices of nanomaterials toxicity in plants are the percentage of germination, root health and elongation, biomass production, and leaf count (Lee et al., 2010). Some studies have also reported the effect of nanomaterials on gene expression, biomass production, shoot length, and seed germination (Dimkpa et al., 2012; El-Temsah and Joner, 2012; Yan et al., 2013; Ghosh et al., 2015).

Nanomaterial toxicity in plants leads to the production of reactive oxygen species and this ROS generation is dependent on the physicochemical attributes of nanomaterials' growth media and environmental conditions (Song et al., 2013; Yang et al., 2017). The effects of nanomaterial toxicity on plants are also dependent on various growth stages and management conditions. To counter this oxidative stress, the antioxidant defense system of plants produces both enzymatic and nonenzymatic antioxidants including glutathione reductase (GR), SOD, CAT, APX, guaiacol peroxidase (GPOX), glutathione *S*-transferases (GST), dehydroascorbate reductase (DHAR), monodehydroascorbate reductase, and glutathione peroxidase (GPX) (Kumar et al., 2018). Additionally, the toxicity can be greatly minimized by the surface coating of nanomaterials as their interaction in the soil plant system will be greatly modified (Hao et al., 2016).

Plants, as primary producers, have an important role for in community build up (McKee and Filser, 2016). Exposure of plants to nanomaterials is a potential pathway for transportation of nanomaterials (Rico et al., 2011). If the transportation of nanomaterials leads to their accumulation in the food chain, then the consumers of higher trophic levels are also exposed to the nanomaterials' toxicity (Zhu et al., 2008; Tolaymat et al., 2015).

5.3 Animals and aquatic organisms

Aquatic ecosystems are a destination of released nanomaterials that exert adverse effects on aquatic life (Exbrayat et al., 2015). These nanomaterials can enter through various direct and indirect routes to water sources. Their toxicology for aquatic organisms is dependent on their bioavailability and bioaccumulation in the aquatic environment. This bioavailability, bioaccumulation, and toxicity is greatly governed by various factors such as total hardness of water, that is, concentration of divalent cations, composition of nanomaterials, pH of media, size and surface area, surface chemistry, and shape (Ovissipour et al., 2013).

In both fresh water and marine environments, a significant number of nanomaterials are present. The findings of recent studies have concluded that nanomaterials serve as pollutants in the water and, depending on their physiochemical properties, their effects are greatly variable. Various laboratory studies have reported that the exposure of nanomaterials is exerting negative influences on invertebrates and fish. The nanomaterials induce oxidative stress in the ecosystem so that negative influences are exerted on these organisms (Hong et al., 2014). Nanomaterials negatively affect various organs of aquatic animals such as brain, intestine, liver, bile, and gills. Other than this, the negative effects are also exerted on the growth and development of aquatic life. Some nanomaterials can also penetrate the chorion, which is the envelope around the developing embryo or egg. In this situation, the nanomaterials can be even more toxic at the juvenile and embryonic stages (Farmen et al., 2012).

The mechanical effects of nanomaterial toxicity include absorption, distribution, metabolic reactions, and the excretions of these substances. After absorption and distribution, the nanomaterials can cause severe threats to the functioning of organs and may even cause death (Shaw and Handy, 2011). Nanomaterial toxicity can lead to decreased Na+/K+ ATPase in the intestine and gills. Depending upon the nanomaterial's toxicity, a change in thiobarbituric acid and variation of lipids in the liver is also observed (Federici et al., 2007). Nanomaterials represent foreign elements along with their own physiochemical properties in the bodies of animals and aquatic life, so they greatly interfere with the normal physiological processes and mechanisms of embryos, growing organisms, and adults. Nanomaterials are responsible for bringing disruption in growth and development and can cause lethal malformations (Handy et al., 2008). Due to their smaller size, they can easily penetrate to the cells without interfering with the defense mechanism. Their penetration to cells causes modification of mitochondria and cell metabolism that may cause cell death. They also interfere with the membrane functions such as and disrupts signal transduction and ionic transport. The physical properties and chemical composition of nanomaterials can cause cytotoxicity. Positively charged nanomaterials can cause complete destruction of the membrane lipid bilayer (Sukhanova et al., 2018).

After inhalation, nanomaterials may be deposited in the lungs and airways and can induce production of reactive oxygen species (ROS) and inflammation (Møller et al., 2010). After the translocation of nanomaterials, inflammatory mediators can be released to the systematic circulation (Erdely et al., 2011). If the transported nanomaterials are transported to the fetal development and pregnancy-related organs, then they can also be taken up by the placental cells and directly interfere with the development of the fetus by inducing inflammation and oxidative stress (Koga and Mor, 2010). Other than this, nanomaterials also affect the genetic material, muscle strength, nervous system, and brain of animals (Brohi et al., 2017).

Due to the frequent use of engineered nanomaterials, their accelerated concentration in the environment is negatively affecting the earthworm population. Possibly, these nanomaterials are carried by the wind or accidentally recycled to the soil. Other than this, the penetration of nanomaterials to the soil also modifies soil solution properties and greatly alters the bioavailability of different nutrients (Lapied et al., 2010).

6. Future prospects of nanotoxicology

Nanotoxicology is a subdivision of the toxicology discipline and is concerned with studying the various toxic effects of nanoscale structures and materials having a diameter of less than 100 nm. The increased toxicity of nanomaterials as compare to the bulk material is because of the fact that nanomaterials are highly reactive as they have high surface area to volume ratio (Recordati et al., 2015). Due to their smaller size and high surface area they can easily diffuse and penetrate to the tissues, biological barriers, and cellular membranes (Barua and Mitragotri, 2014).

Due to major investment in the research and development, nanotechnology has evolved at an unprecedented rate. The use of nanomaterials in replacement and novel technologies has greatly affected consumer, commercial, industrial, and health care products and services (Fadeel, 2019). Although nanotechnological advances provide significant advantages for improving the quality of life and have been provided substantial financial gains, the potential toxicity concerned with the use of nanomaterials should be of primary concern. Sustainable growth and the advantages of nanotechnology can be achieved by using nanomaterials in a safe way (Khan and Shanker, 2015).

However, in the industrial sector, the use of nanomaterials has raised significant concerns due to oral, inhalation, dermal, and parenteral exposure in the environmental, consumer, and occupational settings. Due to widespread use of nanomaterials, diversity of nanomaterials, and sheer numbers, scientists are unable to keep pace by studying nanotoxicology (Stueckle and Roberts, 2019). The 3R principal of replace, reduce, refine could be greatly helpful for the concerns raised about nanotoxicology (Burden et al., 2017). These 3R principals can be made an important part for the regulations related to the involvement of nanomaterials in biocides, medical products, pharmaceuticals, foods, and cosmetics (Rauscher et al., 2017).

Researchers have proposed a risk assessment framework for incorporation of alternative testing for specific nanomaterials. However, most testing models for testing the toxicity of nanomaterials are heavily dependent on animal model testing (Oomen et al., 2018). The issue being that animals are animals, and the mechanism of interaction of nanomaterials in the human body could be different, so more research is required to find a sustainable and efficient solution.

Nano-safety and nanotoxicology has been an intensively researched and discussed topic for more than 20 years. However, there is still a significant gap to understand and harmonize the methods for risk assessment of nanotoxicity (Singh et al., 2019). Despite the valuable ongoing research on nanotoxicology, the underlying mechanisms are still not well understood.

Nanotoxicology is a relatively new discipline, involving a comprehensive and intradisciplinary understanding of nano-bio interactions and their complete underlying mechanisms, which can only be studied by the combination of various perspectives in physics, chemistry, computational sciences, pharmacology, immunology, molecular biology, and nanomedicine, etc. Additionally, continuous and fact-based exchange of communication in the scientific community is critically important. At present there are a good number of studies of nanotoxicology but it is scattered and not systematic; in addition to this, there are some knowledge gaps, so efforts and research activities should be increased to follow a more systematic and comprehensive approach for the better understanding and management of toxicology studies (Singh et al., 2019).

Due to nanotoxicology, negative influences are exerted on potential biological processes, cardiovascular development and the immune, neurological, and reproductive systems (Riediker et al., 2019). Nanotoxicology and the designing and development of safe nanomaterials is a major challenge of the decade, so there is a need to pay critical attention for assessment of nanotoxicological effects by studying the characterization of robust materials, transformations occurring throughout the life cycle of nanomaterials, transformations caused by the nanomaterials in the various biological mediums, determination of appropriate dose metrics, and modeling according to these characteristics (Mattsson and Simkó, 2017).

The future of nanotoxicology is strongly dependent on the efficient research efforts of researchers, policy makers, and students, so the next generation should be trained in such a way that they can be well-equipped and aware of nanotoxicology concerns (Fadeel, 2019).

7. Summary and conclusions

As much as the use of nanoparticles is increasing, the risk graph is also rising rapidly. The more we use it in our food industry, simultaneously the exposure to our bodies increase at the same rate. The potential entry of nanoparticles into the food chain and overall ecosystem is threatening biological systems. The solution of nanotoxicology of engineered and incidental nanoparticles lies only in their safe production, delivery, and ultimate use by the consumer. If we stop repeating past mistakes, track the previous events that led to NPs maximum loss into the environment, and adopt safe approaches toward this technology, then we have hope for safety in the future, which will serve us through this technology rather than cornering us with its deleterious impacts. In the long run, safety standards must be introduced, general awareness regarding this risk must also be given to the public all over the world, safety standards need to be implemented and maintained to minimize interaction of living beings with these particles.

References

Abramenko, N.B., Demidova, T.B., Abkhalimov, E.V., Ershov, B.G., Krysanov, E.Y., Kustov, L.M., 2018. Ecotoxicity of different-shaped silver nanoparticles: case of zebrafish embryos. J. Hazard. Mater. 347, 89–94.

Ahsan, S.M., Rao, C.M., Ahmad, M.F., 2018. Nanoparticle-protein interaction: the significance and role of protein corona. In: Cellular and Molecular Toxicology of Nanoparticles. Springer, Cham, pp. 175–198.

Akter, M., Sikder, M.T., Rahman, M.M., Ullah, A.A., Hossain, K.F.B., Banik, S., et al., 2018. A systematic review on silver nanoparticles-induced cytotoxicity: physicochemical properties and perspectives. J. Adv. Res. 9, 1–16.

Ali, A., Zafar, H., Zia, M., Haq, I.U., Phull, A.R., Ali, J.S., Hussain, A., 2016. Synthesis, characterization, applications, and challenges of iron oxide nanoparticles. Nanotechnol. Sci. Appl. 9, 49.

Anandharamakrishnan, C., Parthasarathi, S., 2019. Food Nanotechnology: Principles and Applications. CRC Press.

Arts, J.H.E., Irfan, M., Keene, A.M., Kreiling, R., Lyon, D., Maier, M., Michel, K., Neubauer, N., Petry, T., Sauer, U.G., Warheit, D., Wiench, K., Wohlleben, W., Landsiedel, R., 2016. Case studies putting the decision-making framework for the grouping and testing of nanomaterials (DF4nanoGrouping) into practice. Regul. Toxicol. Pharmacol. 76, 234–261.
Avramescu, M.L., Rasmussen, P.E., Chénier, M., Gardner, H.D., 2017. Influence of pH, particle size and crystal form on dissolution behaviour of engineered nanomaterials. Environ. Sci. Pollut. Res. 24 (2), 1553–1564.
Barua, S., Mitragotri, S., 2014. Challenges associated with penetration of nanoparticles across cell and tissue barriers: a review of current status and future prospects. Nano Today 9 (2), 223–243.
Brohi, R.D., Wang, L., Talpur, H.S., Wu, D., Khan, F.A., Bhattarai, D., et al., 2017. Toxicity of nanoparticles on the reproductive system in animal models: a review. Front. Pharmacol. 8, 606.
Burden, N., Aschberger, K., Chaudhry, Q., Clift, M.J., Doak, S.H., Fowler, P., et al., 2017. The 3Rs as a framework to support a 21st century approach for nanosafety assessment. Nano Today 12, 10–13.
Bursten, J.R., Roco, M.C., Yang, W., Zhao, Y., Chen, C., Savolainen, K., et al., 2016. Nano on reflection A number of experts from different areas of nanotechnology describe how the field has evolved in the last ten years. Nat. Nanotechnol. 11 (10), 828–834.
Chalupa, D.C., Morrow, P.E., Oberdörster, G., Utell, M.J., Frampton, M.W., 2004. Ultrafine particle deposition in subjects with asthma. Environ. Health Perspect. 112 (8), 879–882.
Clancy, A.J., Bayazit, M.K., Hodge, S.A., Skipper, N.T., Howard, C.A., Shaffer, M.S., 2018. Charged carbon nanomaterials: redox chemistries of fullerenes, carbon nanotubes, and graphenes. Chem. Rev. 118 (16), 7363–7408.
Clavier, A., Carnal, F., Stoll, S., 2016. Effect of surface and salt properties on the ion distribution around spherical nanoparticles: Monte Carlo simulations. J. Phys. Chem. B 120 (32), 7988–7997.
Daima, H.K., Selvakannan, P.R., Kandjani, A.E., Shukla, R., Bhargava, S.K., Bansal, V., 2014. Synergistic influence of polyoxometalate surface corona towards enhancing the antibacterial performance of tyrosine-capped Ag nanoparticles. Nanoscale 6 (2), 758–765.
Dimkpa, C.O., McLean, J.E., Latta, D.E., Manangón, E., Britt, D.W., Johnson, W.P., et al., 2012. CuO and ZnO nanoparticles: phytotoxicity, metal speciation, and induction of oxidative stress in sand-grown wheat. J. Nanopart. Res. 14 (9), 1125.
Donaldson, K., Aitken, R., Tran, L., Stone, V., Duffin, R., Forrest, G., Alexander, A., 2006. Carbon nanotubes: a review of their properties in relation to pulmonary toxicology and workplace safety. Toxicol. Sci. 92 (1), 5–22.
Donia, D.T., Carbone, M., 2019. Fate of the nanoparticles in environmental cycles. Int. J. Environ. Sci. Technol. 16 (1), 583–600.
Ealias, A.M., Saravanakumar, M.P., 2017. A review on the classification, characterisation, synthesis of nanoparticles and their application. IOP Conf. Ser. Mater. Sci. Eng. 263, 032019.
Elsaesser, A., Howard, C.V., 2012. Toxicology of nanoparticles. Adv. Drug Deliv. Rev. 64 (2), 129–137.
El-Temsah, Y.S., Joner, E.J., 2012. Impact of Fe and Ag nanoparticles on seed germination and differences in bioavailability during exposure in aqueous suspension and soil. Environ. Toxicol. 27 (1), 42–49.
Erdely, A., Liston, A., Salmen-Muniz, R., Hulderman, T., Young, S.H., Zeidler-Erdely, P.C., et al., 2011. Identification of systemic markers from a pulmonary carbon nanotube exposure. J. Occup. Environ. Med. 53, S80–S86.
Exbrayat, J.M., Moudilou, E.N., Lapied, E., 2015. Harmful effects of nanoparticles on animals. J. Nanotechnol. 2015, 861092.
Fadeel, B., 2019. The right stuff: on the future of Nanotoxicology. Front. Toxicol. 1, 1.
Farmen, E., Mikkelsen, H.N., Evensen, Ø., Einset, J., Heier, L.S., Rosseland, B.O., et al., 2012. Acute and sub-lethal effects in juvenile Atlantic salmon exposed to low μg/L concentrations of Ag nanoparticles. Aquat. Toxicol. 108, 78–84.
Federici, G., Shaw, B.J., Handy, R.D., 2007. Toxicity of titanium dioxide nanoparticles to rainbow trout (*Oncorhynchus mykiss*): gill injury, oxidative stress, and other physiological effects. Aquat. Toxicol. 84 (4), 415–430.
Firme III, C.P., Bandaru, P.R., 2010. Toxicity issues in the application of carbon nanotubes to biological systems. Nanomedicine 6 (2), 245–256.

Forest, V., Leclerc, L., Hochepied, J.F., Trouvé, A., Sarry, G., Pourchez, J., 2017. Impact of cerium oxide nanoparticles shape on their in vitro cellular toxicity. Toxicol. in Vitro 38, 136–141.

García-Álvarez, R., Hadjidemetriou, M., Sánchez-Iglesias, A., Liz-Marzán, L.M., Kostarelos, K., 2018. In vivo formation of protein corona on gold nanoparticles. The effect of their size and shape. Nanoscale 10 (3), 1256–1264.

Gatoo, M.A., Naseem, S., Arfat, M.Y., Mahmood Dar, A., Qasim, K., Zubair, S., 2014. Physicochemical properties of nanomaterials: implication in associated toxic manifestations. Biomed. Res. Int. 2014, 498420.

Ghosh Chaudhuri, R., Paria, S., 2012. Core/shell nanoparticles: classes, properties, synthesis mechanisms, characterization, and applications. Chem. Rev. 112 (4), 2373–2433.

Ghosh, M., Bhadra, S., Adegoke, A., Bandyopadhyay, M., Mukherjee, A., 2015. MWCNT uptake in Allium cepa root cells induces cytotoxic and genotoxic responses and results in DNA hyper-methylation. Mutat. Res. 774, 49–58.

Gnach, A., Lipinski, T., Bednarkiewicz, A., Rybka, J., Capobianco, J.A., 2015. Upconverting nanoparticles: assessing the toxicity. Chem. Soc. Rev. 44 (6), 1561–1584.

Gottschalk, F., Sonderer, T., Scholz, R.W., Nowack, B., 2009. Modeled environmental concentrations of engineered nanomaterials (TiO2, ZnO, Ag, CNT, fullerenes) for different regions. Environ. Sci. Technol. 43 (24), 9216–9222.

Guerrini, L., Alvarez-Puebla, R.A., Pazos-Perez, N., 2018. Surface modifications of nanoparticles for stability in biological fluids. Materials 11 (7), 1154.

Handy, R.D., Owen, R., Valsami-Jones, E., 2008. The ecotoxicology of nanoparticles and nanomaterials: current status, knowledge gaps, challenges, and future needs. Ecotoxicology 17 (5), 315–325.

Hao, Y., Yu, F., Lv, R., Ma, C., Zhang, Z., Rui, Y., et al., 2016. Carbon nanotubes filled with different ferromagnetic alloys affect the growth and development of rice seedlings by changing the C: N ratio and plant hormones concentrations. PLoS ONE. 11(6).

Helmlinger, J., Sengstock, C., Groß-Heitfeld, C., Mayer, C., Schildhauer, T.A., Köller, M., Epple, M., 2016. Silver nanoparticles with different size and shape: equal cytotoxicity, but different antibacterial effects. RSC Adv. 6 (22), 18490–18501.

Hochella, M.F., Mogk, D.W., Ranville, J., Allen, I.C., Luther, G.W., Marr, L.C., et al., 2019. Natural, incidental, and engineered nanomaterials and their impacts on the Earth system. Science. 363(6434), eaau8299.

Hong, J., Peralta-Videa, J.R., Rico, C., Sahi, S., Viveros, M.N., Bartonjo, J., et al., 2014. Evidence of translocation and physiological impacts of foliar applied CeO_2 nanoparticles on cucumber (*Cucumis sativus*) plants. Environ. Sci. Technol. 48 (8), 4376–4385.

Hou, W.C., Westerhoff, P., Posner, J.D., 2013. Biological accumulation of engineered nanomaterials: a review of current knowledge. Environ Sci Process Impacts 15 (1), 103–122.

Injumpa, W., Ritprajak, P., Insin, N., 2017. Size-dependent cytotoxicity and inflammatory responses of PEGylated silica-iron oxide nanocomposite size series. J. Magn. Magn. Mater. 427, 60–66.

Jeevanandam, J., Barhoum, A., Chan, Y.S., Dufresne, A., Danquah, M.K., 2018. Review on nanoparticles and nanostructured materials: history, sources, toxicity and regulations. Beilstein J. Nanotechnol. 9 (1), 1050–1074.

Jeon, S., Clavadetscher, J., Lee, D.K., Chankeshwara, S.V., Bradley, M., Cho, W.S., 2018. Surface charge-dependent cellular uptake of polystyrene nanoparticles. Nano 8 (12), 1028.

Kaba, S.I., Egorova, E.M., 2015. In vitro studies of the toxic effects of silver nanoparticles on HeLa and U937 cells. Nanotechnol. Sci. Appl. 8, 19.

Kallay, N., Žalac, S., 2002. Stability of nanodispersions: a model for kinetics of aggregation of nanoparticles. J. Colloid Interface Sci. 253 (1), 70–76.

Khan, H.A., Shanker, R., 2015. Toxicity of nanomaterials. Biomed. Res. Int. 2015, 521014.

Khan, I., Saeed, K., Khan, I., 2019. Nanoparticles: properties, applications and toxicities. Arab. J. Chem. 12 (7), 908–931.

Koga, K., Mor, G., 2010. Toll-like receptors at the maternal–fetal interface in normal pregnancy and pregnancy disorders. Am. J. Reprod. Immunol. 63 (6), 587–600.

Krug, H.F., Wick, P., 2011. Nanotoxicology: an interdisciplinary challenge. Angew. Chem. Int. Ed. 50 (6), 1260–1278.

Kumar, V., Sharma, M., Khare, T., Wani, S.H., 2018. Impact of nanoparticles on oxidative stress and responsive antioxidative defense in plants. In: Nanomaterials in Plants, Algae, and Microorganisms. Academic Press, pp. 393–406.

Landau, D.P., Binder, K., 2014. A Guide to Monte Carlo Simulations in Statistical Physics. Cambridge University press.

Lapied, E., Moudilou, E., Exbrayat, J.M., Oughton, D.H., Joner, E.J., 2010. Silver nanoparticle exposure causes apoptotic response in the earthworm Lumbricus terrestris (Oligochaeta). Nanomedicine 5 (6), 975–984.

Lee, C.W., Mahendra, S., Zodrow, K., Li, D., Tsai, Y.C., Braam, J., Alvarez, P.J., 2010. Developmental phytotoxicity of metal oxide nanoparticles to *Arabidopsis thaliana*. Environ. Toxicol. Chem. 29 (3), 669–675.

Leitner, J., Sedmidubský, D., Jankovský, O., 2019. Size and shape-dependent solubility of CuO nanostructures. Materials 12 (20), 3355.

Liu, Y., Li, W., Lao, F., Liu, Y., Wang, L., Bai, R., et al., 2011. Intracellular dynamics of cationic and anionic polystyrene nanoparticles without direct interaction with mitotic spindle and chromosomes. Biomaterials 32 (32), 8291–8303.

Luo, Y.H., Chang, L.W., Lin, P., 2015. Metal-based nanoparticles and the immune system: activation, inflammation, and potential applications. Biomed. Res. Int. 2015, 143720.

Ma, Y., Guo, Y., Wu, S., Lv, Z., Zhang, Q., Ke, Y., 2017. Titanium dioxide nanoparticles induce size-dependent cytotoxicity and genomic DNA hypomethylation in human respiratory cells. RSC Adv. 7 (38), 23560–23572.

Mackevica, A., 2016. Release of Nanomaterials from Consumer Products and Implications for Consumer Exposure Assessment.

Marmiroli, M., Maestri, E., Pagano, L., Robinson, B.H., Ruotolo, R., Marmiroli, N., 2019. Toxicology assessment of engineered nanomaterials: innovation and tradition. In: Exposure to Engineered Nanomaterials in the Environment. Elsevier, pp. 209–234.

Mattsson, M.O., Simkó, M., 2017. The changing face of nanomaterials: risk assessment challenges along the value chain. Regul. Toxicol. Pharmacol. 84, 105–115.

Mc Carthy, D.J., Malhotra, M., O'Mahony, A.M., Cryan, J.F., O'Driscoll, C.M., 2015. Nanoparticles and the blood-brain barrier: advancing from in-vitro models towards therapeutic significance. Pharm. Res. 32 (4), 1161–1185.

McClements, D.J., DeLoid, G., Pyrgiotakis, G., Shatkin, J.A., Xiao, H., Demokritou, P., 2016. The role of the food matrix and gastrointestinal tract in the assessment of biological properties of ingested engineered nanomaterials (iENMs): state of the science and knowledge gaps. NanoImpact 3, 47–57.

McKee, M.S., Filser, J., 2016. Impacts of metal-based engineered nanomaterials on soil communities. Environ. Sci. Nano 3 (3), 506–533.

McNutt, M.K., 2017. Convergence in the geosciences. GeoHealth 1 (1), 2–3.

Meka, A.K., Jenkins, L.J., Dàvalos-Salas, M., Pujara, N., Wong, K.Y., Kumeria, T., et al., 2018. Enhanced solubility, permeability and anticancer activity of vorinostat using tailored mesoporous silica nanoparticles. Pharmaceutics 10 (4), 283.

Mishra, A.R., Zheng, J., Tang, X., Goering, P.L., 2016. Silver nanoparticle-induced autophagic-lysosomal disruption and NLRP3-inflammasome activation in HepG2 cells is size-dependent. Toxicol. Sci. 150 (2), 473–487.

Møller, P., Jacobsen, N.R., Folkmann, J.K., Danielsen, P.H., Mikkelsen, L., Hemmingsen, J.G., et al., 2010. Role of oxidative damage in toxicity of particulates. Free Radic. Res. 44 (1), 1–46.

Monárrez-Cordero, B.E., Amézaga-Madrid, P., Sáenz-Trevizo, A., Pizá-Ruiz, P., Antúnez-Flores, W., Miki-Yoshida, M., 2018. Synthesis and characterization of composite Fe-Ti oxides nanoparticles with high surface area obtained via AACVD. Ceram. Int. 44 (6), 6990–6996.

Mu, Q., David, C.A., Galceran, J., Rey-Castro, C., Krzemiński, Ł., Wallace, R., et al., 2014. Systematic investigation of the physicochemical factors that contribute to the toxicity of ZnO nanoparticles. Chem. Res. Toxicol. 27 (4), 558–567.

Navya, P.N., Daima, H.K., 2016. Rational engineering of physicochemical properties of nanomaterials for biomedical applications with nanotoxicological perspectives. Nano Converg. 3 (1), 1.

Oberbek, P., Kozikowski, P., Czarnecka, K., Sobiech, P., Jakubiak, S., Jankowski, T., 2019. Inhalation exposure to various nanoparticles in work environment—contextual information and results of measurements. J. Nanopart. Res. 21 (11), 222.

Oomen, A.G., Steinhäuser, K.G., Bleeker, E.A., van Broekhuizen, F., Sips, A., Dekkers, S., et al., 2018. Risk assessment frameworks for nanomaterials: scope, link to regulations, applicability, and outline for future directions in view of needed increase in efficiency. NanoImpact 9, 1–13.

O'Shaughnessy, P.T., 2013. Occupational health risk to nanoparticulate exposure. Environ Sci Process Impacts 15 (1), 49–62.

Ovissipour, M., Rasco, B., Sablani, S.S., 2013. Impact of engineered nanoparticles on aquatic organisms. J. Fisheries Livest. Prod. 1, e106.

Pattan, G., Kaul, G., 2014. Health hazards associated with nanomaterials. Toxicol. Ind. Health 30 (6), 499–519.

Peng, S., Li, Z., Zou, L., Liu, W., Liu, C., McClements, D.J., 2018. Improving curcumin solubility and bioavailability by encapsulation in saponin-coated curcumin nanoparticles prepared using a simple pH-driven loading method. Food Funct. 9 (3), 1829–1839.

Piella, J., Bastús, N.G., Puntes, V., 2016. Size-controlled synthesis of sub-10-nanometer citrate-stabilized gold nanoparticles and related optical properties. Chem. Mater. 28 (4), 1066–1075.

Preethi, R., Leena, M.M., Moses, J.A., Anandharamakrishnan, C., 2019. Toxicology Aspects of Nanomaterials.

Rauscher, H., Rasmussen, K., Sokull-Klüttgen, B., 2017. Regulatory aspects of nanomaterials in the EU. Chem. Ing. Tech. 89 (3), 224–231.

Recordati, C., De Maglie, M., Bianchessi, S., Argentiere, S., Cella, C., Mattiello, S., et al., 2015. Tissue distribution and acute toxicity of silver after single intravenous administration in mice: nano-specific and size-dependent effects. Part. Fibre Toxicol. 13 (1), 12.

Rico, C.M., Majumdar, S., Duarte-Gardea, M., Peralta-Videa, J.R., Gardea-Torresdey, J.L., 2011. Interaction of nanoparticles with edible plants and their possible implications in the food chain. J. Agric. Food Chem. 59 (8), 3485–3498.

Riediker, M., Zink, D., Kreyling, W., Oberdörster, G., Elder, A., Graham, U., et al., 2019. Particle toxicology and health-where are we? Part. Fibre Toxicol. 16 (1), 19.

Sajid, M., Ilyas, M., Basheer, C., Tariq, M., Daud, M., Baig, N., Shehzad, F., 2015. Impact of nanoparticles on human and environment: review of toxicity factors, exposures, control strategies, and future prospects. Environ. Sci. Pollut. Res. 22 (6), 4122–4143.

Santner, A., Calderon-Villalobos, L.I.A., Estelle, M., 2009. Plant hormones are versatile chemical regulators of plant growth. Nat. Chem. Biol. 5 (5), 301–307.

Saptarshi, S.R., Duschl, A., Lopata, A.L., 2013. Interaction of nanoparticles with proteins: relation to bio-reactivity of the nanoparticle. J. Nanobiotechnol. 11 (1), 26.

Shaw, B.J., Handy, R.D., 2011. Physiological effects of nanoparticles on fish: a comparison of nanometals versus metal ions. Environ. Int. 37 (6), 1083–1097.

Singh, A.V., Ansari, M.H.D., Laux, P., Luch, A., 2019. Micro-nanorobots: important considerations when developing novel drug delivery platforms. Expert Opin. Drug Deliv. 16, 1259–1275.

Sohail, M.I., Waris, A.A., Ayub, M.A., Usman, M., Zia-ur-Rehman, M., Sabir, M., Faiz, T., 2019. Environmental application of nanomaterials: a promise to sustainable future. In: Engineered Nanomaterials and Phytonanotechnology: Challenges for Plant Sustainability.vol. 87. p. 1.

Song, U., Jun, H., Waldman, B., Roh, J., Kim, Y., Yi, J., Lee, E.J., 2013. Functional analyses of nanoparticle toxicity: a comparative study of the effects of TiO_2 and Ag on tomatoes (*Lycopersicon esculentum*). Ecotoxicol. Environ. Saf. 93, 60–67.

Song, S., Cho, H.B., Kim, H.T., 2018. Surfactant-free synthesis of high surface area silica nanoparticles derived from rice husks by employing the Taguchi approach. J. Ind. Eng. Chem. 61, 281–287.

Stöber, W., Fink, A., Bohn, E., 1968. Controlled growth of monodisperse silica spheres in the micron size range. J. Colloid Interface Sci. 26 (1), 62–69.

Stueckle, T.A., Roberts, J.R., 2019. Perspective on Current Alternatives in Nanotoxicology Research.

Sukhanova, A., Bozrova, S., Sokolov, P., Berestovoy, M., Karaulov, A., Nabiev, I., 2018. Dependence of nanoparticle toxicity on their physical and chemical properties. Nanoscale Res. Lett. 13 (1), 44.
Suttiponparnit, K., Jiang, J., Sahu, M., Suvachittanont, S., Charinpanitkul, T., Biswas, P., 2011. Role of surface area, primary particle size, and crystal phase on titanium dioxide nanoparticle dispersion properties. Nanoscale Res. Lett. 6 (1), 27.
Tolaymat, T., El Badawy, A., Sequeira, R., Genaidy, A., 2015. A system-of-systems approach as a broad and integrated paradigm for sustainable engineered nanomaterials. Sci. Total Environ. 511, 595–607.
Tyagi, I., Gupta, V.K., Sadegh, H., Ghoshekandi, R.S., Makhlouf, A.S.H., 2017. Nanoparticles as adsorbent; a positive approach for removal of noxious metal ions: a review. Sci. Technol. Dev. 34 (3), 195–214.
Van der Merwe, D., Pickrell, J.A., 2018. Toxicity of nanomaterials. In: Veterinary Toxicology. Academic Press, pp. 319–326.
Williams, K.M., Gokulan, K., Cerniglia, C.E., Khare, S., 2016. Size and dose dependent effects of silver nanoparticle exposure on intestinal permeability in an in vitro model of the human gut epithelium. J. Nanobiotechnol. 14 (1), 62.
Yan, S., Zhao, L., Li, H., Zhang, Q., Tan, J., Huang, M., et al., 2013. Single-walled carbon nanotubes selectively influence maize root tissue development accompanied by the change in the related gene expression. J. Hazard. Mater. 246, 110–118.
Yang, J., Cao, W., Rui, Y., 2017. Interactions between nanoparticles and plants: phytotoxicity and defense mechanisms. J. Plant Interact. 12, 158–169.
Yang, Z., Tian, J., Yin, Z., Cui, C., Qian, W., Wei, F., 2019. Carbon nanotube-and graphene-based nanomaterials and applications in high-voltage supercapacitor: a review. Carbon 141, 467–480.
Yin, H., Too, H.P., Chow, G.M., 2005. The effects of particle size and surface coating on the cytotoxicity of nickel ferrite. Biomaterials 26 (29), 5818–5826.
Zhang, W., Lin, W., Pei, Q., Hu, X., Xie, Z., Jing, X., 2016. Redox-hypersensitive organic nanoparticles for selective treatment of cancer cells. Chem. Mater. 28 (12), 4440–4446.
Zhang, L., Wu, L., Si, Y., Shu, K., 2018. Size-dependent cytotoxicity of silver nanoparticles to *Azotobacter vinelandii*: growth inhibition, cell injury, oxidative stress and internalization. PLoS ONE. 13(12), e0209020.
Zhu, H., Han, J., Xiao, J.Q., Jin, Y., 2008. Uptake, translocation, and accumulation of manufactured iron oxide nanoparticles by pumpkin plants. J. Environ. Monit. 10 (6), 713–717.

Further reading

Khanna, P., Ong, C., Bay, B.H., Baeg, G.H., 2015. Nanotoxicity: an interplay of oxidative stress, inflammation and cell death. Nano 5 (3), 1163–1180.
Ryabchikov, Y.V., Al-Kattan, A., Chirvony, V., Sanchez-Royo, J.F., Sentis, M., Timoshenko, V.Y., Kabashin, A.V., 2017, February. Influence of oxidation state on water solubility of Si nanoparticles prepared by laser ablation in water. In: Colloidal Nanoparticles for Biomedical Applications XII. vol. 10078. International Society for Optics and Photonics, p. 100780C.
Soares, S., Sousa, J., Pais, A., Vitorino, C., 2018. Nanomedicine: principles, properties, and regulatory issues. Front. Chem. 6, 360.

CHAPTER 7

Sufficiency and toxicity limits of metallic oxide nanoparticles in the biosphere

Muhammad Irfan Sohail[a], Muhammad Ashar Ayub[a], Muhammad Zia ur Rehman[a], Muhammad Azhar[a], Zia Ur Rahman Farooqi[a], Ayesha Siddiqui[b], Wajid Umar[a,c], Irfan Iftikhar[a], Muhammad Nadeem[a], and Hina Fatima[d]

[a]Institute of Soil and Environmental Sciences, Faculty of Agriculture, University of Agriculture Faisalabad, Faisalabad, Pakistan
[b]Department of Botany, University of Agriculture, Faisalabad, Faisalabad, Pakistan
[c]Doctoral School of Environmental Sciences, Szent Istvan University, Gödöllő, Hungary
[d]School of Applied Biosciences, Kyungpook National University, Daegu, South Korea

1. Introduction

Nanoparticles (NPs) are defined as manufactured, natural, or incidental entities composed of particles, either in free state or in an aggregation, with one or more exterior dimensions in the size of 1–100 nm (McGillicuddy et al., 2017). Nanotechnology is a trillion-dollar industry, which is growing exponentially (Sohail et al., 2019). Nanoscience is transforming every area of life, but this boom has created a challenge to scientists to predict, and thus mitigate, unwanted effects on the ecosystem. Current estimates state that engineered nanomaterials (ENMs) are fundamental components in over 1800 commercial products for diverse applications (Kanwal et al., 2019). According to the recently redeveloped nanotechnology consumer products inventory, metal nanoparticles (MNPs) and metal oxide nanoparticles (MONPs) are widely used ENMs, representing 37% of total NPs-based products globally (Zhang et al., 2018). The synthesis of MONPs is achieved through various routes depending upon the nature of transformation system used. Chemical precipitation, green synthesis via plants, and hydrothermal techniques are widely used methods (Ayub et al., 2019). Further, MONPs possess distinctive structure with unique properties, such as redox potential, high catalytic activity, mechanical stability, biocompatibility, labile surface charges, and high surface area (Tuli et al., 2015). Therefore, MONPs can't be seen as a common inorganic xenobiotics in the biosphere. MONPs are highly reactive compared to their bulk versions. Contemporary studies prove that surface reactivity (Burello and Worth, 2011) and chemical destabilization are involved in potential toxicity (Auffan et al., 2009).

The hazard and risk assessment of MONPs infiltrating the environment is swiftly growing into a decisive and persuasive message for governments, regulatory authorities, and international organizations working on devising the policy frameworks and guidance archives (Sohail et al., 2019). Comprehending and unraveling the ambiguities related to

Nanomaterials: Synthesis, Characterization, Hazards and Safety
https://doi.org/10.1016/B978-0-12-823823-3.00002-1

the hazardous effects of ENMs has been an active area of research since the 1980s (Golbamaki et al., 2018). Since nanotechnology-based industry is accountable for the massive production of MONPs containing products and residues, therefore, the amount of nanowaste is increasing at breakneck speed in the biosphere (Bystrzejewska-Piotrowska et al., 2009). MONPs can enter the biosphere via various routes, i.e., during manufacturing, transportation, usage, and disposal (Wu et al., 2013). It is clear that all EMMs ending up in the aquatic or terrestrial biosphere will interact with aquatic and soil biota and can latently induce toxicity in biological systems (Thiéry et al., 2012). Foregoing studies have reported that different concentrations of MONPs, such as nCuO, $nTiO_2$, and $nCeO_2$, were detected in runoff from sludge from wastewater treatment plants, municipal sewage, rivers and sediments, and soils (Ahmad et al., 2018; Ayub et al., 2019; Markus et al., 2016). However, NPs were an organic part of our biosphere, mainly originating from terrestrial sources, long before the field of nanoscience existed. Among the natural sources, MONPs are largely byproducts of combustion from burning fuels (coal, petroleum, wood, etc.), aerosols from atmospheric phyto-chemistry, and volcanic venting (Bystrzejewska-Piotrowska et al., 2009).

Where rapid and widespread applications of MNPs and MONPs have revolutionized the technology, it has also provoked serious concerns regarding toxicity impacts, predominantly on humans. Thus, nanotechnology is still regarded as a double-edged sword (Kanwal et al., 2019; Solaiman et al., 2019). The exposure of MONPs to biological organisms can be direct and/or indirect. The possible indirect ways include inhalation or digestion of terrestrial and aquatic organisms, and significantly, through plants. The direct passage for aquatic organisms is likely through external surface epithelia or gills. Environmental interaction of MONPs with the surrounding materials and complex organic fluids could result the corona which ultimately depress the functioning of biological systems (Kanwal et al., 2019) and overall ecological sustainability (Rajput et al., 2020).

Globally, assorted nanotoxicological studies have explored the biological impact of core metals, but less emphasis has been placed on ecological risks and human health related to MONPs. Therefore, nanotoxicology of MONPs is an emerging area of scientific research interest. Advancement in this field will also lead to maximized application of MONPs (Wang et al., 2017a,b). However, the question remains whether the toxicity of MONPs emanates from the MONPs themselves or the released metal ions. It is clear that MONPs are more bio-toxic than their bulk versions (Subramaniam et al., 2019). It is widely acknowledged that MONPs have great potential to dissolve in aqueous media, ultimately releasing the toxic core metal ions (Wang et al., 2016a,b,c). A large number of researchers have focused on studying the toxicity of MONPs, i.e., silicon dioxide ($nSiO_2$), titanium dioxide ($nTiO_2$), zinc oxide (nZnO), cerium dioxide ($nCeO_2$), chromium dioxide ($nCrO_2$), and copper oxide (nCuO). The general trend of reported toxicities to biological compartments is as follows: $nCuO > nZnO > nTiO_2 > nSiO_2$

(Deline and Nason, 2019; Dizaj et al., 2014; Wu et al., 2013). The MONPs of iron oxides (nFe_xO_x), CeO_2, and aluminum oxide (nAl_2O_3) are relatively regarded as nontoxic or subject to media concentrations (Deline and Nason, 2019). Under multiple environmental factors, MONPs inevitably undergo chemical, physical, and biological transformations. Therefore, it is vital to understand thoroughly the environmental transformation processes and related toxicity properties of MONPs in order to predict better their effects on the environment (Zhang et al., 2018). The knowledge that the scientific world has gained up to now is not enough to draw conclusions on the fate and actual release of MONPs in the natural environment (Lekamge et al., 2020).

Nano-toxicological testing is carried out to characterize the potentially adverse effects of NPs on biotic components of the environment, with a view to ensuring the safest possible use of NPs. Environmental nanotoxicology places great emphasis on ecological interactions at population, community, and ecosystem levels, while accounting for underlying organismal and cellular effects. Ecological nanotoxicology powerfully links exposure and ENMs properties, biochemical mechanisms, and ecological and physical processes. Mechanistic models—experimentally measured effects with underlying mechanisms—are required to integrate information from multiple scales and predicting the outcomes for realistic exposure regimes. Predictive models for MONPs toxicity are difficult to develop, as the relevant particle properties are heavily dependent on the surrounding environment (Fernández-García and Rodriguez, 2011). It is therefore vital that MONPs' behavior is studied in matrices that reflect real-world complexity (Cox, 1989; Deline and Nason, 2019). Prevalent studies on exploring the release of NPs are still incomplete, thus it is vital to procure more accurate appraisals of ENMs entering the environment at various phases of their life cycle (Gottschalk and Nowack, 2011). To achieve prevention of ecotoxicity, it is necessary to elaborate the toxicity mechanism. Despite the fact that several reports are already present in the literature, exegesis of actual toxicity mechanisms seems a convoluted subject, as contrasting results are common. Researchers have also endeavored to predict the toxicity of MONPs related to properties of MONPs (Burello and Worth, 2011). However, these predictions have mainly focused on developing the relationship between an NP's energy band structure to redox potentials of radical-forming ENMs (Zhang et al., 2012a,b), whereas surface thermodynamics of ENMs have not received enough attention. Understanding the dissolution process of release of core metal ions is also fundamental to shed light on toxicity mechanisms (Barnard, 2010). Computational ventures to forewarn of the toxicity originating from ROS production has also been studied (Sharifi et al., 2012). Nonetheless, such predictions can only be plausible if ROS production can precisely be quantified and correlated to presence of MONPs. Quantitative structure-activity relationships (QSAR) models have been implicated to foresee the toxicity but the implementation of QSAR is also hampered by the dearth of scientific evidence of the explicit and definite mechanism of toxicity (Burello and Worth, 2011; Djurišić et al., 2014, 2015).

Currently, no complete package of safety guidelines exists regarding the release of MONPs into biospheres. The United Kingdom and United States have already published environmental quality standards for the aquatic biosphere but most models focus only on core metal ions. However, the paramount contrast between MONPs and metal ions is the charge (Baker et al., 2014). Therefore, considerable efforts are needed to assess the impacts of MONPs. Although numerous tools have been developed for gauging and managing ENM exposure to the environment, the EU Commission relies on life cycle-based models (Hischier, 2014; UNEP (United Nations Environment Programme), 2011). Life cycle assessment (LCA) is administered by the international standards ISO-14040 series (ISO (International Organization for Standardization), 2006a,b). Yet many researchers suggest that studies based on LCA assessment do not represent the whole life cycle (Salieri et al., 2015; Miseljic and Olsen, 2014). The existing knowledge gaps regarding the release of ENMs (life cycle catalogue) and budding toxicity of these ENMs on the environment and humans (impact assessment) are hampering understanding of the true implications of LCA. In addition, fate and transport models (Salieri et al., 2015)used to evaluate the concentration of ENMs in the biosphere are progressively emerging (Liu and Cohen, 2014). However, scores of challenges concerning MONPs utilization in consumer products—like inappropriate and illegitimate labeling and packaging thus violating the regulatory guidelines—exist (Subramaniam et al., 2019). Therefore, under the umbrella of the above discussion, this chapter is designed to help manufacturers, regulation agencies, environmental engineers, and scientists in the field to understand the developments in MONPs, their interactions with environmental segments, and potential toxicity limits in commercial products. The knowledge gap and future research needs are also identified.

2. MONPs categories and characteristics

2.1 Categories of MONPs

MONPs can be categorized with respect to their origin, dimensions, chemical composition, and level of toxicity (Pietroiusti and Magrini, 2015). The details are discussed below.

2.1.1 Categories based on origin

Generally, MONPs can come from two sources, i.e., natural and anthropogenic. The MONPs that originate from natural sources are also known as ultrafine particles. Natural sources can be soils, dust storms, breaking sea waves, volcanic eruptions, and forest fires (Kumar et al., 2014). Living organisms also contain some NPs naturally, like biogenic magnetite, calcium hydroxyapatite, etc. (Hochella Jr et al., 2012). Nowadays, several types of MONPs are being engineered with distinctive and exclusive characteristics for divergent applications. Engineered MONPs can come in several shapes and forms like, plates, shells, fibers, rods, needles, sphere, tubes, rings, etc. Engineered MONPs have controlled composition, dimensions, and shape compared to natural MONPs (Oberdörster et al., 2005).

2.1.2 Categories based on dimensions

MONPs can also be categorized based on their dimensions (Dolez, 2015). MONPs that have all their dimensions in the 1–100 nm scale are known as zero-dimensional NPs (0D), i.e., nMgO, nZnO-rings, $nTiO_2$, etc. MONPs that have one external dimension in micrometer scale and two dimensions in nanoscale are known as one-dimensional MONPs (1D). This category includes nAl_2O_3, $nSiO_2$, $nZrO_2$, etc. (Brown and Stevens, 2007). MONPs that have only one external dimension in nanometer scale and two external dimensions in micrometer scale are known as two-dimensional NPs (2D). Mostly this category includes coatings and films of MONPs, which are predominantly utilized in electronics and physics for electronic components (Dolez, 2015). The last category includes NPs or nanocomposites that have all their external dimensions in micrometer scale but display internal nanoscale features, known as three-dimensional (3D).

2.1.3 Categories based on chemical composition

MONPs can be classified based on the type of inherent core metal. Oxides of several metals are synthesized and used in various sectors. Every metal has different characteristics, and based on those characteristics, nanoparticles of metal oxides have different applications in different fields. The metals whose oxides have been used extensively in the generation of MONPs are Al, Zn, Zr, Mg, Fe, Cu, Co, Mn, and rare earth metals (Fig. 1).

2.1.4 Categories based on toxicity level

Recent developments in nanotechnology have increased the application of NPs in various sectors. Despite their benefits, the environmental fate and toxicity of different NPs are major concerns. The potential risk of NPs in health terms was first observed in 1990 (Oberdörster et al., 2007). Since then, research has been carried out on this aspect. In the case of MONPs, the bulk metal oxides differ from the MONPs in physical, chemical, mechanical, and electrical properties. These changes in properties change the reactivity of compounds. Several scientists have compared the toxicity potential of bulk metal oxides and MONPs. Srivastav et al. (2016) carried out an experiment to evaluate the toxicity potential of nZnO and bulk ZnO in Wistar rats. Results showed that nZnO showed more toxic effects than bulk ZnO. This might be due to the higher reactivity and higher penetration of nanosized particles than for bulk metal oxide. Similarly, Alves et al. (2019) recently evaluated the toxicity potential of nZnO and bulk ZnO on earthworms and springtails. He reported that nZnO significantly affected the reproduction of both earthworms and springtails at $4000\,mg\,kg^{-1}$ compared to bulk ZnO. Lai et al. (2018a,b)

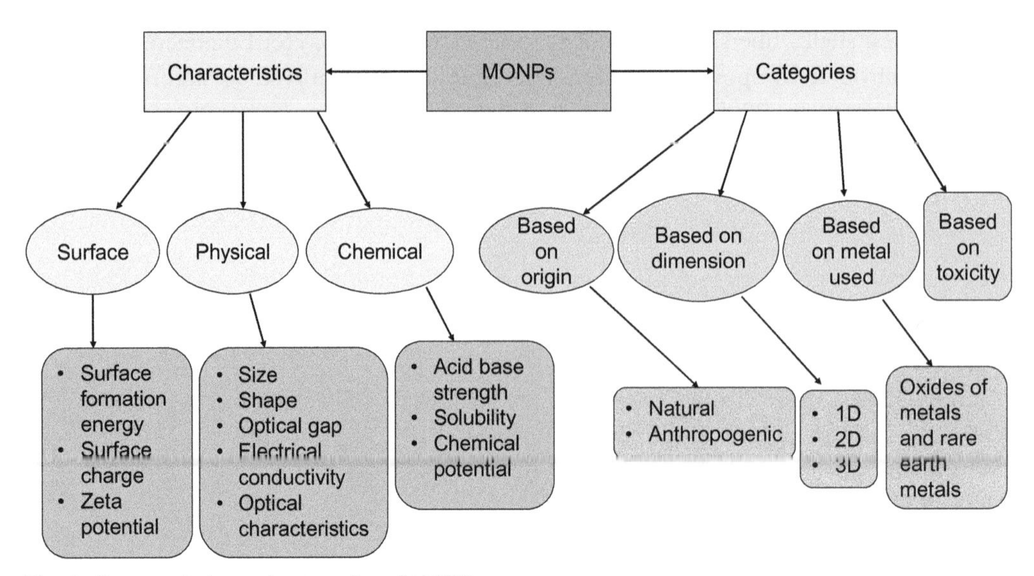

Fig. 1 Characteristics and categories of MONPs.

conducted a study to evaluate the toxicity of $_{n}$CuO in mice (C57BL/6). Results showed that the exposure of $_{n}$CuO induced pulmonary fibrosis in mice (57BL/6). A study conducted by Chattopadhyay et al. (2015) suggested that $_{n}$CoO is toxic to the primary immune cells of humans. The toxicity of $nTiO_2$ on human astrocyte and neuronal cells was evaluated by Coccini et al. (2015). The study reported that exposure to $nTiO_2$, even at low dose ($\geq 0.1\,\mu g\,mL^{-1}$), significantly affected the human cerebral SY5Y and cell lines D384. Apart from humans and animals, MONPs showed toxic effect in aquatic animals and plants as well. The toxicity of iron oxide NPs was evaluated by Alkhatib et al. (2019). The results showed drastic effects on growth. Now, with the efforts and dedication of several research groups, exposure limits to NPs in the workplace have been proposed (Pietroiusti and Magrini, 2015). Therefore, MONPs can be categorized based on their toxicity level (Table 1).

2.2 Characteristics

The inherent characteristics of MONPs play a significant part in their toxic behavior. Interaction of MONPs with living organisms largely depends on their surface area and size (Sukhanova et al., 2018). MONPs possess a large surface area, which increases their catalytic activity and reaction capacity. The size of NPs is comparable to DNA helix diameter (2 nm), protein globules (2–10 nm), and cell membrane thickness (10 nm), which shows that NPs can easily penetrate cells and organelles. Pan et al. (2007) reported that gold NPs of 1.4 nm were 60 times more toxic than the MONPs of 15 nm size. Additionally, size of MONPs also determines the interaction of NPs with defense and

Table 1 Exposure limits of different MONPs.

Sr. no.	MONPs categories	Exposure limits	References
1	Fiber-like MONPs	0.007–0.03 mg m^{-3} or 10^4–10^5 fibers m^{-3}	Pietroiusti and Magrini (2015)
2	MONPs (<6000 kg m^{-3}), i.e., $_n$TiO$_2$, $_n$Al$_2$O$_3$, $_n$ZnO	2×10^7 and 4×10^7 particles m^{-3}, between 0.066 and $0.1 \times$ WEL (work exposure limit) and between 0.3 and 3.5 mg m^{-3}	Pietroiusti and Magrini (2015) and Dolez (2015)
3	MONPs (>6000 kg m^{-3}), i.e., $_n$Fe$_2$O$_3$, $_n$SnO, $_n$Sb$_2$O$_3$, and nCeO$_2$	2×10^7 particle m^{-3} and $0.066 \times$ WEL	Dolez (2015)
	The value for insoluble MONPs that have no WEL is 0.3 mg m^{-3}		Dolez (2015)

transport systems of the body and cells (Sukhanova et al., 2018). Like size, the toxicity of MONPs largely depends on shape, i.e., rod, cylinder, cube, ellipsoid, sheet, or sphere. For example, spherical MONPs caused more endocytosis than nanotubes did (Champion and Mitragotri, 2006).

Chemical composition and crystal structure of NPs also influence the NPs' toxic potential. A toxicity comparison of nZnO and nSiO$_2$ of 20 nm size was carried out on mouse fibroblasts. The two types of NPs showed different toxicity mechanisms. Silica NPs caused DNA structure alteration, while nZnO caused oxidative stress (Yang et al., 2009). Gurr et al. (2005) evaluated the effect of crystal structure of NPs on toxicity potential and reported that rutile-like crystal structure NPs (nTiO$_2$) caused lipid peroxidation, DNA damage, and production of micronuclei, while the NPs of anatase-like crystal structures of the same size were nontoxic. By determining the interaction of NPs with biological systems, it has become clear that surface charge also plays an important role in nanotoxicity (Schaeublin et al., 2011; El Badawy et al., 2011). It was reported that positively charged NPs enter cells easily and cause more toxic effects than negatively charged and neutral NPs do (Sukhanova et al., 2018). Positively charged NPs bind to the negatively charged DNA more strongly and prolong the G0/G1 phase of the cell cycle. On the other hand, negatively charged NPs did not show any effect on the cell cycle (Liu et al., 2011a,b). Physical, chemical, and surface characteristics of NPs are addressed below and characteristics are also summarized in Fig. 1.

2.2.1 Surface characteristics

The surrounding environment directly reacts with the surface of MONPs. According to Organization for Economic Co-operation and Development (OECD) guidelines, the surface properties of MONPs are considered an important component when used in risk assessment applications (Hunt and Watkiss, 2007). Along with external layers of atoms, some of the internal atomic layers also influence the surface properties of MONPs (Stelmakh et al., 2017).

2.2.1.1 Surface formation energy

One of the major differences between macro and nano scale is that, on the surface of nanoparticles, a significant increase in atoms is noted compared to the total number of atoms forming the whole material. This property of MONPs significantly influences their surface reactivity compared to microparticles. Atoms on the surface of MONPs are not fully occupied and show more reactivity because of high free energy (Escorihuela et al., 2018a,b).

2.2.1.2 Surface charge

Surface charge controls the aggregation and dispersion of MONPs and ultimately affects the cellular uptake of MONPs (Ying et al., 2015). Surface charge depends on the partially filled p-orbitals of oxygen, d-orbital of metals, and partially saturated surface bonds. This characteristic may affect the ROS generation. MONPs that have redox potential may show toxicity to cells, while stable MONPs do not show toxicity (Saliner et al., 2008).

2.2.1.3 Zeta potential

Measurement of electrostatic interaction between dispersed NPs is known as zeta potential. It is considered as the universal characteristic of NPs. Cell adhesion and the biological inactivation of cells depends on the zeta potential. Zhang et al. (2008a,b) reported the interaction of iron oxide NPs with normal and cancer-infected epithelial cells of the breast. In the case of NPs the surface charge, which includes pKa of the particle and electrical charge in the double interfacial layer, is termed zeta potential (Ying et al., 2015). According to the OECD, zeta potential is a good characteristic to evaluate the toxicity of less soluble MONPs (Zhang et al., 2008a,b). The results of the above study showed that zeta potential was not only affected by conditions in suspension, such as temperature, type of ion, pH and ionic strength (Wang et al., 2013a,b), but was also affected by other properties.

2.2.2 Physical characteristics

2.2.2.1 Size of MONPs

The size of NPs has a significant effect on their chemical, electrical, and thermodynamic properties (Liu, 2006). Chavali and Nikolova (2019) reported that the properties of magnetic oxide NPs changed according to their size. Jun et al. (2008) detected change in the behavior of -Fe_2O_3 NPs. It was reported that at 12 nm size they showed superparamagnetic behavior while at 55 nm size they showed ferromagnetic behavior. Change in the properties with the change in size significantly influences the toxicity potential of NPs. Wongrakpanich et al. (2016) evaluated the size-dependent toxicity of $_n$CuO in lung epithelial cells. Results revealed that $_{n}CuO$ with a size of 24 nm exhibited more cytotoxicity than the NPs of 4 nm size. This behavior might be due to the change in internal dissolution of Cu^{+2}. In another study, Sanders et al. (2012) reported that 25 nm TiO_2 anatase showed greater phototoxic effect to human epithelial cells, compared to 142 nm anatase.

2.2.2.2 Shape of MONPs

MONPs are synthesized in different shapes like cubes, spheres, rods, and cylinders, depending on the purpose. The shape of the MONPs can also affect different properties (Misra et al., 2014; Yang and Ma, 2010). It was reported that spherical-shaped MONPs are more readily taken up by cells than rod-shaped ones (Chithrani et al., 2006). Similarly, Heng et al. (2011) reported that spherical-shaped ZnO NPs were taken up by cells more readily than ZnO sheets were. Toxicity potential of MONPs varies according to their shape. Liu et al. (2017) reported that spherical-shaped TiO_2 NPs penetrated the blood brain barrier more quickly than rod-shaped ones. In another study, Dong et al. (2019) reported that rod-shaped Al NPs showed strong toxicity effects on rat astrocytes compared to Al NPs in flakes. Similarly, shape related toxicity behavior of ceria NPs was studied by Forest et al. (2017), showing that rod-shaped ceria NPs were more lethal to cells than octahedron/cube-shaped NPs were.

2.2.2.3 Optical characteristics of MONPs

Optical properties of MONPs play an important role during development of technological devices. Optical properties can be determined by light scattering in solution as well as by electronic transitions in the material, which can form electron-hole pairs in the electronic structure (Mistrik et al., 2017). Due to Coulomb's interaction, excitons are formed by the holes and electrons existing in a material. To understand the optical nature of semiconductors, these excitons play an important role. To develop high performance optoelectronic devices such as laser diodes, semiconductors, and light-emitting diodes, NPs are a good candidate, when the size is decreased, as in the case of nZnO (Li et al., 2009). The band-gap property of MONPs is exclusive and important for various technologies. UV-visible spectroscopy can be used to determine the optical band gap. It can be estimated by the Taucs model (Kahouli et al., 2015; Khan et al., 2014).

2.2.3 Chemical characteristics

2.2.3.1 Electrical conductivity

In terms of conductivity, MONPs can be covalent or ionic in nature. According to Boltzmann, electronic charge carriers are a function of temperature and energy band gap, given the electronic conduction. There are two types of electronic conduction—p-type and n-type—which depend on the principle charge carrier, i.e., hole or electron (Petty, 2019). By creating more spaces, the number of free charges can be increased, creating a nonstoichiometric effect. On increasing the electronic contribution, an increase in conductivity was reported (Dhumal et al., 2019; Lavik et al., 1997). In gas-sensing devices, such as WO_3, SnO_2, a strong dependence on size was reported (Franke et al., 2006).

2.2.3.2 Acid/base strength

Redox potential of the metal center and deprotonation and protonation of the oxygen atoms on the surface influence the acid/base properties of MONPs (Reddy et al., 2013). Cations and anions act as acidic and basic, respectively, while the hydroxyl groups on the surface can be either acidic or basic. The acid base properties of MONPs are due to the movement of valence electrons. Aluminum oxide and Ga_2O_3 are examples of strong acidic MONPs (Glinka et al., 2001). The base or acid character of the material significantly influences the isoelectric point (IEP) of NPs. It was reported that antibacterial activities of $_{n}TiO_2$ were affected by IEP (Djurišić et al., 2015).

2.2.3.3 Solubility

Toxicity and bioavailability of MONPs depends on the solubility in an aqueous medium. It is considered an important parameter by the OECD to evaluate the toxicity and fate of MONPs (OECD, 2016). Dimensions also affect the solubility of the compound in solvent. Escorihuela et al. (2018a,b) developed a method to evaluate the solubility of MONPs in aqueous solutions, which was applied to nZnO in water. In the case of partially soluble MONPs, toxicity depends on the release of free metal ions in the solution, i.e., CuO and ZnO (Robinson et al., 2015).

2.2.3.4 Chemical potential

The driving force responsible for the formation of MONPs into their end shape and composition is known as chemical potential. The chemical potential of the surface atoms of the NPs depends on the curvature radius of the surface (Thorat and Bauer, 2020; Guozhong, 2004). On the surface of NPs, the chemical potential of the atoms showed higher values at convex surfaces than flat surfaces. The chemical potential of NPs tells us about the crystallographic structure, shape, and composition. At nanoscale, stress and free energy can induce changes in cell parameters and alter the thermodynamic stability in NPs compared to bulk materials. This phenomenon is observed in MONPs such as Al_2O_3 (Rodríguez and Fernández-García, 2007), TiO_2 (Zhang and Banfield, 2005), MoO_x (Song et al., 2003), and VO_x (Rodríguez and Fernández-García, 2007).

3. Synthesis and characterization of MONPs

Preparation of MONPs with controlled features is the priority for the commercialization of MONPs. Basically, there are two major approaches for the synthesis of nanoparticles: (1) bottom-up approach, which involves the assembling of molecules and atoms to generate a wide range of MONPs; and (2) top-down approach, where bulk materials are converted to fine NPs by leaching out bit by bit. Based on these two approaches there are three ways to synthesize MONPs, which include chemical, physical, and biological methods.

3.1 Chemical methods

Purposeful execution of different chemical reactions to obtain a product or several products is known as chemical synthesis. Several techniques are discussed here.

3.1.1 Sol-gel method

There are two components in sol-gel: sol, which represents the colloidal suspension of chemicals in liquid, and gel, which is the liquid containing the polymers. In this method, sol of different types is mixed to form a network of polymers (gel) (Brinker and Scherer, 2013). There are two major phenomena in this method: hydrolysis and condensation. After condensation the extra water is removed, and the resulting material is either xerogel or aerogel, depending on the method of drying. There are two types of sol-gel method: aqueous sol-gel and nonaqueous sol-gel. If the reaction medium is water, than it is known as aqueous sol-gel and if the reaction medium used is an organic solvent than it would be nonaqueous sol-gel method. Novel cone-shaped nZnO NPs were obtained by the decomposition of Zn $(OAc)_2$-TOPO complex by this method (Joo et al., 2005). Vijayalakshmi and Rajendran (2012) reported the preparation of $nTiO_2$ with a size around 9 nm and highly crystalline, using titanium VI isopropoxide and ethanol. They added HNO_3 to assist the hydrolysis process. Apart from that, several other MONPs have been prepared by different research groups, e.g., nZnO (Zavar, 2017), $nSnO_2$ (Al-Hada et al., 2018), and $_nWO_3$ (Wang et al., 2019).

3.1.2 Microemulsion method

A thermodynamically stable system consisting of two immiscible liquids is called a microemulsion. In a microemulsion, droplets are formed by the dispersed phase, which is in lower volume (Sharma and Shah, 1985). This system greatly depends on the dispersion phase. The dispersion phase might be water in oil (W/O), in which water is dispersed in oil or oil in water (O/W); in this type, oil is dispersed in a bulk of water. There is also a possibility of oil in oil emulsion (O/O), in which two oils (nonpolar and polar) are mixed (Tadros, 2009). Generally, microemulsion systems consist of three components, i.e., nonpolar, polar, and surfactant (Malik et al., 2012; Hussain and Batool, 2017). In this method, surfactants are used as emulsifiers. Surfactants can be classified as cationic (cetyltrimethyle ammonium bromide), anionic (dioctylsulfosuccinate sodium), nonanionic (Triton X-100), and zwitterionic (phosphatidylcholine) (Kronberg et al., 2014; Marchand et al., 2003). Boutonnet used a chemical reduction method to synthesize some noble colloidal metals (Kanwar et al., 2018; Ganguli et al., 2010). Specific structure, high stability, and very small droplet size are important properties of microemulsions (Hussain and Batool, 2017; Eriksson et al., 2004). Housaindokht and Nakhaei Pour (2013) synthesized controlled sized iron oxide NPs using a microemulsion technique. Pemartin-Biernath et al. (2016) also reported the preparation of cerium oxide NPs using oil in water reaction of microemulsion.

3.1.3 Hydrothermal method

Hydrothermal or solvothermal is one of the most effective and common methods to synthesize MONPs with a variety of morphologies. In this method, an autoclave is filled with an organic compound or water, the reactants placed into it, and the reaction carried out under high pressure and high temperature conditions. If the preparation is carried out in water, than the process is known as hydrothermal, while if the preparation is carried out in a nonaqueous solvent then it is known as solvothermal method (Wu et al., 2002; Cushing et al., 2004). Several types of autoclave have been discussed in the literature (Hakuta et al., 2005). Normally, Teflon-lined autoclaves are preferred because of their ability to bear high pressure and temperature. One of the main factors in synthesizing nanostructured inorganic materials is the precise control of the hydrothermal method (Shi et al., 2013a,b). Yan et al. (2008) reported the preparation of ceria NPs using $Na_3PO_4{\cdot}6H_2O$, Ce $(NO_3)_2{\cdot}6H_2O$, and NaOH as a precursor by a simple hydrothermal technique. Preparation of other MONPs, i.e., CuO, NiO, and ZnO was also reported by this method (Liu et al., 2015; Sue et al., 2011; Maryanti et al., 2014).

3.1.4 Polyol method

Usually, multivalent and high-boiling alcohols are used to synthesize MONPs. In this method, polyols can act as a solvent as well as a reducing agent, and can also control the growth of MONPs. Different polyols can be used in this method, i.e., butylene glycol, ethylene glycol, diethylene glycol, triethylene glycol, tetraethyleneglycol, and propylene glycol, etc. (Carroll et al., 2011). Kim et al. (2008) reported the synthesis of $_nCu_2O$ by using ethylene glycol (reducing agent), copper nitrate (precursor), and poly (vinyl pyrrolidone) as a stabilizing agent. Lee et al. (2008a,b) reported the preparation of nZnO. Several other MONPs were also prepared by using this method, like TiO_2 (Dong et al., 2015), CeO_2 (Zheng et al., 2015), Fe_2O_3 (Ma et al., 2014), and Al_2O_3 (Itoh et al., 2015).

3.1.5 Chemical vapor synthesis

Chemical vapor synthesis (CVS) is a relatively new method, which is a modification of the chemical vapor deposition (CVD) method used to produce thin films or coatings. In CVS, particle formation and homogeneous nucleation are favored. Short reaction time, low particle size, high purity, and low agglomeration are the advantages of CVS (Srdić et al., 2000). Stijepovic et al. (2015) synthesized gallium oxide NPs by the CVS method.

Similarly, Polarz et al. (2005) synthesized nZnO by CVS method using a Zn compound alkyl-alkoxy as precursor material, converted to ZnO chemically by gas-phase.

3.2 Physical methods

In this approach thermal energy, mechanical pressure, and radiation of high energy are used to synthesize MONPs. These methods come under the top-down category. The MONPs generated by this approach are monodisperse and free of solvent contamination. This approach is not as economical as the bottom-up approach due to generation of huge amounts of waste. Several methods come under the physical approach to generate MONPs.

3.2.1 High energy ball milling

John Benjamin first developed this method in 1970 to generate alloys with high strength, which can bear high pressure and temperature. This method is used to generate MONPs of different shapes and dimensions by using energy efficiently (Grumezescu, 2016; Xing et al., 2013). In this method, moving balls transfer their kinetic energy to mill the material. The kinetic energy is used to break the chemical bonds and to convert the required material into very fine particles with newly developed surfaces. This method is also known as the mechanochemical method, because sometimes high pressure and temperature (>1000°C) are involved. nZnO NPs were synthesized by this method using ZnO microcrystalline powder with a crystal size of ~30 nm (Ul-Islam and Butola, 2018; Salah et al., 2011; De Carvalho et al., 2013).

3.2.2 Physical vapor deposition

A collective set of several processes to generate MONPs of very small size is known as physical vapor deposition (PVD). There are three general steps in this eco-friendly process including: (1) solid source vaporization; (2) vapor transportation; and (3) growth and nucleation to synthesize MONPs (Dhand et al., 2015). There are different types of PVD methods to synthesize MONPs: sputtering, pulsed laser deposition, and vacuum arc.

3.2.2.1 Sputtering

Sputtering is a commonly used vacuum-based process to generate MONPs. The principle of momentum transfer is used in this method. The atoms from the required material are removed/ejected with bombardment of ions. The material is deposited by using radio frequency, DC, and pulse DC. There are several steps involved in deposition:

(a) Between two electrodes, a plasma of neutral gas, i.e., Ar is generated.
(b) A potential is applied between electrodes to accelerate the ions in the plasma.
(c) The energetic ions hit the target material to remove the material.
(d) The removed material is transported to the substrate.

There are two types of sputtering: simple and magnetron. The difference is only the magnet, which is attached behind the target material in magnetron sputtering. Bouchat et al. (2013) reported the synthesis of $nTiO_2$. Aubry et al. (2019) and Rydosz et al. (2019) synthesized Fe, Cu, Sn, and Ti-oxide NPs using pulsed-DC magnetron sputtering.

3.2.2.2 Pulse laser deposition/laser ablation method (LA)

A strong laser beam is used to evaporate the solid source (Chen and Yeh, 2002). In this method, a continuous or pulsed laser can be used. Synthesis of nZnO was reported by this method with a size range of 10–32 nm; the material was vaporized by an Nd:YAG laser of 532 nm and an aqueous SDS solution was used for deposition (Al-Dahash et al., 2018; Singh and Gopal, 2007). Harano et al. (2002) and Sakiyama et al. (2004) also reported the production of $nTiO_2$ and nNiO by the LA method.

3.2.2.3 Vacuum arc (VA)

Like sputtering, this is also a vacuum-based method to synthesize MONPs. In this method, the arc is used to vaporize the target material and then the vaporized material is deposited. Synthesis of $_nCuO$ was reported by the solid liquid arc discharge process (Tharchanaa et al., 2020; Yao et al., 2005). Synthesis of other MONPs was also reported by this method, i.e., ZnO (Ashkarran et al., 2009), TiO_2 (Ashkarran et al., 2010a,b,c), and ZrO_2 (Ashkarran et al., 2010a,b,c). Lepeshev et al. (2016) also reported the synthesis of $nTiO_2$ with an average particle size of 6 nm by the vacuum arc method, using a cluster deposition source.

3.2.3 Laser pyrolysis

Haggerty and his colleagues first developed the CO_2 laser pyrolysis technique (a vapor phase synthesis technique) in the early 1980s (Daraio and Jin, 2012; D'Amato et al., 2013). A large variety of MONPs can be synthesized using this technique (Fe_2O_3, SiO_2, TiO_2, Al_2O_3). In this method, laser-induced chemical reactions generate condensable products. In this process, one of the reactants should be able to absorb through the resonant vibrational mode of infrared laser radiations of CO_2. An additional chemical, such as ethylene gas, ammonia, or sulfur hexafluoride, is used to carry out the process of energy transfer between precursor and laser light (Dhand et al., 2015). This is a very useful method to generate MONPs of 5–60 nm. D'Amato et al. (2013) reported the generation of $nTiO_2$ and $nSiO_2$ by laser pyrolysis technique using titanium (VI) isopropoxide and tetraethyl orthosilicate, respectively, as precursor materials. Popovici et al. (2007) and Arias et al. (2018) synthesized iron oxide NPs by the CO_2 laser pyrolysis technique. Similarly, synthesis of $_nFe_2O_3$ was also reported with an average size of 5 nm by using Fe $(CO)_5$ vapors (Veintemillas-Verdaguer et al., 2004).

3.2.4 Flame spray pyrolysis

This is the latest in flame aerosol technology (Teoh et al., 2010). It is a one-step process of combustion, in which an organic solvent of high enthalpy is used in liquid form. High temperature and self-sustaining flames, high temperature gradient, proven scalability, usage of liquid feeds, and less volatile precursor are some elements of this process. Flame spray pyrolysis (FSP) is one of the most exploited methods for the synthesis of MONPs. There are several steps to the combustion process: (1) evaporation of metal; (2) nucleation; (3) growth by coalescence and sintering; and (4) particle generation. Sokolowski et al. (1977) first synthesized nAl_2O_3 by combustion of aluminum acetylacetonate in a benzene-ethanol solvent mixture as precursor. Wallace et al. (2013) and Tricoli et al. (2016) reported the synthesis of nZnO by this method. Many other researchers also reported the production of other MONPs like Mn_2O_3, ZrO_2, WO_3, Fe_2O_3, SnO_2, Al_2O_3, and V_2O_5 (Laine et al., 2006; Sel et al., 2014; Tricoli et al., 2008; Pokhrel et al., 2010).

3.2.5 Electrospraying technique

Electrospraying is like electrospinning, and both techniques are used to fabricate MONPs (Koštáková et al., 2012). The electromechanical device used in this method contains the mixture of solvent and the selected polymer, which is then taken up by syringe and high voltage is applied to the capillary tip, which results in the generation of charged droplets. After evaporation of the solvent, particles or fibers are collected at the counter electrode. It was reported that $nSiO_2$ of 20 nm size were generated using ethylene glycol as solvent (Jaworek and Sobczyk, 2008; Tang and Cheng, 2013). Similarly, synthesis of other MONPs was also reported by this method, such as Al_2O_3 (Jaworek and Sobczyk, 2008; Witharana et al., 2012), TiO_2 (Jaworek and Sobczyk, 2008; Shi et al. 2013), and ZrO_2 (Jaworek and Sobczyk, 2008; Jamal, 2013; Eshed et al., 2011).

3.3 Biological methods

Biological synthesis of MONPs is eco-friendly, cost-effective, less toxic, and efficient then other methods. Biological entities like bacteria, fungi, viruses, plants, and yeast have been deployed to synthesize MONPs. There are three categories of biological methods: plant extracts, microorganisms, and biomolecule templates. These methods are described below.

3.3.1 Biological synthesis of MONPs using plant extracts

Environmentally friendly synthesis of MONPs using the extracts of plants is considered a reliable method (Akhtar et al., 2013; Kaur et al., 2016). Chemicals such as polyphenols, phenolic acid, alkaloids, tannins, polysaccharides, and terpenoids present in the extracts of flowers, leaves, and stems act as a bio-reductant (Jeevanandam et al., 2016; Silveira et al., 2018). The characteristics of MONPs, such as, size, shape, and quantity produced are

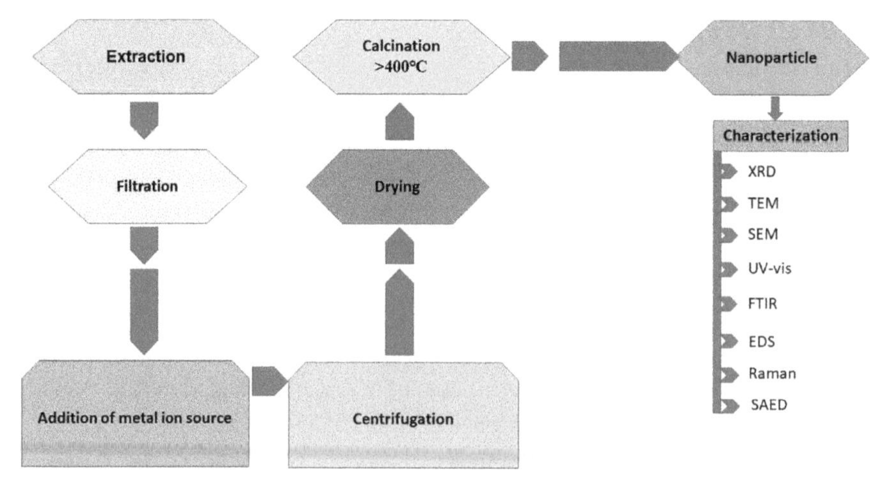

Fig. 2 Plant-mediated synthesis and characterization of MONPs.

significantly affected by pH, concentration of metal salt, and the nature of the extract (Mittal et al., 2013). Rufus et al. (2017) developed an ecofriendly and efficient method to synthesize α-Fe_2O_3 NPs by using the leaf extract of cashew tree as a bio-reductant. The synthesized NPs ranged in the size from 29 to 50 nm. Sundrarajan et al. (2017) reported the synthesis of tetragonal nTiO_2 with an average particle size of 10 nm by using *Morinda citrifolia* leaf extracts following the hydrothermal method. Similarly, Nasrollahzadeh and Sajadi (2015) synthesized nTiO_2 by using titanyl hydroxide and *Euphorbia heteradena Jaub* root extract as a bio-reductant. Madan et al. (2016) reported the synthesis of nZnO with wurtzite hexagonal structure. They used *Azadirachta indica* (neem) leaf extract as a template followed by combustion of solution at low temperature. Similarly, Ali et al. (2016) synthesized nZnO following this biogenic, efficient, and cost-effective technique using *Aloe vera* leaf extract (*Aloe barbadensis* Miller). Synthesis of Mn_3O_4 was carried out by using neem extract as a reducing agent (Sharma et al., 2016). Green synthesis of iron oxide NPs was carried out by Al-Ruqeishi et al. (2019) using the polyphenols present in the leaves of Omani mango. A graphical scheme of plant-mediated synthesis of MONPs is shown in Fig. 2.

3.3.2 Biological synthesis of MONPs using microorganisms

Among microorganisms, prokaryotic bacteria are extensively utilized for the synthesis of MONPs. Bacteria are preferred because they are easy to manipulate (Thakkar et al., 2010). Bacteria use several reducing enzymes, like nitrate-dependent reductase or NADH-dependent reductase, to reduce the size of metal ions (Boroumand Moghaddam et al., 2015). There are several bacterial species used to synthesize MONPs such as *Pseudomonas* sp., *Actinobacter* sp., *Lactobacillus* sp., *Escherichia coli*, *Cyanobacterium* sp., *Klebsiella pneumonia*, and *Bacillus cereus* (Shah et al., 2015; Velusamy et al., 2016). Bacteria can use extracellular or intracellular mechanisms to synthesize MONPs (Ghosh et al., 2017). Mostly less toxic MONPs such as Fe_xO_x, ZnO, CuO, and TiO_2 are synthesized by bacterial-mediated processes (Jeevanandam et al., 2016). Kirthi et al. (2011) reported the synthesis of nTiO_2 using *Bacillus subtilis* and TiO $(OH)_2$ as a reducing agent and precursor, respectively. Synthesized NPs ranged in size from 66 to 77 nm and were spherical in shape. Khan and Fulekar (2016) used *Bacillus amyloliquefaciens* to synthesize nTiO_2 and the resultant NPs ranged in size from 22 to 97 nm and were spherical in shape. Similarly, synthesis of nZnO was also reported by using *Aeromonas hydrophila*. The generated NPs were spherical in shape and the average particle size was 58 nm (Jayaseelan et al., 2012). Kundu et al. (2014) also synthesized nZnO by $ZnSO_4$, using H_2O as a precursor and *Rhodococcus pyridinivorans* NT2 as a reducing agent. The size of the NPs ranged from 100 to 120 nm and they were hexagonal in shape.

Fungi have also been utilized to synthesize MONPs. Differentiation, growth, and metabolism of fungi depends on metals directly or indirectly, making it interesting to synthesize MONPs using different strains of fungi. Enzymes released by fungi play an important role in biosynthesis of MONPs (Salaheldin et al., 2016; Khandel and Shahi, 2018). It was reported that nTiO_2 were synthesized by using *Aspergillus flavus* as a capping and reducing agent. It was observed that the synthesized NPs were in aggregated form with an average particle size ranging from 62 to 74 nm. Rajakumar et al. (2012) synthesized titanium dioxide NPs by using different salt precursors (chloride, nitrate, oxide, and sulfate salts). It was reported that incubation of 0.1 mM solution of precursor salt for 72 h at 28°C resulted in the production of NPs in larger amounts. Release of a greater number of enzymes per unit of biomass and higher ability to take up metals are the main advantages of fungal biosynthesis of MONPs. Synthesis of nZnO by using *Aspergillus fumigatus* TFR was also reported (Raliya and Tarafdar, 2013). Similarly, Ibrahem et al. (2017) synthesized nZnO by using the filtrate of *Aspergillus niger*. The resultant NPs were spherical in shape and ranged in size from 41 to 75 nm. Synthesis of bismuth oxide NPs was carried out by using the spores of *Fusarium oxysporum* (Jeevanandam et al., 2016).

4. Methods used to analyze and characterize the MONPs

It is crucial to analyze and characterize the MONPs in plants and animal tissues as well as in different consumer products to study the uptake patterns, interactions, biokinetics, and biotransformation of these particles in the cells. Various methods can be used to detect and characterize the MONPs in different cells and materials. Methods and techniques used to detect and characterize MONPs in different materials are discussed below and shown in Fig. 2.

4.1 Microscopy-based techniques

A light microscope or compound microscope is not suitable for detection and characterizing of nanoparticles in any type of specimen because of their limited resolution, i.e., resolution power up to 200 nm. However, the near field scanning optical microscope (NSOM), which has a minimum resolution of 12 nm, can be used to some extent. In this method, the nanometric tip of the optical probe is positioned in such a way that it is close to the sample and interacting with it. This technique is also suitable for detection of fluorescent nano-objects. However, fluorescence of NPs can also be detected by the confocal laser scanning microscope (CLSM), which is capable of sectioning across the specimen with a focused beam of light (Lewinski et al., 2010). Atomic force microscopy (AFM) and chemical force microscopy (CFM) are also used for NPS detection and characterization.

Scanning electron microscopy (SEM) and transmission electron microscopy (TEM) are the most common microscopy-based techniques for characterization and detection of NPs in different types of samples. Both sectioned and intact samples can be analyzed, and resolution less than 1 nm can be obtained under SEM (Montes et al., 2017). In SEM, a focused beam of electrons scans the surface of the sample and information about the composition and topography of the sample can be obtained by the signals produced by the surface of the sample due to interactions between electrons and atoms. In TEM, a beam of electrons is transmitted though the cell and an image produced by the interaction of electrons with the sample. TEM can produce images in high resolution at less than a nanometer (0.07 nm), visualizing a single column of atoms. High lateral resolution of TEM enables it to detect the individual NPs dynamics in a living animal or plant cell. Visualizing the NPs positions in the cell with SEM and TEM can provide precise information about the size, shape, structure, morphology, and aggregation or dispersion state of these particles, which can be helpful in understanding uptake and localization of these MONPs within the cell or tissue (Yan and Chen, 2018).

Variable-pressure scanning electron microscopy (VP-SEM) is a modified form of SEM with extended capabilities and complexity with the addition of variable pressure. This can be used for high spatial resolution for conducting and analyzing in situ experiments (Thiel, 2019). Very little or no sample preparation is required, compared to conventional SEM, in VP-SEM application. VP-SEM implies imaging in the gas environment for charge control while environmental SEM (ESEM) deals with imaging of water or hydrated samples, often with a temperature-controlled stage (Joubert, 2017). SEM can also be coupled with a focused ion beam of electrons (FIB-SEM) for sample sectioning and 3D imaging. Scanning transmission electron microscope (STEM) is also coupled with a focused beam of electrons for specimen scanning, receives signal just like TEM through the ultrathin specimen, and its resolving power can reach up to atomic level (Dudkiewicz et al., 2012). STEM electron tomography can be used for characterization of 3D morphology of NPs for stable as well as embedded NPs (Mourdikoudis et al., 2018).

Conventional EM operates under vacuum conditions. NPs in liquid media cannot be studied with conventional microscopy. Cryo- and liquid electron microscopes are modified forms of SEM, TEM, and FIB-SEM. Cryo-EM images a frozen sample and allows the visualization of specimens in their native frozen environment by vitrifying at cryogenic temperature (Danino, 2012). Samples are usually frozen with liquid nitrogen. Cryo-EM is used to study NPs' growth, complex aggregation mechanisms, and complex growth mechanisms, and to visualize the molecular template (Pallares et al., 2016). It can achieve resolution below nanometer or subnanomolar scale. Liquid- and cryo-EM techniques can be used for hydrated food and environmental and cosmetic samples. The resolving power of these EM are not better than conventional SEM and TEM. Environmental SEMs and TEMs (ESEM, ETEM) are used as alternative approaches for detecting and characterizing NPs in hydrated samples. This technique allows the imaging of a sample in its original state and this imaging process takes place under lower pressure than the conventional EM (Dudkiewicz et al., 2012).

4.2 X-ray-based techniques

X-ray scattering techniques are the most widely used techniques for determination of composition, pore size, and crystalline structure of specific single and compound elements. X-ray diffraction (XRD) is the most commonly used for characterization of NPs on the basis of crystalline structure, grain size, lattice parameters, and the nature of the phase. Other parameters can also be estimated in a specific sample by comparison of the broadening of the most intense peak in XRD measurement by using Scherrer's equation. Samples can be fed in powder form in XRD. It is not suitable for amorphous material and size below than 3 nm due to over-broadening of peaks (Mourdikoudis et al., 2018). X-ray absorption spectroscopy (XAS) measures materials' X-ray absorption coefficient as a function of energy. It includes X-ray absorption near edge structure (XANES) and extended X-ray absorption fine structure (EXAFS). In situ XAS is used to examine the mechanism of formation and kinetics of NPs. Every element has a set of characteristic absorption edges corresponding to different binding energies of its electrons, giving XAS element selectivity. It can be used for determination of the chemical state of species, Debye-Waller forces, and interatomic distances, and can also be used for noncrystalline NPs (Srabionyan et al., 2016).

X-ray scattering techniques, such as small angle X-ray scattering (SAXS), are extensively used for characterization of MONPs, i.e., ZnO and FeO NPs. Scattering of X-rays at low angles contains information of size and shape of particles. This technique determines the growth kinetics, size distribution, and particle size (Ramallo-Lopez et al., 2007). X-ray photoelectron spectroscopy (XPS) is mostly used for surface chemical analysis. XPS is very useful for coating and surface characterization of solid-state NPs. It is used to determine the elemental composition, electronic structure, ligand binding, and oxidation states of nanostructures. The X-ray fluorescence (XRF) technique can be used for identification and quantification of elements in solid, liquid, and powdered samples (Dudkiewicz et al., 2012).

4.3 Spectroscopy-based techniques

Scattering spectroscopy is a common technique used for particle analysis. It is based on the measurement of the amount of light scattered by a substance—incident as well as polarizing angles—at certain a wavelength. Particle size and molecular weight measurement of the substance can be done by detecting the interference patterns of light. In liquids, dynamic light scattering (DLS) is the main scattering technique used for measurement of size of NPS (Brar and Verma, 2011). Nanoparticle tracking analysis (NTA) is a similar approach to DLS in which Brownian movement of an individual particle is analyzed by video, where positional changes of this individual particle is tracked step by step in two directions by a light scattering center (Kendall et al., 2009). X-ray scattering techniques are also used for particle size analysis, composition, and other characterizations of nanomaterials, as discussed earlier. Raman spectroscopy is based on the Raman effect and can be used for molecular identification, sizing, structure, and oxidation states analysis. It is categorized under both light scattering and laser-based methods. When the laser light impinges upon a molecule, a type of scattering occurs, which results in shifting up and down in the energy of laser photons and this information can be used for analysis (Wong et al., 2009). Laser-induced fluorescence (LIF) and laser-induced breakdown detection (LIBD) are other laser-based techniques for quantification of number concentration and size of NPs.

Mass spectroscopy allows molecular mass and chemical characterization of a sample compound and allows detection at minimum levels of parts per trillion. Matrix-assisted laser desorption ionization (MALDI), electrospray ionization (ESI), and inductively coupled plasma (ICP) are used in MS for NPs detection and characterization (Tiede et al., 2010). ICP-based techniques are used for a wide range of solid, liquid, and suspension samples. ICP optical emission spectroscopy (ICP-OES), ICP mass spectroscopy (ICP-MS), and single particle ICP-MS (SP-ICP-MS) are mainly adopted for NPs analysis. ICP-OES is used for measurement of elemental composition and NPs number concentration. ICP-MS is widely used for detection, characterization, and quantification of NPs. For sizing of a single particle, SP-ICP-MS is adopted and is used for obtaining concentration and size of metal-based NPs by running in single particle mode and achieving single particle spikes (Yan and Chen, 2018).

5. Application of MONPs: Current status and future trends in industry

In recent years, nanotechnology has emerged as an important tool for synthesis of new materials with novel and desirable properties. Nanotechnology is a rapidly growing trillion-dollar industry. The largest share of manufacture and application among different NMs belongs to MONPs (Djurišić et al., 2014; Zhang et al., 2018). The global market for metal oxide nanoparticles indicates that MONPs' current production is estimated at ~1.66 million tons, compared to 0.27 million tons in 2012 (Vance et al., 2015; Rajput et al., 2018a,b,c,d). Among the existing ENMs, MONPs are identified in approximately 37% of the total products containing NPs (Kumar et al., 2018; López-Muñoz et al., 2019). For example, global production of ZnO nanoparticles is estimated at 0.1–1.2 million tons/year (Kumar and Dhawan, 2013; Swain et al., 2016; Singh, 2019). Similarly, it is estimated in a number of studies that the current median global consumption of Ag NPs is 55 tons/year with other estimates from future markets suggesting Ag NP consumption is as high as 360 tons to 450 tons/year (Piccinno et al., 2012; Lazareva and Keller, 2014; Zhang et al., 2016; McGillicuddy et al., 2017). More than 400 products have been advertised as containing Ag NPs (Zhang et al., 2018).

Owing to exclusive and useful physio-chemical properties like magnetism, optics, surface energy, shape, size, etc., the MONPs have revalorized various technology trends (some examples of which are listed in Table 2) such as commercial products (e.g., fuel cells, plastics, consumer products, agriculture, food safety) (Van Devener and Anderson, 2006; Lee et al., 2010a,b; Lagaron and Lopez-Rubio, 2011; Wang et al., 2016a,b,c; Prasad et al., 2017; Rizwan et al., 2017), environmental applications (e.g., environmental analysis, sensing, remediation) (Bai and Zhou, 2014; Trujillo-Reyes et al., 2014a,b; Mahdavi et al., 2015; Amde et al., 2015, 2016; Gan et al., 2016; Ju-Nam and Lead, 2016; Lee et al., 2016; Mirzaei et al., 2016; Vuong et al., 2016; Ghasemi et al., 2017; Rahman et al., 2017; Singhal et al., 2017), sustainable chemistry (e.g., catalysis) (Almukhlifi and Burns, 2016; Gawande et al., 2016; Montini et al., 2016), and health (e.g., antimicrobials, cancer treatment) (Salata, 2004; Applerot et al., 2009; Amde et al., 2015; Ansari et al., 2015; Chauhan et al., 2015; Tuli et al., 2015; Al-Ajmi et al., 2016; Yang et al., 2016a,b; Saba and Amini, 2017). The current applications of MONPs in different eras of technologies are summarized in Table 2.

5.1 Applications in electronics and related industries

MONPs have significantly contributed to innovations in applied nanotechnology. For instance, MONPs have been effectively used in boosting the performance of batteries, trace gas sensors, solar cells, magnetic storage devices, microstrip and patch-type transparent antennas, energy conversion, and many more due to their ability to increase sensitivity and thereby enhance the analytical precision of instruments down to single atomic and molecular level detection; e.g., cobalt oxide NPs and carbon nanotubes have several

Table 2 Trending applications of MONPs in various industries.

Industrial sectors	Applications	MONPs	References
Agriculture	Nanofertilizer, improved nutrients uptake by plants	$_{n}ZnO$, $_{n}CuO$, $_{n}Fe_xO_x$, $_{n}\alpha\text{-}Fe_2O_3$	Bisquera et al. (2017), Palchoudhury et al. (2018), Wang et al. (2019), Kumar et al. (2020), and Younes et al. (2020)
	Oxidative stress reduction and plant growth improvements	$_{n}ZnO$, $_{n}Fe_xO_x$	Rizwan et al. (2019)
	Nanopesticides	$_{n}CuO$, $_{n}MgO$, $_{n}ZnO$, $_{n}CeO_2$	Malaikozhundan and Vinodhini (2018), Pandey et al. (2018), and Tiwari et al. (2020)
	Nanosensors for monitoring pathogen attack, pesticides detection and soil conditions	$_{n}ZnO$	Fraceto et al. (2018) and Kaushal and Wani (2017)
Paint industry	Photocatalytic property against air pollutants and UV protection	$_{n}TiO_2$	Varma et al. (2019) and West et al. (2019)
	Prevention of metal corrosion	$_{n}TiO_2$	Yousefi and Mahmoudian (2019)
Biomedicine sciences	Targeted drug delivery and drug nanocarriers	$_{n}ZnO$, superparamagnetic $_{n}Fe_3O_4$	Mishra et al. (2017), Karimzadeh et al. (2017), and Mirza et al. (2020)
	Anticancer, antibacterial, antidiabetic, antiinflammatory potential	$_{n}ZnO$, $_{n}CdO$-ZnO, and nZnO/kaoline composites	Mirzaei and Darroudi (2017) and Fahimmunisha et al. (2020)
	Bioimaging	$_{n}ZnO$, magnetic $_{n}Fe_3O_4$	Rosenberg et al. (2018), Wierzbinski et al. (2018), and Eixenberger et al. (2019)
	Protein adsorption, antioxidant potential	$_{n}ZnO$, $_{n}CeO_2$	Jiang et al. (2018)

Table 2 Trending applications of MONPs in various industries—cont'd

Industrial sectors	Applications	MONPs	References
Cosmetics industry	Skincare products; sun-blocks, and plasmonic enhancement for sunscreen products, moisturizers, lotions	$_nZnO$, $_nTiO_2$	Trivedi and Murase (2017), Schneider and Lim (2019), and Gershon et al. (2017)
	Hair care products; shampoos	$_nZnO$, $_nTiO_2$	Dréno et al. (2019)
	Makeup; concealers, correctors	$_nZnO$, $_nTiO_2$	Fang et al. (2017) and Hosny et al. (2017)
Food packaging and safety	Nano-biosensors for monitoring food quality and safety	$_nZnO$, $_nTiO_2$	Srivastava et al. (2018)
	Antimicrobial activities	$_nZnO$, $_nTiO_2$, $_nAl_2O_3$	Garcia et al. (2018)
	Zinc supplement	$_nZnO$	Agarwal et al. (2017)
Electronics and information technology-related industries	Lithium ion batteries, solar cells	$_nZnO$, $_nTiO_2$	Gu et al. (2018)
	Semiconductors gas sensors	$_nZnO$, $_nTiO_2$, $_n(\alpha\text{-}Fe_2O_3)$, $_nNiO$, $_nCu_2O$	Korotcenkov and Cho (2017) and Gao and Zhang (2018)
	Fluorescent lights and refrigerators	$_nTiO_2$	Haider et al. (2019)
Environmental remediation	Economic exploration, refining and processing of oil and gas	$_nCeO_2$, $_nNiO$, $_nAl_2O_3$, $_nMnO_2$, $_nPdO$, $_nCo_3O_4$, $_nNi\text{-}Pd/TiO_2$	Khalil et al. (2017)
	Removal of pollutants dyes, and sensing of heavy metal ions	$_nCuO$, $_nZnO$	Singh et al. (2018)
	Photocatalytic activity	$_nTiO_2$	Haider et al. (2019)
	Biodegradable metal oxide plastic polymer	$_nZnO\text{-}CuO$	Varaprasad et al. (2017)
	Water and air purification	$_nTiO_2$	Haider et al. (2019)

applications in the manufacturing of highly integrated devices (Fu et al., 2005). Similarly, nanocomposites of MONPs with polymers provide optical properties in the visible region and photoluminescence along with enhanced processability resulting in wide applications in electronics and related fields for manufacturing of light-emitting diodes, solar cells, transistors, and optoelectronic coatings. For instance, ZnO NPs with polyamide nanocomposites are used for enhanced photoluminescence spectroscopy (Hajibeygi et al., 2017) and tungsten trioxide NPs are being used for their photochromic properties under UV radiation (Soytaş et al., 2019). In addition, in laser-driven fabrication related industries, perhaps the most promising application of MONPs is in different catalytic processes including photocatalysis and electrocatalysis (Wang and Gao, 2019). Electromagnetic properties of oxide nanoparticles allow them to be applicable in specific areas. For instance, based on piezoelectric properties, ZnO NPs have been utilized for nanogenerators that can harvest energy from low-frequency motion to operate human-wearable gadgets (Gao et al., 2009; Xu and Wang, 2011). MONPs are also significantly used in dye-sensitized solar cells as an electron transport layer (Akin and Sonmezoglu, 2018). Specially, titanium dioxide (TiO_2) NPs are used in dye-sensitized solar cells due to its electronic bandgap of ~3.2 eV and ideal position of its electron conduction band (Mishra et al., 2009). Likewise, organic coatings on the surface of steel with ZnO nanofillers make it highly resistant to corrosion (Pruna, 2019).

5.2 Applications for environmental remediations

The desire for new technologies for environmental remediations that exhibit improvements in terms of eco-compatibility and affordability over conventional methods provided inspiration for using nanotechnology for environmental pollution remediation (Noubactep et al., 2012). Effluents from agriculture, industry, landfills, and urban waste can have a severe global impact (Schwarzenbach et al., 2010). Adsorption has been widely employed as an effective strategy for decontamination of wastewater; on a commercial scale, MONPs such as ZnO NPs in combination with other materials have been successfully used as nanoadsorbents (Kecili and Hussain, 2018). Synergistic effects of ZnO-chitosan nanoparticles have proved useful in removal of 96% of the pesticide permethrin from wastewater (Kyzas and Matis, 2015). Similarly, modified ZnO NPs have been used on an industrial scale for the adsorption of maximum arsenic (III) and cadmium (II) ions from aqueous solutions (Gnanasangeetha and Thambavani, 2013; Salmani et al., 2014). MONPs have a wide range applications in mitigation of air pollution, e.g., nanosurfaces of calcium carbonates coated with titanium dioxide have been used as adsorbents of carbon dioxide—an important outdoor air pollutant (Ibrahim et al., 2016). Likewise, nanofibers of titanium dioxide coated with nano-silver particles have shown the capacity to decompose air pollutants such as VOCs (volatile organic compounds) by 21% and oxides of nitrogen (NO_x) by 30%, as reported by Srisitthiratkul et al. (2011). MONPs have promising applications in remediation of environmental pollution due to their robust, less toxic, and advanced pollution monitoring and remedial capacity.

5.3 Applications in the textile industry

Nanoparticles have several potential uses in clothing industries. Engineered nanofibers, also known as metal oxide fabrics, for protective clothing with the capability of removing toxic chemicals are highly in demand (Lu et al., 2017). One such industrial application includes surface coatings of fabric with polyacrylamidoxime, which results in dissolution of diisopropyl fluorophosphate in aqueous solution (Turaga et al., 2012). MONPs have also been widely used as nanofillers due to good mechanical properties and resistance against corrosion and oxidation. For instance, coelectrodeposition of TiO_2, CeO_2, SiO_2, and Al_2O_3 to the Zn matrix has been used for its enhanced anticorrosion and tribological properties in the textile industry. Similarly, electroplated Ni with TiO_2 nanocoatings on clothing exhibited less wear and high hardness of the engineered clothing.

5.4 MONPs applications in biomedical engineering and sciences

In recent years, as demand for the development of new drugs is pressing and given the inherent nanoscale processes of the biological compartments, nanotechnology has promising applications in diverse medical fields such as cardiovascular medicine and oncology (Boisseau and Loubaton, 2011). As therapeutic and diagnostic agents; nanoparticles, nanocages, and nanoshells have been used for the discovery and delivery of nanotechnology-based drugs. In fact, the term "nanomedicine" implies development and combination of atoms and molecules to yield novel molecular assemblies or nanoparticles for provision of a personalized medicine on the scale of the individual cell, organelle, or even smaller component (McNeil, 2005). Many MONPs, such as zinc oxide (ZnO) and iron oxide NPs, are important remedies for acute diseases, i.e., hepatitis, HIV, malaria, and cancer (Surendiran et al., 2009). Numerous MONPs have been prospective candidates for the detection of various biomolecules. Graphene has wide electroanalytical applications due to good optical, thermal, electrical, and mechanical properties. However, deposition of MONPs such as Co_3O_4, MnO_2, TiO_2, Fe_2O_3, ZnO, and NiO onto graphene sheets exhibits biocompatibility and increased electrochemical activity that makes it a competent transducer for biomolecules detection (Immanuel et al., 2019).

Based on particle size and surface attributes of MONPs, both active and passive drug targeting can be accomplished after oral, parenteral, nasal, or intraocular administration. Furthermore, site-specific targeting of drug delivery can be achieved by either using magnetic guidance or attaching specific ligands. For instance, surface coatings can be reformed and conjugated with nucleotides, nucleic acids, enzymes, peptides, antibodies, proteins, and small molecules (Mudshinge et al., 2011), and therefore can be efficiently targeted to a specific tissue, cell, or tumor, without being destroyed by the body's immune system. MONPs are engineered to deliver drugs to targeted tumor cells, which permits direct treatment of these cells (Rasmussen et al., 2010). Targeted delivery ensures more efficient drug allocation and decreased drug toxicity. For instance, a biocompatible iron oxide nanoprobe known as chlorotoxin (CTX) coated with polyethylene glycol

(PEG) is capable of targeting and accumulating in glioma tumor cells (Sun et al., 2008). It can also transport engineered drugs across the blood brain barrier.

Transition MONPs are of potential use in biomedicine due to the capacity to manipulate the particles with an external magnetic field (Qiao et al., 2009). These magnetic MONPs can be used with strong noninvasive tools such as MRI despite low signal sensitivity problems (Xie et al., 2010). Similarly, a magnetic field is used to heat up magnetic nanoparticles for application in the treatment of hyperthermia (Laurent et al., 2011). Furthermore, superparamagnetic MONPs have come to light as promising biomedical tools for early diagnosis of tumors, inflammation, diabetes, and atherosclerosis, due to biocompatibility and infinitesimal size. Their magnetic properties have ensured potential applications of these superparamagnetic MONPs for gene therapy, molecular, and stem cell tracking, target-specific drug delivery, rapid DNA sequencing using magnetic separation technologies, and as enhanced-resolution contrast agents for imaging (Chertok et al., 2008; Lodhia et al., 2010; Mahmoudi et al., 2011; Li et al., 2013).

Antimicrobial medicines are widely used for the prevention and control of infectious agents such as pseudomonas, staphylococcus, listeria, and vibrio (Jindal et al., 2015). However, indiscriminate use of these drugs has caused antibiotic resistance leading to adverse health risks (World Health Organization, 2018). Therefore, attention has been focused on the use of MONPs as a substitute to conventional antibiotics due to their selectivity, specificity, robustness, minuscule size, high resistance, and broad-spectrum antimicrobial activities against bacteria, viruses, and fungi. MONPs such as ZnO, AgO, and TiO_2 can kill pathogens by inhibiting synthesis of biomolecules and/or blocking normal cellular functions and processes. Therefore, these particles are used as coatings for the treatments of burns and wounds (Graves, 2014).

5.5 Applications in food safety and food packaging

Foodborne infections and diseases are a global community health issue. Every year, foodborne pathogens are the cause of millions of hospitalizations, illnesses, and even deaths (Jevšnik et al., 2013). One of the possibilities is the manufacturing of antimicrobial packaging that can directly interact with the food product to reduce, inhibit, or retard the growth of microorganisms and their spores present on food surfaces (Suppakul et al., 2003). MONPs such as zinc oxide (ZnO) possess antimicrobial efficacy against microbes such as *Staphylococcus aureus*, *Pseudomonas aeruginosa*, *Campylobacter jejuni*, *Bacillus subtilis*, *Escherichia coli*, and *Lactobacillus plantarum* (Hernández-Sierra et al., 2008; Llorens et al., 2012). ZnO has also been commercially used for active food packaging because of its white appearance and high flexibility in addition to its ability to hamper UV radiation (Espitia et al., 2012). In addition, being a supplement for micronutrient zinc it plays a significant role in the growth and development of humans and animals. Commercially available ZnO food packaging has improved mechanical strength, stability, and barrier properties, in addition to its ability to retain food as fresh with longer shelf lives, safety, convenience, and high cost-to-benefit ratios (Al-Naamani et al., 2016). Other MONPs such as Cu_2O, CuO, TiO_2, and MgO are also actively used for food packaging. These MONPs have the capability of scavenging oxygen and oxidizing ethylene, which has also proved useful in extending food shelf life (Stoimenov et al., 2002; Garcia et al., 2018).

5.6 Applications in precision agriculture

To address the rising global challenges of sustainable agriculture and food security, several innovative technologies and advancements have been made in recent years. Nanotechnology has potential in dealing with multiple problems in the agriculture sector like negative effects of climate change, nutrient deficiency, pest control, low plant growth, and yield (Ditta, 2012; Prasad et al., 2017). MONPs have a wide range of applications in precision agriculture, for example, NPs encapsulated with micronutrients have been used for the constant supply of nutrients and water to crops in nutrient-poor soils. Similarly, pest-resistant cultivars and varieties of crops have been biogenetically engineered using nanoparticle-mediated transfer of desired DNA or genes into crops. Remote sensing devices have been manufactured through using metal-oxide based nanosensors for monitoring and maintaining plant health and soil quality for precision agriculture (Sharon et al., 2010). Examples include the use of TiO_2 NPs as nanopesticides for tomato crops because of its ability to produce ROS to kill bacterial leaf spot. CuO, MnO, and ZnO NPs are effective in boosting plant production and protecting against pathogenic attack when cultivated in fusarium wilt fungus-infested soil (Chhipa and Joshi, 2016). Similarly, MnO, CuO, and ZnO NPs have been using as nanofertilizers to boost vigor and yield of tomatoes and eggplants grown in disease-infested soils (Elmer and White, 2016; Wang et al., 2016a,b,c). These MONPs have also been effective in providing nutritional benefits to the plant organs using a smart delivery system, thereby increasing agronomic production (Mousavi and Rezaei, 2011). ZnO NPs translocate to all plant organs via both apoplast and symplast pathways. Although translocation and accumulation of MONPs have been noted in plant seedlings received foliar applications, the fate of NPs in mature plants is largely undetermined (Rai et al., 2014).

5.7 Applications in the cosmetics industry

MONPs are widely used in the cosmetics industry, especially in manufacturing of sunscreens, hair care products, facial and topical creams, and toothpaste (Julia and Li, 2011; Katz et al., 2015). Recently, MONPs, particularly ZnO and TiO_2, have increasingly been used in the cosmetics industry as an inorganic constituent of sunscreens, because of their ability to reflect and scatter harmful sun radiation, particularly UV-A and UV-B, compared to typical chemical-based UV filters and sunscreens that may cause disturbances in the endocrine system and also skin rashes and irritation (Lewicka et al., 2011; Smijs and Pavel, 2011; Huang et al., 2013; Lu et al., 2015). In addition, MONPs such as Fe_xO_x, TiO_2, and Al_2O_3 are actively used as concealer in makeup products by companies like Alusion and Soltan. The hair care industry has exhibited a remarkable input of ENMs. The intrinsic properties and minuscule size of MONPs can directly target hair shaft and follicle cells, thereby making it possible to gain maximum benefits from active ingredients. $nTiO_2$ has been engineered into hair care treatments effectively (Mihranyan et al., 2012; Steckiewicz et al., 2019). Extensive research and probable applications of MONPs in cosmetology and dermatology are increasing day by day to improve hair and skin cosmesis and cure related diseases (Mu and Sprando, 2010).

6. Potential exposure routes of MONPs to humans and fate in biosphere

Nanotechnology is a novel concept and is used in many fields including cosmetics (Morganti, 2010), food (Siegrist et al., 2008; Handford et al., 2014), medicine (Kahan et al., 2009), and agriculture (Handford et al., 2014; Mukhopadhyay, 2014) for breakthroughs. This technology is a blessing on one side, but on the other hand it increases the chances of exposure of the consumers—mainly humans. The MONPs in the human environment have been widely studied and categorized for their life cycles assessments and exposure potential during life cycle stages. These NPs are disposed or released directly or indirectly to surface or ground water, rivers, canals, lakes, and finally the seas. These NPs can also be released in air and agricultural soil. Therefore, these MONPs can accumulate in aquatic creatures, animals, humans, and agricultural plants, leading to potential toxicity, and it is of key public interest to explore the various routes of exposure and their effects.

The main route of MONPs' entry into the human body is through primary ingestion, i.e., either through food or indirectly from food container dissolution. Secondary ingestion routes include the inhalation of airborne nanoparticles (Yang et al., 2016a,b; Bergin and Witzmann, 2013; Borm et al., 2006) or they can penetrate the body through skin (Wang et al., 2007a,b, Borm et al., 2006). Once MONPs enter the body, they can easily translocate from one part to another through blood circulation. The distribution of MONPs in the body is possibly dependent on physiochemical properties, e.g., size, polarity, hydrophilicity, lipophilicity, and catalytic activity (Yang et al., 2016a,b; Wang et al., 2007a,b). The size of MONPs is inversely proportional to surface area, i.e., as size decreases surface area and biological activity increase (Agnihotri et al., 2014; Ershov et al., 2016). Compared to larger particles, it is hypothesized that smaller-sized particles are more toxic and easily taken up by cells due to higher activity. Recent studies revealed that orally taken MONPs can translocate from the gastrointestinal tract to other body parts, i.e., liver, kidney, spleen, lungs, heart, and brain (Hillyer and Albrecht, 2001),

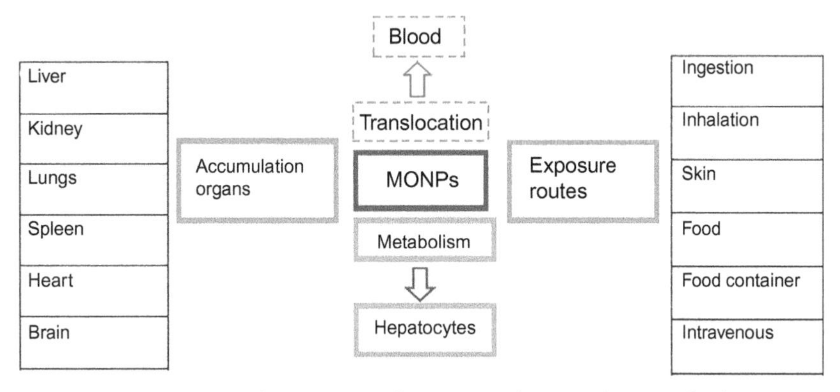

Fig. 3 MONP exposure routes to humans, translocation, and accumulation in body organs, and their metabolism.

while the highest accumulation was reported in liver of mice following oral and intravenous exposures (Kermanizadeh et al., 2014; Wang et al., 2007a,b). van der Zande et al. (2012) also reported the highest accumulation in liver and spleen of rats. Hepatocyte blood cells play a vital role in MONPs' metabolism (Wallace Hayes and Kruger, 2014). Liver, lungs, and kidney are primary accumulation and distribution sites for MONPs (Wang et al., 2013a,b). Potential exposure routes to humans, translocation, and accumulation are shown in Fig. 3.

7. Biokinetics of MONPs in multicellular organisms

The concerns and theories regarding the biokinetics of MONPs in different organisms (i.e., aquatic and terrestrial animals, terrestrial plants, etc.) are still questionable. Extensive use of MONPs in consumer products increased the risk of environmental and human exposure to toxic effect of these MONPs (Ramskov et al., 2015). MONPs exposure into environment and to human body can be potential toxic. Determining the biokinetics of MONPs in biosphere is important in order to determine the potential risks for human health (Zhang et al., 2008a,b).

7.1 Biokinetics of MONPs in aquatic animals

Effluents from wastewater treatment plants, terrestrial water run-off, direct release from consumer products, and the air compartment are some important sources of MONPs in aquatic ecosystems (Vale et al., 2016). In aqueous environments, characteristics and behavior of MONPs depend on the size, shape, surface charge, and chemical composition of these particles. Aggregation, agglomeration, and stabilization of particles depends on the aquatic environmental conditions, i.e., temperature, pH, ionic strength, composition, etc. (Corsi et al., 2014). Mobility of these particles also increases in water, increasing the exposure of aquatic animals to these particles. Biokinetics of MONPs in aquatic animals include different processes such as aqueous or dietary uptake, dietary assimilation and elimination, etc. (Tsui and Wang, 2007). MONPs enter the body of aquatic animals by either aqueous uptake or dietary uptake, but the relative importance and rate of uptake remains to be determined (Li and Wang, 2013). The difference in tissue concentration of bio-accumulated metal NPs and aqueous concentration suggests that ingestion and aqueous uptake are the main routes of entry of MONPs in aquatic animals. Animal cells contain cytoplasm, in which cell organelles are present, and which is enclosed by a lipid plasma membrane. The plasma membrane has many systems and mechanisms for restricting movement of materials into and out of the cell (e.g., transporters, channels, pores, etc.), and is also known as a selectively permeable membrane (Fabrega et al., 2011).

Endocytosis is a process by which MONPs enter the cell of animals and are further classified as phagocytosis and pinocytosis. Materials outside the cell ranging from 1 to 100 nm are taken in through invaginations of the plasma membrane in the form of vesicles and transported into the cell. MONPs enter aquatic animal cells by three major endocytosis pathways, i.e., clathrin-mediated endocytosis, clathrin-independent endocytosis, and macro pinocytosis and phagocytosis (Luoma et al., 2014). These NPs may also enter the cell by diffusion through the plasma membrane or by disrupting it. MONPs are taken up by channels and transporters by dissolving into ions (Rocha et al., 2016). Differences between the functional ecology and physiological traits greatly alter the MONPs' bioaccumulation in aquatic animals living in the same ecosystem and environment. Processes that involve bioaccumulation as well as bioavailability of MONPs are dependent on the nature and concentration of these particles, routes of exposure, nature of aqueous environment, and biology of subjected aquatic organisms (Châtel and Mouneyrac, 2017). Low exposure of cells to MONPs leads to disturbance of the biochemical processes of organisms, but high-dose chronic exposure results in oxidative stress and necrosis (Klaper et al., 2014). MONPs-based cytotoxicity of aquatic animals needs to be studied at different biological organization levels, including a need to explore the behavior and fate of MONPs in aquatic animals.

7.2 Biokinetics of MONPs in terrestrial animals/humans

Skin, lungs, ingestion, and gastrointestinal paths are the main entry points of MONPs in terrestrial animals including human beings. However, injections, implants, and inhaled particles that absorb into the blood stream are the other sources of MONPs in the body. Once they have entered the circulatory system through any source, these MONPs can be transported all over the body, depositing in and causing damage to numerous body organs including the brain (Oberdörster et al., 2005; Janrao et al., 2014). MONPs can enter through undamaged skin, but the rate of nanoparticle entry is accelerated when the skin is broken due to acne, wounds, sunburn, or other abnormalities (Oberdörster et al., 2005). Potential for MONPs' entry into the body through micropores present in the outer layer of the skin, i.e., epidermis for sweat, follicle sites, and sebaceous glands, is also revealed. Entry of these materials can be recognized by the body immune system because of the presence of nerve endings, dendritic cells, lymph vessels, blood, and macrophages in the derma (Desai et al., 2010). Inhaled MONPs can be deposited throughout the respiratory tract, including the nose, pharynx, and lungs. Larger particles can settle in the lungs' surface, while smaller diameter particles enter the alveoli and smaller airways via diffusion (Fröhlich and Salar-Behzadi, 2014). The gastrointestinal tract is the most important pathway of macromolecules as well as MONPs intake through food, medication, and water also. Endogenous sources, i.e., intestinal calcium and phosphate secretions are also causes. Some contaminated food products increase the intake of MONPs via the gastrointestinal tract (Di Gioacchino et al., 2009). These materials absorb in the gastrointestinal tract and penetrate to mucus depending on the size, dose, chemistry, and characteristics (Yuan et al., 2013).

Different types of cells are vulnerable to MONPs and these materials can interact and make complexes with subcellular structures. However, shape, size, type, and chemistry determine the cellular uptake, localization, and ability to boost oxidative processes (Kettler et al., 2014). Cellular uptake is mainly due to two endocytic pathways: phagocytosis and pinocytosis. Phagocytosis is commonly found in macrophages, dendritic cells, and neutrophils, while pinocytosis is found in nearly all types of cells. Pinocytosis is further classified as micropinocytosis, caveolae-mediated, clathrin-mediated, and caveolae/clathrin-independent endocytosis (Yameen et al., 2014). MONPs can also be taken up by the cells by passive uptake mechanisms and adhesive interactions initiated by steric interaction, electrostatic charges, interfacial tension, or van der Waals forces (Peters et al., 2006). These MONPs can be found in numerous locations within the cell, i.e., cell membrane, cytoplasm, lipid vesicles, mitochondria, and nuclear membrane, as well as the nucleus of the cell (Xia et al., 2006). MONPs, apart from the nervous system, can cause toxic consequences in skin, gastrointestinal tract, and lungs, as these are the organs in direct contact with the environment and the major entry points of the nanoparticles (Janrao et al., 2014; Yameen et al., 2014).

7.3 Biokinetics of MONPs in food crops

Wide applications of MONPs ensure their availability in soil and agricultural systems. They exist in the environment surrounding plants and, due to strong interactions of plants with their surroundings, the chances of hazardous material uptake including MONPs increases (Anjum et al., 2016). MONPs may be included in the food chain by plant-based food products, ultimately posing a threat to animals as well as humans (Deng et al., 2014). Plants may uptake MONPs through the leaves, but roots are the main entry point into plant systems, and hence, the food chain and ultimately, humans' and animals' systems. Where plants are exposed to MONPs-contaminated soils, different processes of biokinetics, i.e., absorption, translocation, and accumulation of these material in plant tissues, may be modulated by the type of vegetation and size, characteristics, type, or chemical composition of nanoparticles (Rico et al., 2011). Root surface area and surface reactivity can affect the adsorption of nanoparticles by roots. Although adsorption of MONPs on root surface cannot be designated as uptake by the plants, use of surfactants and ions can compete with adsorption of these NPs on plant root surfaces (Zhou et al., 2011). These NPs can penetrate the epidermis and reach the endodermis, passing through the cortex (Zhao et al., 2012). Moreover, lateral roots can provide extra surface area for penetration of MONPs and ultimately transport to the stele or central cylinder (Peng et al., 2015).

MONPs usually interact with the cell wall containing cellulose micro fibrils, hemicellulose, lignin, and pectin, which restrict the movement of larger particles into the cell (Deng et al., 2014). MONPs enter the apoplastic or symplastic pathways after passing through nanopores of the cell wall by diffusion. In addition, facilitated diffusion with carrier proteins, ion channels, aquaporens, and some organic compounds can facilitate MONPs entry into most regulated symplastic pathways in major crops for entrance into the cell (Rico et al., 2011). In the apoplastic pathway, particles absorbed by the root access the endodermis by the spaces between the cell walls and plasma membrane as well as epidermal and cortical cells (Deng et al., 2014). The Casparian strip is the last apoplastic barrier for these MONPs to enter the vascular tissues and beyond. In many food crop species, it hinders the translocation of these nanoparticles, and in some species, it is proven that the aggregated nanomaterials accumulate in the endodermis. Gaps present in the Casparian strip can facilitate entrance of MONPs. The absorbed MONPs in the vascular tissues are translocated into the plant shoot system and aerial parts of the plant by means of water and nutrients absorption and patterns as well as the water transpiration stream (Lee et al., 2010a,b; Schwabe et al., 2015). Mature leaves, compared to younger leaves, are closer to the roots and exposure time of mature leaves is more than that of younger leaves. Thus, mature leaves can be designated as major sites of MONPs storage (Peng et al., 2015). These MONPs can further translocate into flowers and other edible portions of plants.

8. Ecotoxicity of MONPs in environmental components

The biosphere/ecosphere is the zone of the planet Earth where life exists and is basically the sum of all the ecosystems. An ecosystem is a community where biotic (living) and abiotic (nonliving) components interact with each other. However, sometimes biological, chemical, and physical factors disrupt the biochemistry of the natural ecosystem. MONPs are one of the factors responsible for ecological toxicity. Further, increased consumption of MONPs can cause unprecedented threats to the ecosystem (Abramenko et al., 2018; Bellani et al., 2016). Since the majority of MONPs are being applied imprecisely without any concern about their fate and ecological threats (Rajput et al., 2020), the production, consumption and dumping process of MONPs is adding thousands of tons of ENMs to the soil every year (Bundschuh et al., 2018; McGillicuddy et al., 2017). Not only soil biota, but aquatic life, plants, and humans are also being severely affected. MONPs have been observed to cause impairment in the development of animals (Sun et al., 2016) and human beings (Gliga et al., 2018). Hence, it is imperative to focus on the ecological risk assessment and safety level evaluation of MONPs. However, there are numerous challenges, i.e., inconstant NPs exposure, surface modification over time leading to changing physicochemical properties, and lack of specialized methods and equipment to characterize and quantify MONPs in different medium (Selck et al., 2016). The overall ecological impacts of MONPs toxicity are summarized in Table 3.

8.1 Effect of MONPs on soil and soil biota

Owing to their small size, MONPs can be translocated from one place to other, eventually ending up accumulating in soils for long periods of time (Tolaymat et al., 2017; Strambeanu et al., 2015; Keller et al., 2017). Once NPs gain entry into the terrestrial system, they behave differently in relation to the properties of the soil with which they interact. In the terrestrial system, MONPs undergo various chemical (redox reactions and dissolution) and physical transformations, i.e., changes in shape, size, and aggregation (Peijnenburg et al., 2016). However, it is very difficult to track these transformations in real systems owing to the absence of developed techniques (Ju-Nam and Lead, 2016; Cornelis et al., 2014). MONPs' toxicity becomes more severe once NPs get attached to the inorganic and organic substances present in the soil system. For instance, organic matter contents can be adsorbed on the surface of MONPs and can imply certain changes to their ionic characteristics and eventually play a role in their stability (Ghosh et al., 2008). There are very few reports published focusing on the effects of MONPs on soil's physical, chemical, and biological properties. For instance, Ben-Moshe et al. (2013) revealed that lower concentrations of MONPs posed insignificant effects on the macroscopic properties of the soil whereas a significant change was observed in the reaction and activity of bacterial communities of that soil. In addition, the effect was stronger in the case of CuO nanoparticles compared to Fe_3O_4 nanoparticles. Soil with lower clay and organic matter contents was noted to be comparatively more affected by MONPs. For instance, CuO NPs exhibited stronger effects on hydrolytic activity, oxidative potential, and community composition of bacteria in sandy loam soil compared to clay loam soil (Frenk et al., 2013). Another study reported that Ag NPs caused negative effects on the metabolic activity, nitrification potential, and the abundances of bacteria, whereas the reverse was observed for FeO NPs (He et al., 2016; McGillicuddy et al., 2017). However, certain benefits from NPs have also been reported. For example, their role as

Table 3 Ecotoxicity of MONPs on major environmental components.

MONPs	Target organism	Observed effects	References
Ag NPs	Bacteria, eukaryotes, and ammonia-oxidizing bacteria	Decreased soil microbial metabolic activity and metabolic efficiency.	He et al. (2016)
FeO NPs	Bacteria, eukaryotes, and ammonia-oxidizing bacteria	Increased soil microbial metabolic activity and metabolic efficiency.	He et al. (2016)
Ag NPs	Acidobacteria, actinobacteria, cyanobacteria, nitrospirae	Ag^+ caused serious damage to the cell wall of *N. europaea*, and reduced the number of acido-bacteria, actinobacteria, and cyanobacteria.	Juan et al. (2017)
Ag NPs	N-fixers, siderophore producers, and P-solubilizers bacteria	AgNPs modified the soil bacterial diversity and different concentrations of AgNPs had different effects on the functional bacterial diversity.	Mehta et al. (2016)
TiO_2 NPs	Bacteria, archaea, and ammonia-oxidizing clades	Strong negative impacts on nitrification enzyme activities and the abundances of ammonia-oxidizing microorganism.	Simonin et al. (2016)
CuO NPs	*Raphanus sativus*, *Lolium perenne*, *Lolium rigidum*	Reduction in plant growth was recorded.	Atha et al. (2012)
TiO_2 NPs	*Oryza sativa* L.	Disturbed metabolic flux. Starch, sucrose, glyoxylate, and dicarboxylate metabolism were inhibited significantly.	Wu et al. (2017)
Ag NPs	*Triticum aestivum* L.	Increased nuclear erosion, elongation, and decrease in the number of cells undergoing mitosis were recorded.	Abdelsalam et al. (2018)
Cu NPs	*Oryza sativa* L.	Seed germination was reduced.	Shaw and Hossain (2013)
CeO_2 NPs	*Oryza sativa* L.	Rice grains with less glutelin, S, Fe, valeric acids, prolamin, lauric, and starch were observed.	Rico et al. (2013)
CeO_2 NPs	*Triticum aestivum* L.	Modified S and Mn storage in grains, increase in amino acid and linolenic acid concentration.	Rico et al. (2014)
CuO NPs	*Hordeum sativum* distichum	Impairment in root morphology, cell wall, epidermis, cortical layers, and vascular bundles was observed.	Rajput et al. (2018a)

Table 3 Ecotoxicity of MONPs on major environmental components—cont'd

MONPs	Target organism	Observed effects	References
CuO NPs	*H. diversicolor* and *S. plana*	Behavioral impairments and activation of defense biomarkers were recorded.	Buffet et al. (2011)
ZnO NPs	*Oryzias melastigma*	Increased mortality and heart rate.	Cong et al. (2017)
FeO NPs	Mussel	Augmentation in the level of ROS.	Taze et al. (2016)
NiO and CoO NPs	*Artemia salina*	Oxidative stress was reported.	Ates et al. (2016)
Fe_3O_4 NPs	*Artemia salina* larvae	Mitochondrial morphology was disrupted, and cysts and larvae of *Artemia salina* were impacted.	Zhu et al. (2017)

antimicrobial agents and carriers of drugs is exceptionally important in the treatment of infectious diseases (Raghunath and Perumal, 2017).

8.2 Effect of MONPs on aquatic animals

Aquatic sediments or oceans and terrestrial lands are the ultimate disposal sites for released MONPs (Tolaymat et al., 2017). Once present in aquatic systems, MOPs not only render the water unsuitable for human consumption but also pose lethal effects on marine fauna and flora. Augmented MONPs tend to decrease the dissolved oxygen concentration, causing negative effects on the quality of water and marine life. This eventually leads to an aquatic system with pollution-tolerant organisms (Sharma and Rawat, 2009). Further, MONPs have the capability to perdure and bioaccumulate in aquatic animals, eventually becoming part of the food chain (Kahlon et al., 2018). MONPs can enter the body of a living organism in various ways, i.e., through cell membranes, ion transport chain, or by endocytosis. Once inside, MONPs can facilitate the production of ROS by interfering with the electron transport chain, ultimately disrupting the functions of organelles (Liu and Hurt, 2010). Various studies have reported the effects of MONPs on rotifers (Snell and Hicks, 2011), arthropods (Hanna et al., 2013), annelids (Cong et al., 2011), and mollusks (García-Negrete et al., 2013; Baker et al., 2014). Moreover, MONPs were observed to be responsible for specific behavioral and biochemical changes in response of the marine organism. For instance, addition of $10\,\mu g\,L^{-1}$ of CuO Nps in the seawater in laboratory experiments activated several defense biomarkers, i.e., catalase (CAT), glutathione-*S*-transferase (GST), and superoxide dismutase (SOD) in *S. plana* and CAT and GST in *H. diversicolor*. *S. plana* exposed to CuO NPs and soluble Cu exhibited certain behavioral impairments. However, in the case of *H. diversicolor*, a decrease in burrowing behavior was observed only when they were exposed to soluble Cu (Buffet et al., 2011). In addition, ZnO NPs caused certain negative effects in the marine medaka *Oryzias melastigma*, i.e., increased mortality and heart rate. In addition, a substantial decrease in the total percent hatching success and an increased malformation percentage were observed (Cong et al., 2017). Further, Taze et al. (2016) showed that iron oxide nanoparticles can cause modification in the physiology of animals by causing oxidative stress in their hemocytes. Mussels exposed to iron oxide NPs displayed a substantial augmentation in the level of ROS, prooxidant-antioxidant balance (PAB), protein carbonylation, and lipid peroxidation.

8.3 Effect of MONPs on food crops

Augmented use of ENMs in almost every sector of life has led researchers to consider seriously the consequences of MONPs on food crops. Although the mechanism for their entry into the plant body is not fully understood, it is known that, owing to their small size, they can ingress the plant body through cell membrane carrier proteins, root exudates, and ion channels, and also by endocytosis or by plasmodesmata transport (Rajput et al., 2018a,b,c,d). NPs' toxicity varies with respect to exposure duration, concentration size, and the type of NP. For instance, Cu NPs facilitate plant growth at lower concentration as an essential micronutrient; however, at higher concentrations, they have been reported to hinder plant growth due to an antagonistic relationship with other nutrients such as B, Mo, Mn, Mg, and Zn (Nair and Chung, 2014).

Furthermore, a number of studies have mentioned the negative effects of MONPs, like inhibited growth, reduced seed germination rate, oxidative stress, diminished root and shoot growth, less biomass, and changes in photosynthesis and transpiration rate of plants exposed to MONPs (Atha et al., 2012; Kasana et al., 2017; Shams et al., 2018). For instance, Shaw and Hossain (2013) reported that treatment of rice seedlings with <0.5 mM concentration of Cu NPs triggered a significant inhibition in a crop's seed germination. Similarly, another study demonstrated that application of $10\,\mu g\,L^{-1}$ of Cu NPs to barley (*Hordeum sativum* distichum) caused a substantial reduction in germination rate and root and shoot elongation of the crop (Rajput et al., 2018a,b,c,d). CuO NPs toxicity is not only limited to seed germination and root and shoot elongation, their effects have also been reported on the morphology and overall growth of plants. For instance, CuO NPs deformed stomatal aperture (Rajput et al., 2015), blocked stomata (Singh and Kumar, 2016), and disturbed ultrastructure of leaves (Olchowik et al., 2017) and root morphology of wheat (Shaw et al., 2014). In addition, Cu, CeO_2 NPs displayed potential to alter physiology and food quality of cereal crops. Two experiments were conducted in order to identify the effects of CeO_2 NPs on nutrition, yield, and antioxidant properties of rice (Rico et al., 2013) and wheat (Rico et al., 2014). It was observed that CeO_2 NPs altered the amino acid composition, the fatty acid content, and the sulfur and manganese distribution in the grains. However, contrasting effects have been reported for MONPs in plants and well summarized by Ahmad et al. (2018).

8.4 Effect of MONPs on air quality

MONPs can get into the atmosphere both directly (emission from urban combustion sources, i.e., power generation plants) and indirectly by gaseous emission from the same sources, which eventually leads to atmospheric MONPs nucleation. MONPs coming out from combustion sources are usually of three major types: primary, delayed primary, and secondary MONPs (Rönkkö and Timonen, 2019). Since poor air quality can lead to various fatal diseases including cancer, respiratory disorders, and cardiovascular diseases (Ibrahim et al., 2016), it is crucial to set standards and regulations for their permissible levels in the air in order to reduce MONPs' associated risk. However, as yet no air quality regulations have been introduced by any country owing to the absence of adequate knowledge about the safe levels of NPs in the air and lack of standard measurement techniques (Djurisic et al., 2013; Durán and Seabra, 2012; Ingale and Chaudhari, 2013; Kahru and Dubourguier, 2010; Kahru and Ivask, 2013; Kumar et al., 2011, 2014; Lei et al., 2014; Liang et al., 2016).

A double-edged sword phenomenon is evident for MONPs. On the one hand, they cause death, especially when exposure is for longer duration and exposed concentration is relatively high. For instance, Song et al. (2009) reported a case of a group of women (age group 18–47) working in a printing factory. Two women out of seven died after exposure to ENMs for 5–13 months. Later, pathological examination revealed the presence of particles of about 30 nm in diameter in the lung tissues of the patients. It has been suggested that long-term exposure to the NPs especially in combination with the organic compounds can cause serious damage to health. On the other hand, various studies emphasize the importance of the MONPs as air cleaners. For instance, ZnO NPs revealed substantial potential to remove nitrogen oxide and sulfur oxides from the air by chemosorption (Singh et al., 2012). Ag/SBA-15 nanocomposites exhibited a tendency to remove significant amounts of carbon monoxide by CO oxidation at pH=5 (Zhang et al., 2011). In addition, Baltrusaitis et al. (2011) reported the unique ability of TiO_2 NPs to adsorb SO_2. This adsorption was light-dependent, i.e., in the presence of light irradiation the oxidation resulted in the formation of SO_4^{2-}; however, in the absence of light irradiation, SO_2 was adsorbed in the form of SO_3.

9. Endpoint toxicity studies of MONPs in multicellular organisms

9.1 Cytotoxicity/phytotoxicity

Cytotoxicity is related to cells, and cytotoxicity of MONPs was first discovered in microbes. Nanoparticles of Fe_3O_4, TiO_2, CuO, and ZnO proved to be toxic for different microbial species. These MONPs are antibacterially efficient by inducing toxicity to the bacterial cells and causing their mortality (Stankic et al., 2016; Jeong et al., 2018). To study the effects of MONPs on human health, quantitative structure-activity relationship (QSAR) and other models were used to predict the cytotoxicity potential. It was observed that MONPs can affect human health by introducing cytotoxicity in cells (Pan et al., 2016; Simeone and Costa, 2019). Results in another study indicated that MONPs like ZnO, CuO, and MgO produce cytotoxicity in the human vascular endothelial cells, if the exposure to these is increased, by producing intracellular ROS and affecting permeability of plasma membranes (Sun et al., 2011). MONPs have also been evaluated for their cytotoxic potential to human epithelial cells and it was revealed that MONPs were able to generate ROS (Fahmy and Cormier, 2009); the same type of experiment was done on bronchial epithelial cells and resulted in cytotoxicity (Liu et al., 2011a,b). In another study, human and catfish cells were exposed to MONPs (ZnO, TiO_2, CuO, and Co_3O_4) and resulted in cytotoxicity, which caused ROS production and destruction of cell and mitochondrial membranes (Wang et al., 2011a,b). Among all the MONPs, nZnO are the most dangerous to human cells and have proved lethal (Lozano et al., 2011). To benefit from the cytotoxic potential of MONPs, scientists are trying to destroy cancerous cells using MONPs (Rasmussen et al., 2010). When applied in plants, MONPs have proved beneficial for removing heavy metal uptake in plants at optimum concentrations but proved lethal when these concentrations are exceeded (Siddiqi and Husen, 2017; Kim et al., 2012).

9.2 Genotoxicity and carcinogenicity

Due to the small size of MONPs, they are easily transported into biological systems. Genotoxic and carcinogenicity potential of Co, Ni, and Cu-oxide NPs have been tested by Magaye et al. (2012) through in vivo mammalian cells testing and proved to be a potential genotoxic and carcinogenic source via abnormal apoptosis, oxidative stress, and pro-inflammatory effects. Nickel-based nanoparticles are also known to be carcinogenic in nature, initiating inflammation, DNA damage, micronucleus formation, and causing sarcoma and other activities that accelerate cancer (Magaye and Zhao, 2012). The same effects are summarized in another study, i.e., MONPs induce genotoxicity through damaging DNA, chromosomal aberrations, DNA strand breaks, oxidative DNA damage, and mutations (Kwon et al., 2014).

9.3 Allergenicity

Inhalation of MONPs leads to pulmonary toxicity; i.e., ZnO-nanoparticles induce pulmonary oxidative stress as well as inflammation in mice. It has been shown that the release of Zn^{2+} from ZnO nanoparticles causes allergies (Horie et al., 2015; Huang et al., 2015; Vennemann et al., 2017). Bronchocentric interstitial inflammation was observed after inhalation of AgO NPs. AgO NPs are accumulated in the lungs and induce inflammatory responses in the peritoneum (Chuang et al., 2013).

9.4 Neurotoxicity

In vitro neurotoxicity studies of $nCeO_2$ showed dose-dependent neurotoxicity (Darroudi et al., 2013). Significant evidence indicates that MONPs elicit toxicity in humans exposed by introducing blood-brain barrier, oxidative stress, and inflammatory mechanisms (Karmakar et al., 2014). In another study, nZnO increased the production of pro-inflammatory cytokines in the serum and the brain of mice, increased oxidative stress level, impaired learning and memory abilities, and caused hippocampal pathological changes (Tian et al., 2015). Other laboratory experiments showed that Ag, Cu, and Al oxide NPs induced neurotoxicity by altering sensory, motor, and cognitive functions at developmental stages (Shanker Sharma and Sharma, 2012). In another study, NPs-induced brain damage was studied in mice indicating that MONPs can even cause death (Sharma et al., 2013).

9.5 Immunotoxicity

Immunotoxicity refers to reduced immune system efficiency against pathogenic/microbial diseases and abiotic stresses. CeO, ZnO, and FeONPs have been widely explored for their immunotoxicity potential. The toxic effects of MONPs on lymphocytes were reported by (Easo and Mohanan, 2015). Similarly, the lysosomal destabilization, phagocytosis, and changes in the functions of the phagocytic cells, which ultimately decreases the ability of animals to defend themselves against pathogens and infectious diseases have been studied (Jovanović and Palić, 2012). In a ex-situ study, MONPs had induced mortality in fish via bioaccumulation (Gagnon et al., 2018). Therefore, MONPs have significant potential to react with biological systems to reduce immune system response when surrounded by a protein corona (Corbo et al., 2016).

9.6 Hepatotoxicity

MONPs commonly lead to induction of hepatotoxicity in mammals, especially humans. nZnO NPs exhibited inflammation in cells, necrosis, hydropic degeneration, hepatocytes apoptosis, anisokaryosis, karyolysis, nuclear membrane irregularity, glycogen content depletion, and hemosiderosis, which indicates that ZnO NPs can cause oxidative stress in liver and other cells and thus affect the functions of many organs (Almansour et al., 2017). Disturbances to antioxidant enzymes (Volkovova et al., 2015), damage to liver cells, and increases in oxidative stress and inflammation in mice have been demonstrated by Wei et al. (2016) and Gao et al. (2017a,b).

9.7 Teratogenicity/reproductive toxicity

Several reports have been published indicating MONPs' potential to alter the reproductive systems of terrestrial and aquatic animals. nCuO NPs cause a decrease in antioxidant enzymes, induce oxidative stress, apoptosis and developmental abnormalities (Ganesan et al., 2016), and even transgenerational toxicity in rats (Kadammattil et al., 2018). In zebrafish, effects on embryo and larvae survival, delay in hatching rate, tissue ulceration and malformation, and mortality have also been observed (Zhu et al., 2008a,b, 2012).

9.8 Methods for analyzing the biological effects of MONPs

MONPs have dissimilar physical and chemical properties from each other including surface, thermal, electrical, and optical properties (Auffan et al., 2009). These properties also differ from their native material (core metal) (Santos et al., 2015). To understand completely MONPs' potential for environmental toxicity, it is imperative to explore the direct and indirect effects for achieving regulatory guidelines for environmental and human safety. As discussed in the introduction to this chapter, the assessment of various toxic effects on biological systems is a difficult job. Different methods and techniques are used to analyze the toxicity effect of MONPs on organisms. The methods of analyzing are classified into two main groups: in vivo and in vitro. The in vitro methods have more importance in the assessment of toxicity than in vivo methods do because of advantages such as being fast, cheap, and causing only minor ethical concerns. The proliferation assay, necrosis assay, apoptosis assay, DNA damage assay, and oxidative stress assay are the subdivision of assessment.

10. Methods for analyzing the biological effects MONPs

10.1 In vitro toxicity assessment methods

10.1.1 Proliferation assay

The proliferation assay is typically used to detect changes in the metabolism of cells and number of cells in division. The 3-(4,5-dimethylthiazol-2-yl)-2,5-diphenyltetrazolium bromide (MTT) is a frequently used tetrazolium salt for in vitro evaluation methods (Sayes et al., 2007). This technique has various advantages, including reproducible results, quick yield, and less manipulation of model cells (Molinari et al., 2003). Culture media changes such as cholesterol, pH, and ascorbate (Natarajan et al., 2000; Abe and Saito, 1999) in MTT assay can lead to the altered measurement of tetrazolium salt (Molinari et al., 2003). Production of formazan is also a disadvantage of MTT assay. Other types of assays that produce soluble dyes are preferred, such as WST-1 or STT. Another method used to measure the redox potential of cells is Almar Blue; this method is much better than MTT assay due to the easy and simple preparation of samples (Punshon et al., 2005). Visual counting of proliferating cells after NPs exposure was carried out in a colo-genic assay (Casey et al., 2007).

10.1.2 Necrosis assay

This method is used to analyze the toxicity of NPs by the viability and integrity of the cell membrane. The integrity of the membrane is evaluated by the uptake of a dye like trypan blue or natural red (Monteiro-Riviere et al., 2005; Huang et al., 2004). It gives reliable, inexpensive, rapid, and reproducible results for analyzing the toxicity effects of MONPs. In a study, Miranda et al. (2017) reported that Ag NPs reduced the stability of the lysosome-endosome system by up to 30%, which was measured by neutral red assay. Trypan blue is also used to evaluate cell deaths, as it enters in the dead cells (Strober, 2015). Kononenko et al. (2017) studied the stability of cell membranes in a medium of nZnO (13–737 μM) and it was reported that nZnO at concentrations of >369 μM showed cytotoxic effects.

10.1.3 Apoptosis assay

Apoptosis assay is one of the best in vitro assay techniques for the evaluation of MONPs' toxicity. The results of many experiments indicate that MONPs produce free ions or radicles, which then cause DNA damage and apoptosis that induce oxidative stress in the cell (Ryter et al., 2007; Li and Osborne, 2008). It was reported that DNA damage and apoptosis can be caused by oxidative stress in the cellular system (Ryter et al., 2007). Several researchers reported the role of NPs in apoptosis. Ahamed et al. (2008) reported that Ag NPs were responsible for apoptosis in stem cells of mice. There are several methods to assess the apoptosis, such as inspection of morphological changes, comet assay, TUNEL assay, and annexin-V assay (Lee et al., 2007; Jin et al., 2007; Mo and Lim, 2005; Pan et al., 2007). In toxicity assessment, the typically used death markers are propidium iodide (PI) and annexin-V—the principle of this assay is that florescence is enhanced when phosphatidylserine is bound to annexin-V, which is the indication of plasma membrane externalization. At late stages of apoptosis, PI, which acts as a dye, stains the nucleus, which only happens when cell membrane integrity is lost (Berghe et al., 2013; Silva, 2010). Breakage of DNA strands can be detected by comet assay (Kumar et al., 2013; Bajpayee et al., 2013). TUNEL assay is used to detect apoptotic cells and DNA damage (Gavrieli et al., 1992).

10.1.4 Oxidative stress assay

The oxidative stress assay technique is a very important technique for plants and other organisms in which ROS and reactive nitrogen species (RNS) are measured. X-band electron paramagnetic resonance (EPR) is used to detect the measurement of ROs and RNS (Magder, 2006). This technique is more expensive than others and it has some limitations, due to the fluorescent probe reacting with other reactive species, which sometimes causes misleading results (Halliwell and Whiteman, 2004). The above problems can be solved by using a nonfluorescent probe: DCFDA. There are several other assays used to detect lipid peroxidation such as TBA assay, Amplex red assay, Nitro blue tetrazolium assay, and C11-BIODIPY assay (Hussain et al., 2006).

10.2 In vivo assessment methods

The in vivo assessment methods are less important and tedious because these tests are based on radiolabels. Different in vivo toxicity assessment techniques are used such as hematology, biodistribution, histopathology, and serum chemistry. These studies are used to localize the route of NPs, evaluate the metabolism and excretion of NPs, and check the level of toxicity caused by NPs. Microfluidics and micro-electrochemistry were also used to examine the toxicity assessment of MONPs (Guo et al., 2012; Baker et al., 2008; Lei et al., 2008; Zhu et al., 2008a,b). After exposure to NPs, the in vivo toxicity assessment can be carried out by observing the changes in cell types and in serum chemistry (Baker et al., 2008). To evaluate the level of toxicity after NPs exposure, histopathology of organs, cells, and tissues is used (Lei et al., 2008). Histopathological examination is carried out using the main tissues exposed to NPs such as kidney, brain, lungs, spleen, liver, eyes, and heart (Baker et al., 2008; Zhu et al., 2008a,b). Microfluidics and micro-electrochemistry are the most advanced techniques used for in vivo toxicity assessment (Ewing et al., 1983).

11. Permissible limits of MONPs in environmental compartments and consumer products

MONPs have become widespread in almost every sector of technology and thus, MONPs release in the environment is not only a scientific concern but also related to societies and sustainability of the ecosphere. At certain concentrations, MONPs are highly toxic to the environment.

These NPs affect plant growth, the microbial community, and human health. CuO and ZnO have damaged the soil ecosystem because they have greater ability to transform and solubilize. These properties play a vital role in evaluating the concentration of adequacy and toxicity of MONPs. The threshold level of MONPs also depends upon the level of stress and the chemical composition, which includes size, shape, and surface area (Rajput et al., 2018a,b,c, 2020). Currently, no complete package of safety guidelines exists regarding the release of MONPs into biospheres. The United Kingdom and United States have already published environmental quality standards for the aquatic biosphere, but most models focus only on core metal ions. However, the paramount contrast between MONPs and metal ions is the charge or valency of metal ion (Baker et al., 2014). Therefore, considerable efforts are needed to assess the impacts of MONPs. Although numerous tools have been developed for gauging and managing ENMs' exposure to the environment, the EU Commission relies on life cycle-based models (Hischier, 2014; UNEP (United Nations Environment Programme), 2011). Life cycle assessment

Table 4 Effects of different MONPs on soil microorganisms.

Sr. no.	MONPs	Concentration in soil (mg kg^{-1})	Effect on soil microbial community	References
1.	ZnO, Fe_3O_4	500–1000, 2000	Bacterial communities of black and saline-alkali soils and the activities of the enzymes, i.e., catalase, invertase, phosphatase, and urease were decreased.	You et al. (2018)
2.	Fe_3O_4	0.1–10	Reduced bacterial population.	Cao et al. (2016)
3.	ZnO	1000	Affected plate counts of *Azotobacter* P-solubilizing and K-solubilizing and inhibited enzyme activities.	Chai et al. (2015)
4.	CuO	0–1000	Decreased soil microbial biomass and community structure, enzymatic activities.	Xu et al. (2015)
5.	Fe	550	Individual analysis showed effect on bacterial group.	Shah et al. (2014)
6.	CuO	10	Affected soil microbial community.	Ben-Moshe et al. (2013)
7.	Cu	220	Reduced the biomass of N and C, disturbed structure of microbial community.	Kumar et al. (2012)
8.	ZnO	500–2000	Altered soil bacterial community structure and decreased bacterial diversity.	Ge et al. (2011)

(LCA) is administered by the international standards ISO-14040 series (ISO (International Organization for Standardization), 2006a,b). Yet many researchers suggest that studies based on LCA assessment do not cover the absolute life cycle (Salieri et al., 2015; Miseljic and Olsen, 2014). The existing knowledge gaps regarding the release of ENMs (life cycle catalogue) and growing toxicity of these ENMs on the environment and humans (impact assessment) are hampering the true implications of LCA. Into the bargain, fate and transport models used to evaluate the concentration of ENMs in the biosphere are progressively emerging (Liu and Cohen, 2014; Salieri et al., 2015). In addition, scores of challenges concerning MONPs utilization in consumer products, like inappropriate and illegitimate labeling and packaging, thus violating the regulatory guidelines, exist (Subramaniam et al., 2019). Some MONPs have been reviewed in Table 3 and Tables 4 and 5 explaining the toxic effects at various concentrations in soil biota and plants, respectively.

Table 5 Effects of different MONPs on plant growth.

Crop	Scientific name	MONPs ($mg L^{-1}$)	Effect on crop	References
Iron oxide nanoparticles				
Sunflower	*Helianthus annuus* L.	50–100	Reduced nutrient uptake, translocation, and root hydraulic conductivity	Martínez-Fernández et al. (2016)
Tomato	*Solanum lycopersicum* L.	50–100	Inhibited root hydraulic conductivity	Martínez-Fernández et al. (2016)
Lettuce	*Lactuca sativa* L.	1–50	Reduced the root elongation	Liu et al. (2016)
Rice	*Oryza sativa* L.	2.200	Inhibited phyto-hormones	Gui et al. (2015)
Lettuce	*Lactuca sativa* L.	10–20	Increased antioxidant enzyme activity, reduced root size, reduced chlorophylls content, affected hydraulic conductance	Trujillo-Reyes et al. (2014a,b)
Ryegrass, pumpkin	*Lolium perenne* L. *Cucurbita pepo* L.	30–500	Increased root elongation, blocked aquaporins, oxidative stress	Wang et al. (2011a,b)
Soybean	*Glycine max* L.	0–1	Increased leaf and pod dry weight, 48% increase in grain yield	Sheykhbaglou et al. (2010)
Zinc oxide nanoparticles				
Mustard	*Brassica nigra* L.	500–1500	Affected seed germination and seedling growth	Zafar et al. (2017)
Soybean	*Glycine max* L.	0.05–0.5	Necrosis, leaf chlorosis, and affected photo system-II quantum efficiency	Priester et al. (2017)
Spinach	*Spinacia oleracea* L.	1000	Shoot and root length, total weight, chlorophyll, and carotenoid contents reduced	Singh and Kumar (2016)
Alfalfa	*Medicago sativa* L.	50–750	Reduced root and shoot biomass by 80%	Bandyopadhyay et al. (2015)

Table 5 Effects of different MONPs on plant growth—cont'd

Crop	Scientific name	MONPs (mg L^{-1})	Effect on crop	References
Bean	*Phaseolus vulgaris* L.	100–500	Inhibited growth, imbalanced nutrients in shoots, Na increased, Fe, Mn, Zb, and Ca decreased	Dimkpa et al. (2015)
Maize, rice	*Zea mays* L. *Oryza sativa* L.	2000	Decreased elongation of root	Yang et al. (2015)
Pea	*Pisum sativum* L.	25–500	Reduced chlorophyll and CAT content in leaves and APOX in root and leaves	Mukherjee et al. (2014)
Soybean	*Glycine max* L.	500	Seed formation affected, growth of shoot and root was reduced	Yoon et al. (2014)
Radish, grasses	*Raphan sativus* L.	10–1000	Inhibited growth, damaged genetic material	Atha et al. (2012)
Wheat	*Triticum aestivum* L.	500	Reduced root growth, increased lipid peroxidation and oxidized glutathione in root, deceased chlorophyll content in shoot, increased ROS production	Dimkpa et al. (2012)
Maize	*Zea mays* L.	20	Aggregates penetrated the root epidermis, cortex, and accumulated in xylem vessels	Zhao et al. (2012)
Zucchini	*Cucurbita pepo* L.	1000	Reduced biomass (78%–90%)	Stampoulis et al. (2009)
Radish, rapeseed, ryegrass, corn	*Raphan sativus* L., *Brassica napus* L., *Lolium perenne* L., *Zea mays* L.	2000	Reduced root growth and elongation	Lin and Xing (2007)
Copper oxide nanoparticles				
Oak		0.50	Disturbed shape, plastoglobules, and starch contents of leaf cells	Olchowik et al. (2017)

Continued

Table 5 Effects of different MONPs on plant growth—cont'd

Crop	Scientific name	MONPs (mg L^{-1})	Effect on crop	References
Spring barley	*Hoedeum vulgare* L.	10,000	Affected germination rate, retarded root and shoot length	Rajput et al. (2017)
Spinach	*Spinacia oleracea* L.	1000	Reduced shoot and root length, carotenoid content, total weight, and chlorophyll	Singh and Kumar (2016)
Cotton	*Gossypium hirsutum* L.	1000	Height and root length decreased after 10-day exposure in vitro	Van Nhan et al. (2016)
Carrot	*Daucus carota* L.	1–1000	Biomass decreased significantly	Ebbs et al. (2016)
Cotton	*Gossypium hirsutum* L.	10–1000	Reduced the uptake of minerals (B, Mo, Mn, Mg, Zn, Fe)	Le Van et al. (2016)
Onion	*Allium cepa* L.	0–80	Reduced/stopped growth of root length and showed deformation on root cap surface and meristematic zone	Deng et al. (2016)
Mustard	*Brassica nigra* L.	0–500	Reduced shoot growth, root shortening	Rao and Shekhawat (2016)
Lettuce, alfalfa	*Lactuca sativa* L., *Medicago sativa* L.	0–1000	Reduced root growth, decreased P, S, Cu, and Fe in shoots, increased APX in roots	Hong et al. (2015)
Bean	*Phaseolus vulgaris* L.	100–500	Inhibited growth, imbalanced metal nutrients in shoots, increased Na, decreased Fe, Mn, Zn, and Ca	Dimkpa et al. (2015)
Maize, rice	*Zea mays* L.	2000, 40–80	Shoot and root length was reduced	Yang et al. (2015)
Cilantro	*Coriandrum sativa* L.	0–100	Affected germination, reduced shoot elongation	Zuverza-Mena et al. (2015)

Table 5 Effects of different MONPs on plant growth—cont'd

Crop	Scientific name	MONPs ($mg\,L^{-1}$)	Effect on crop	References
Mustard	*Brassica nigra* L.	0–1500	Reduction in root growth, affected antioxidant activity	Nair and Chung (2015)
Soybean	*Glycine max* L.	50–500	Reduced shoot/root growth, weight, and total chlorophyll content	Nair and Chung (2014)
Mungbean	*Vigna radiate* L.	200–500	Reduced root length and total chlorophyll contents	Nair et al. (2014)
Barley	*Hoedeum vulgare* L.	0–1.5	Induced ROS, damaged membrane	Shaw et al. (2014)
Cucumber	*Cucumis sativus* L.	100–600	Reduced the germination of seed and elongation of root	Moon et al. (2014)
Maize	*Zea mays* L.	2–100	Chlorotic symptoms, reduced biomass and root elongation	Wang et al. (2012)
Wheat	*Triticum aestivum* L.	500	Shoot length reduced by 13% and root length reduced 59%, brown necrotic lesions on roots, decreased chlorophyll levels	Dimkpa et al. (2012)
Radish, grasses	*Raphan sativus* L.	10–1000	Affected growth and damaged nucleic acid	Atha et al. (2012)
Mungbean, wheat	*Vigna radiate* L., *Triticum aestivum* L.	200–1000	Reduced plant biomass	Lee et al. (2008a)

12. Conclusions

In this chapter we emphasized the contemporary advancements in the field of nanotoxicology, mainly focused on abundant usage of MONPs in industry. This chapter has elucidated the significance of transformations-related toxicological effects of MONPs with the goal to assess more effectively the impacts on natural environments. As nanotechnology-based industry is enhancing its dimension into various fields, the contribution of MONPs is shown by the fact that more than 1800 products contain MONPs. From production to disposal, these MONPs are intentionally or unintentionally released into the environment and ultimately get dumped into terrestrial and aquatic environments. Among various routes, wastewater discharge is the main pathway. MONPs undergo many transformations in air, terrestrial, and aquatic environments, and the resulting toxic properties of transformed particles are distinct from those of the source particles.

The main transformation processes, i.e., aggregation/agglomeration, adsorption, deposition, dissolution, redox reactions, and interaction with macromolecules via various mechanisms, are interlinked to MONPs' toxicity in the environment. The processes are affected by behaviors of the MONPs (e.g., size, morphology, surface charge), and conditions of the medium such as pH, temperature, and the presence of various components (proteins, polysaccharides, radicals, surfactants). MONPs can also remain bioavailable in their nano-form to the biota. However, most of the MONPs transform to other forms (their corresponding metal ions), which are more bioavailable than their sources. The stabilized MONPs are less bioavailable than the ones which undergo various transformations in the environment. The use of coating/stabilizing agents or doping can alter the morphological properties of MONPs. Knowledge of toxicity mechanisms for a given MONPs can inform how it can be redesigned to reduce toxicity. Many MONPs exhibit toxicity that can be attributed to more than one mechanism. Development of novel instruments and methods that can explore transformed MONPs from the natural and biological environments is crucial for the comprehensive characterization of the properties. Additional studies are needed for investigating slow transformation and its related toxicity at environmentally relevant concentrations despite the limitations in instrumental technology and test methods. International regulatory authorities such as the Working Group on Regulations for Nanotechnology and US Food and Drug Administration (FDA), Scientific Network for Risk Assessment of Nanotechnologies in Food and Feed (Nano Network), and Australia New Zealand Food Standards Code, etc. must be proactive in devising the regulatory standards.

13. Prospects

The magnitude of the challenges in predicting the environmental behaviors and potential effects of MONPs is daunting. As nanotechnology progresses, ever-increasing variations in the chemical and physical composition of nanomaterials can be envisioned. MONPs have been shown to act as a double-edged sword, and therefore, to assess the potential for risk from MONPs, scientists need to explore more thoroughly the potential for exposure to MONPs to occur, the likely range of exposure concentrations, and the form of the materials to which organisms will be exposed. Exposure science and nanotoxicology is emerging as a mature field that can readily address the current issues. The regulatory guidelines based on toxicological studies must be devised immediately, but also be monitored by governments and industrial stakeholders. The rise in implementation of MONPs in cosmetics and other consumer products emphasizes the requirement for a safe and less toxic alternative to synthetic nanomaterials. Moreover, the prospect of biogenic synthesis of nanomaterials must be extensively explored to produce MONPs with hazard-free properties. An eco-friendly approach to the synthesis procedure of MONPs can significantly affect its outcome and help to overcome the hazardous effects associated with MONPs. Synthesis of green nanoparticles is rapidly opening new avenues in pharmaceutical applications as well as the cosmetics industry. Numerous biological sources including crude plant extracts and microorganisms have been recognized as potential precursors for the nonhazardous preparation of various MONPs. Stringent guidelines and regulations must be standardized to prevent the mentioned health hazards associated with the excessive use of MONPs in every industry. Moreover, controlling the dissolution of MONPs is biggest challenge in the field. For improved nanoparticle stability to reduce overall dissolution through use of a shell material or use of dopants and other materials to tune band gap properties or antioxidant addition, chelating agents can revolutionize the MONPs application with safety as priority parameter. Nanotechnology-based industry is becoming central to each area of life, and MONPs form the backbone to nanoscience; therefore, research domains must be prioritized to address the hazard impacts of MONPs in the biosphere.

References

Abdelsalam, N.R., Abdel-Megeed, A., Ali, H.M., Salem, M.Z., Al-Hayali, M.F., Elshikh, M.S., 2018. Genotoxicity effects of silver nanoparticles on wheat (*Triticum aestivum* L.) root tip cells. Ecotoxicol. Environ. Saf. 155, 76–85.

Abe, K., Saito, H., 1999. Cholesterol does not affect the toxicity of amyloid β fragment but mimics its effect on MTT formazan exocytosis in cultured rat hippocampal neurons. Neurosci. Res. 35 (3), 165–174.

Abramenko, N.B., Demidova, T.B., Abkhalimov, E.V., Ershov, B.G., Krysanov, E.Y., Kustov, L.M., 2018. Ecotoxicity of different-shaped silver nanoparticles: case of zebrafish embryos. J. Hazard. Mater. 347, 89–94.

Agarwal, H., Kumar, S.V., Rajeshkumar, S., 2017. A review on green synthesis of zinc oxide nanoparticles—an eco-friendly approach. Resour. Eff. Technol. 3 (4), 406–413.

Agnihotri, S., Mukherji, S., Mukherji, S., 2014. Size-controlled silver nanoparticles synthesized over the range 5–100 nm using the same protocol and their antibacterial efficacy. RSC Adv. 4 (8), 3974–3983.

Ahamed, M., Karns, M., Goodson, M., Rowe, J., Hussain, S.M., Schlager, J.J., Hong, Y., 2008. DNA damage response to different surface chemistry of silver nanoparticles in mammalian cells. Toxicol. Appl. Pharmacol. 233 (3), 404–410.

Ahmad, H.R., Zia-ur-Rehman, M., Sohail, M.I., ul Haq, M.A., Khalid, H., Ayub, M.A., Ishaq, G., 2018. Effects of rare earth oxide nanoparticles on plants. In: *Nanomaterials in Plants, Algae, and Microorganisms*. Academic Press, pp. 239–275.

Akhtar, M.S., Panwar, J., Yun, Y.S., 2013. Biogenic synthesis of metallic nanoparticles by plant extracts. ACS Sustain. Chem. Eng. 1 (6), 591–602.

Akin, S., Sonmezoglu, S., 2018. Metal oxide nanoparticles as electron transport layer for highly efficient dye-sensitized solar cells. In: Emerging Materials for Energy Conversion and Storage. Elsevier, pp. 39–79.

Al-Ajmi, M.F., Hussain, A., Ahmed, F., 2016. Novel synthesis of ZnO nanoparticles and their enhanced anticancer activity: role of ZnO as a drug carrier. Ceram. Int. 42 (3), 4462–4469.

Al-Dahash, G., Mubder Khilkala, W., Abd Alwahid, S.N., 2018. Preparation and characterization of ZnO nanoparticles by laser ablation in NaOH aqueous solution. Iran. J. Chem. Chem. Eng. 37 (1), 11–16.

Al-Hada, N.M., Kamari, H.M., Baqer, A.A., Shaari, A.H., Saion, E., 2018. Thermal calcination-based production of SnO_2 nanopowder: an analysis of SnO_2 nanoparticle characteristics and antibacterial activities. Nanomaterials 8 (4), 250.

Ali, K., Dwivedi, S., Azam, A., Saquib, Q., Al-Said, M.S., Alkhedhairy, A.A., Musarrat, J., 2016. Aloe vera extract functionalized zinc oxide nanoparticles as nanoantibiotics against multi-drug resistant clinical bacterial isolates. J. Colloid Interface Sci. 472, 145–156.

Alkhatib, R., Alkhatib, B., Abdo, N., Laith, A.E., Creamer, R., 2019. Physio-biochemical and ultrastructural impact of (Fe_3O_4) nanoparticles on tobacco. BMC Plant Biol. 19 (1), 253.

Almansour, M.I., Alferah, M.A., Shraideh, Z.A., Jarrar, B.M., 2017. Zinc oxide nanoparticles hepatotoxicity: histological and histochemical study. Environ. Toxicol. Pharmacol. 51, 124–130.

Almukhlifi, H.A., Burns, R.C., 2016. The complete oxidation of isobutane over CeO_2 and Au/CeO_2, and the composite catalysts MO_x/CeO_2 and Au/MO_x/CeO_2 (Mn^+= Mn, Fe, Co and Ni): the effects of gold nanoparticles obtained from n-hexanethiolate-stabilized gold nanoparticles. J. Mol. Catal. A Chem. 415, 131–143.

Al-Naamani, L., Dobretsov, S., Dutta, J., 2016. Chitosan-zinc oxide nanoparticle composite coating for active food packaging applications. Innovative Food Sci. Emerg. Technol. 38, 231–237.

Al-Ruqeishi, M.S., Mohiuddin, T., Al-Saadi, L.K., 2019. Green synthesis of iron oxide nanorods from deciduous Omani mango tree leaves for heavy oil viscosity treatment. Arab. J. Chem. 12 (8), 4084–4090.

Alves, M.L., Oliveira Filho, L.C.I.D., Nogueira, P., Ogliari, A.J., Fiori, M.A., Baretta, D., Baretta, C.R.D.M., 2019. Influence of ZnO nanoparticles and a non-nano ZnO on survival and reproduction of earthworm and springtail in tropical natural soil. Rev. Bras. Cienc. Solo. 43.

Amde, M., Tan, Z.Q., Liu, R., Liu, J.F., 2015. Nanofluid of zinc oxide nanoparticles in ionic liquid for single drop liquid microextraction of fungicides in environmental waters prior to high performance liquid chromatographic analysis. J. Chromatogr. A 1395, 7–15.

Amde, M., Liu, J.F., Tan, Z.Q., Bekana, D., 2016. Ionic liquid-based zinc oxide nanofluid for vortex assisted liquid liquid microextraction of inorganic mercury in environmental waters prior to cold vapor atomic fluorescence spectroscopic detection. Talanta 149, 341–346.

Anjum, N.A., Rodrigo, M.A.M., Moulick, A., Heger, Z., Kopel, P., Zítka, O., Adam, V., Lukatkin, A.S., Duarte, A.C., Pereira, E., Kizek, R., 2016. Transport phenomena of nanoparticles in plants and animals/humans. Environ. Res. 151, 233–243.

Ansari, M.A., Khan, H.M., Alzohairy, M.A., Jalal, M., Ali, S.G., Pal, R., Musarrat, J., 2015. Green synthesis of Al_2O_3 nanoparticles and their bactericidal potential against clinical isolates of multi-drug resistant *Pseudomonas aeruginosa*. World J. Microbiol. Biotechnol. 31 (1), 153–164.

Applerot, G., Lipovsky, A., Dror, R., Perkas, N., Nitzan, Y., Lubart, R., Gedanken, A., 2009. Enhanced antibacterial activity of nanocrystalline ZnO due to increased ROS-mediated cell injury. Adv. Funct. Mater. 19 (6), 842–852.

Arias, L.S., Pessan, J.P., Vieira, A.P.M., Lima, T.M.T.D., Delbem, A.C.B., Monteiro, D.R., 2018. Iron oxide nanoparticles for biomedical applications: a perspective on synthesis, drugs, antimicrobial activity, and toxicity. Antibiotics 7 (2), 46.

Ashkarran, A.A., Mahdavi, S.M., Ahadian, M.M., 2009. ZnO nanoparticles prepared by electrical arc discharge method in water. Mater. Chem. Phys. 118 (1), 6–8.

Ashkarran, A.A., Afshar, S.A.A., Aghigh, S.M., 2010a. Photocatalytic activity of ZrO_2 nanoparticles prepared by electrical arc discharge method in water. Polyhedron 29 (4), 1370–1374.

Ashkarran, A.A., Kavianipour, M., Aghigh, S.M., Afshar, S.A., Saviz, S., Zad, A.I., 2010b. On the formation of TiO_2 nanoparticles via submerged arc discharge technique: synthesis, characterization and photocatalytic properties. J. Clust. Sci. 21 (4), 753–766.

Ashkarran, A.A., Mahdavi, S.M., Ahadian, M.M., 2010c. Photocatalytic activity of ZnO nanoparticles prepared via submerged arc discharge method. Appl. Phys. A 100 (4), 1097–1102.

Ates, M., Demir, V., Arslan, Z., Camas, M., Celik, F., 2016. Toxicity of engineered nickel oxide and cobalt oxide nanoparticles to *Artemia salina* in seawater. Water Air Soil Pollut. 227 (3), 70.

Atha, D.H., Wang, H., Petersen, E.J., Cleveland, D., Holbrook, R.D., Jaruga, P., Dizdaroglu, M., Xing, B., Nelson, B.C., 2012. Copper oxide nanoparticle mediated DNA damage in terrestrial plant models. Environ. Sci. Technol. 46 (3), 1819–1827.

Aubry, E., Liu, T., Dekens, A., Perry, F., Mangin, S., Hauet, T., Billard, A., 2019. Synthesis of iron oxide films by reactive magnetron sputtering assisted by plasma emission monitoring. Mater. Chem. Phys. 223, 360–365.

Auffan, M., Rose, J., Bottero, J.Y., Lowry, G.V., Jolivet, J.P., Wiesner, M.R., 2009. Towards a definition of inorganic nanoparticles from an environmental, health and safety perspective. Nat. Nanotechnol. 4 (10), 634–641.

Ayub, M.A., Sohail, M.I., Umair, M., ur Rehman, M.Z., Usman, M., Sabir, M., Rizwan, M., Ali, S., Ahmad, Z., 2019. Cerium oxide nanoparticles: Advances in synthesis, prospects and application in agro-ecosystem. In: Comprehensive Analytical Chemistry. vol. 87. Elsevier, pp. 209–250.

Bai, J., Zhou, B., 2014. Titanium dioxide nanomaterials for sensor applications. Chem. Rev. 114 (19), 10131–10176.

Bajpayee, M., Kumar, A., Dhawan, A., 2013. The comet assay: assessment of in vitro and in vivo DNA damage. In: Genotoxicity Assessment. Humana Press, Totowa, NJ, pp. 325–345.

Baker, G.L., Gupta, A., Clark, M.L., Valenzuela, B.R., Staska, L.M., Harbo, S.J., Pierce, J.T., Dill, J.A., 2008. Inhalation toxicity and lung toxicokinetics of C60 fullerene nanoparticles and microparticles. Toxicol. Sci. 101 (1), 122–131.

Baker, T.J., Tyler, C.R., Galloway, T.S., 2014. Impacts of metal and metal oxide nanoparticles on marine organisms. Environ. Pollut. 186, 257–271.

Baltrusaitis, J., Jayaweera, P.M., Grassian, V.H., 2011. Sulfur dioxide adsorption on TiO_2 nanoparticles: influence of particle size, coadsorbates, sample pretreatment, and light on surface speciation and surface coverage. J. Phys. Chem. C 115 (2), 492–500.

Bandyopadhyay, S., Plascencia-Villa, G., Mukherjee, A., Rico, C.M., José-Yacamán, M., Peralta-Videa, J.R., Gardea-Torresdey, J.L., 2015. Comparative phytotoxicity of ZnO NPs, bulk ZnO, and ionic zinc onto the alfalfa plants symbiotically associated with Sinorhizobium meliloti in soil. Sci. Total Environ. 515, 60–69.

Barnard, A.S., 2010. One-to-one comparison of sunscreen efficacy, aesthetics and potential nanotoxicity. Nat. Nanotechnol. 5 (4), 271–274.

Bellani, L., Giorgetti, L., Riela, S., Lazzara, G., Scialabba, A., Massaro, M., 2016. Ecotoxicity of halloysite nanotube–supported palladium nanoparticles in *Raphanus sativus* L. Environ. Toxicol. Chem. 35 (10), 2503–2510.

Ben-Moshe, T., Frenk, S., Dror, I., Minz, D., Berkowitz, B., 2013. Effects of metal oxide nanoparticles on soil properties. Chemosphere 90 (2), 640–646.

Berghe, T.V., Grootjans, S., Goossens, V., Dondelinger, Y., Krysko, D.V., Takahashi, N., Vandenabeele, P., 2013. Determination of apoptotic and necrotic cell death in vitro and in vivo. Methods 61 (2), 117–129.

Bergin, I.L., Witzmann, F.A., 2013. Nanoparticle toxicity by the gastrointestinal route: evidence and knowledge gaps. Int. J. Biomed. Nanosci. Nanotechnol. 3(1–2).

Bisquera, K.P.P., Salazar, J.R., Romero, E.S., Mar, L.L., Lopez, A., Monserate, J.J., 2017. Synthesis and characterization as zinc oxide nanoparticles as a source of zinc micronutrient in organic fertilizer. Int. J. Agric. Technol. 13 (7.2), 1695–1706.

Boisseau, P., Loubaton, B., 2011. Nanomedicine, nanotechnology in medicine. C.R. Phys. 12 (7), 620–636.

Borm, P.J., Robbins, D., Haubold, S., Kuhlbusch, T., Fissan, H., Donaldson, K., Schins, R., Stone, V., Kreyling, W., Lademann, J., Krutmann, J., 2006. The potential risks of nanomaterials: a review carried out for ECETOC. Part. Fibre Toxicol. 3 (1), 11.

Boroumand Moghaddam, A., Namvar, F., Moniri, M., Azizi, S., Mohamad, R., 2015. Nanoparticles biosynthesized by fungi and yeast: a review of their preparation, properties, and medical applications. Molecules 20 (9), 16540–16565.

Bouchat, V., Moreau, N., Colomer, J.F., Lucas, S., 2013. On Some Applications of Nanoparticles Synthesized in the Gas Phase by Magnetron Discharges.

Brar, S.K., Verma, M., 2011. Measurement of nanoparticles by light-scattering techniques. TrAC Trends Anal. Chem. 30 (1), 4–17.

Brinker, C.J., Scherer, G.W., 2013. Sol-Gel Science: The Physics and Chemistry of Sol-Gel Processing. Academic press.

Brown, P., Stevens, K. (Eds.), 2007. Nanofibers and Nanotechnology in Textiles. Elsevier.

Buffet, P.E., Tankoua, O.F., Pan, J.F., Berhanu, D., Herrenknecht, C., Poirier, L., Amiard-Triquet, C., Amiard, J.C., Bérard, J.B., Risso, C., Guibbolini, M., 2011. Behavioural and biochemical responses of two marine invertebrates *Scrobicularia plana* and *Hediste diversicolor* to copper oxide nanoparticles. Chemosphere 84 (1), 166–174.

Bundschuh, M., Filser, J., Lüderwald, S., McKee, M.S., Metreveli, G., Schaumann, G.E., Schulz, R., Wagner, S., 2018. Nanoparticles in the environment: where do we come from, where do we go to? Environ. Sci. Eur. 30 (1), 1–17.

Burello, E., Worth, A.P., 2011. QSAR modeling of nanomaterials. Wiley Interdiscip. Rev. Nanomed. Nanobiotechnol. 3 (3), 298–306.

Bystrzejewska-Piotrowska, G., Golimowski, J., Urban, P.L., 2009. Nanoparticles: their potential toxicity, waste and environmental management. Waste Manag. 29 (9), 2587–2595.

Cao, J., Feng, Y., Lin, X., Wang, J., 2016. Arbuscular mycorrhizal fungi alleviate the negative effects of iron oxide nanoparticles on bacterial community in rhizospheric soils. Front. Environ. Sci. 4, 10.

Carroll, K.J., Reveles, J.U., Shultz, M.D., Khanna, S.N., Carpenter, E.E., 2011. Preparation of elemental Cu and Ni nanoparticles by the polyol method: an experimental and theoretical approach. J. Phys. Chem. C 115 (6), 2656–2664.

Casey, A., Herzog, E., Davoren, M., Lyng, F.M., Byrne, H.J., Chambers, G., 2007. Spectroscopic analysis confirms the interactions between single walled carbon nanotubes and various dyes commonly used to assess cytotoxicity. Carbon 45 (7), 1425–1432.

Chai, H., Yao, J., Sun, J., Zhang, C., Liu, W., Zhu, M., Ceccanti, B., 2015. The effect of metal oxide nanoparticles on functional bacteria and metabolic profiles in agricultural soil. Bull. Environ. Contam. Toxicol. 94 (4), 490–495.

Champion, J.A., Mitragotri, S., 2006. Role of target geometry in phagocytosis. Proc. Natl. Acad. Sci. 103 (13), 4930–4934.

Châtel, A., Mouneyrac, C., 2017. Signaling pathways involved in metal-based nanomaterial toxicity towards aquatic organisms. Comp. Biochem. Physiol. C Toxicol. Pharmacol. 196, 61–70.

Chattopadhyay, S., Dash, S.K., Tripathy, S., Das, B., Mandal, D., Pramanik, P., Roy, S., 2015. Toxicity of cobalt oxide nanoparticles to normal cells; an in vitro and in vivo study. Chem. Biol. Interact. 226, 58–71.

Chauhan, I., Aggrawal, S., Mohanty, P., 2015. ZnO nanowire-immobilized paper matrices for visible light-induced antibacterial activity against *Escherichia coli*. Environ. Sci. Nano 2 (3), 273–279.

Chavali, M.S., Nikolova, M.P., 2019. Metal oxide nanoparticles and their applications in nanotechnology. SN Appl. Sci. 1 (6), 607.

Chen, Y.H., Yeh, C.S., 2002. Laser ablation method: use of surfactants to form the dispersed Ag nanoparticles. Colloids Surf. A Physicochem. Eng. Asp. 197 (1–3), 133–139.

Chertok, B., Moffat, B.A., David, A.E., Yu, F., Bergemann, C., Ross, B.D., Yang, V.C., 2008. Iron oxide nanoparticles as a drug delivery vehicle for MRI monitored magnetic targeting of brain tumors. Biomaterials 29 (4), 487–496.

Chhipa, H., Joshi, P., 2016. Nanofertilisers, nanopesticides and nanosensors in agriculture. In: Nanoscience in Food and Agriculture 1. Springer, Cham, pp. 247–282.

Chithrani, B.D., Ghazani, A.A., Chan, W.C., 2006. Determining the size and shape dependence of gold nanoparticle uptake into mammalian cells. Nano Lett. 6 (4), 662–668.

Chuang, H.C., Hsiao, T.C., Wu, C.K., Chang, H.H., Lee, C.H., Chang, C.C., Cheng, T.J., 2013. Allergenicity and toxicology of inhaled silver nanoparticles in allergen-provocation mice models. Int. J. Nanomedicine 8, 4495.

Coccini, T., Grandi, S., Lonati, D., Locatelli, C., De Simone, U., 2015. Comparative cellular toxicity of titanium dioxide nanoparticles on human astrocyte and neuronal cells after acute and prolonged exposure. Neurotoxicology 48, 77–89.

Cong, Y., Banta, G.T., Selck, H., Berhanu, D., Valsami-Jones, E., Forbes, V.E., 2011. Toxic effects and bioaccumulation of nano-, micron-and ionic-Ag in the polychaete, *Nereis diversicolor*. Aquat. Toxicol. 105 (3–4), 403–411.

Cong, Y., Jin, F., Wang, J., Mu, J., 2017. The embryotoxicity of ZnO nanoparticles to marine medaka, *Oryzias melastigma*. Aquat. Toxicol. 185, 11–18.

Corbo, C., Molinaro, R., Parodi, A., Toledano Furman, N.E., Salvatore, F., Tasciotti, E., 2016. The impact of nanoparticle protein corona on cytotoxicity, immunotoxicity and target drug delivery. Nanomedicine 11 (1), 81–100.

Cornelis, G., Hund-Rinke, K., Kuhlbusch, T., Van den Brink, N., Nickel, C., 2014. Fate and bioavailability of engineered nanoparticles in soils: a review. Crit. Rev. Environ. Sci. Technol. 44 (24), 2720–2764.

Corsi, I., Cherr, G.N., Lenihan, H.S., Labille, J., Hassellov, M., Canesi, L., Dondero, F., Frenzilli, G., Hristozov, D., Puntes, V., Della Torre, C., 2014. Common Strategies and Technologies for the Ecosafety Assessment and Design of Nanomaterials Entering the Marine Environment.

Cox, P.A., 1989. The Elements. Their Origin, Abundance, and Distribution.

Cushing, B.L., Kolesnichenko, V.L., O'Connor, C.J., 2004. Recent advances in the liquid-phase syntheses of inorganic nanoparticles. Chem. Rev. 104 (9), 3893–3946.

D'Amato, R., Falconieri, M., Gagliardi, S., Popovici, E., Serra, E., Terranova, G., Borsella, E., 2013. Synthesis of ceramic nanoparticles by laser pyrolysis: from research to applications. J. Anal. Appl. Pyrolysis 104, 461–469.

Danino, D., 2012. Cryo-TEM of soft molecular assemblies. Curr. Opin. Colloid Interface Sci. 17 (6), 316–329.

Daraio, C., Jin, S., 2012. Synthesis and patterning methods for nanostructures useful for biological applications. In: Nanotechnology for Biology and Medicine. Springer, New York, NY, pp. 27–44.

Darroudi, M., Hakimi, M., Sarani, M., Oskuee, R.K., Zak, A.K., Gholami, L., 2013. Facile synthesis, characterization, and evaluation of neurotoxicity effect of cerium oxide nanoparticles. Ceram. Int. 39 (6), 6917–6921.

De Carvalho, J.F., De Medeiros, S.N., Morales, M.A., Dantas, A.L., Carriço, A.S., 2013. Synthesis of magnetite nanoparticles by high energy ball milling. Appl. Surf. Sci. 275, 84–87.

Deline, A.R., Nason, J.A., 2019. Evaluation of labeling methods used for investigating the environmental behavior and toxicity of metal oxide nanoparticles. Environ. Sci. Nano 6 (4), 1043–1066.

Deng, Y.Q., White, J.C., Xing, B.S., 2014. Interactions between engineered nanomaterials and agricultural crops: implications for food safety. J. Zhejiang Univ. Sci. A 15 (8), 552–572.

Deng, F., Wang, S., Xin, H., 2016. Toxicity of CuO nanoparticles to structure and metabolic activity of *Allium cepa* root tips. Bull. Environ. Contam. Toxicol. 97 (5), 702–708.

Desai, P., Patlolla, R.R., Singh, M., 2010. Interaction of nanoparticles and cell-penetrating peptides with skin for transdermal drug delivery. Mol. Membr. Biol. 27 (7), 247–259.

Dhand, C., Dwivedi, N., Loh, X.J., Ying, A.N.J., Verma, N.K., Beuerman, R.W., Lakshminarayanan, R., Ramakrishna, S., 2015. Methods and strategies for the synthesis of diverse nanoparticles and their applications: a comprehensive overview. RSC Adv. 5 (127), 105003–105037.

Dhumal, R.S., Bommidi, D., Salehinia, I., 2019. Thermal conductivity of metal-coated tri-walled carbon nanotubes in the presence of vacancies-molecular dynamics simulations. Nanomaterials 9 (6), 809.

Di Gioacchino, M., Verna, N., Gornati, R., Sabbioni, E., Bernardini, G., 2009. Metal nanoparticle health risk assessment. Nanotoxicity 519–542.

Dimkpa, C.O., McLean, J.E., Britt, D.W., Anderson, A.J., 2015. Nano-CuO and interaction with nano-ZnO or soil bacterium provide evidence for the interference of nanoparticles in metal nutrition of plants. Ecotoxicology 24 (1), 119–129.

Dimkpa, C.O., McLean, J.E., Latta, D.E., Manangón, E., Britt, D.W., Johnson, W.P., Boyanov, M.I., Anderson, A.J., 2012. CuO and ZnO nanoparticles: phytotoxicity, metal speciation, and induction of oxidative stress in sand-grown wheat. J. Nanoparticle Res. 14 (9), 1125.

Ditta, A., 2012. How helpful is nanotechnology in agriculture? Adv. Nat. Sci. Nanosci. Nanotechnol. 3 (3), 033002.

Dizaj, S.M., Lotfipour, F., Barzegar-Jalali, M., Zarrintan, M.H., Adibkia, K., 2014. Antimicrobial activity of the metals and metal oxide nanoparticles. Mater. Sci. Eng. C Mater. Biol. Appl. 44 (1), 278.

Djurisic, M., Vidal, G.S., Mann, M., Aharon, A., Kim, T., Santos, A.F., Zuo, Y., Hübener, M., Shatz, C.J., 2013. PirB regulates a structural substrate for cortical plasticity. Proc. Natl. Acad. Sci. 110 (51), 20771–20776.

Djurišić, A.B., Liu, X., Leung, Y.H., 2014. Zinc oxide films and nanomaterials for photovoltaic applications. *physica status solidi (RRL)–Rapid*. Res. Lett. 8 (2), 123–132.

Djurišić, A.B., Leung, Y.H., Ng, A.M., Xu, X.Y., Lee, P.K., Degger, N., Wu, R.S.S., 2015. Toxicity of metal oxide nanoparticles: mechanisms, characterization, and avoiding experimental artefacts. Small 11 (1), 26–44.

Dolez, P.I., 2015. Nanomaterials definitions, classifications, and applications. In: Nanoengineering. Elsevier, pp. 3–40.

Dong, H., Chen, Y.C., Feldmann, C., 2015. Polyol synthesis of nanoparticles: status and options regarding metals, oxides, chalcogenides, and non-metal elements. Green Chem. 17 (8), 4107–4132.

Dong, L., Tang, S., Deng, F., Gong, Y., Zhao, K., Zhou, J., Liang, D., Fang, J., Hecker, M., Giesy, J.P., Bai, X., 2019. Shape-dependent toxicity of alumina nanoparticles in rat astrocytes. Sci. Total Environ. 690, 158–166.

Dréno, B., Alexis, A., Chuberre, B., Marinovich, M., 2019. Safety of titanium dioxide nanoparticles in cosmetics. J. Eur. Acad. Dermatol. Venereol. 33, 34–46.

Dudkiewicz, A., Luo, P., Tiede, K., Boxall, A., 2012. Detecting and characterizing nanoparticles in food, beverages and nutraceuticals. In: Nanotechnology in the Food, Beverage and Nutraceutical Industries. Woodhead Publishing, pp. 53–81.

Durán, N., Seabra, A.B., 2012. Metallic oxide nanoparticles: state of the art in biogenic syntheses and their mechanisms. Appl. Microbiol. Biotechnol. 95 (2), 275–288.

Easo, S.L., Mohanan, P.V., 2015. In vitro hematological and in vivo immunotoxicity assessment of dextran stabilized iron oxide nanoparticles. Colloids Surf. B: Biointerfaces 134, 122–130.

Ebbs, S.D., Bradfield, S.J., Kumar, P., White, J.C., Musante, C., Ma, X., 2016. Accumulation of zinc, copper, or cerium in carrot (*Daucus carota*) exposed to metal oxide nanoparticles and metal ions. Environ. Sci. Nano 3 (1), 114–126.

Eixenberger, J.E., Anders, C.B., Wada, K., Reddy, K.M., Brown, R.J., Moreno-Ramirez, J., Weltner, A.E., Karthik, C., Tenne, D.A., Fologea, D., Wingett, D.G., 2019. Defect engineering of ZnO nanoparticles for bioimaging applications. ACS Appl. Mater. Interfaces 11 (28), 24933–24944.

El Badawy, A.M., Silva, R.G., Morris, B., Scheckel, K.G., Suidan, M.T., Tolaymat, T.M., 2011. Surface charge-dependent toxicity of silver nanoparticles. Environ. Sci. Technol. 45 (1), 283–287.

Elmer, W.H., White, J.C., 2016. The use of metallic oxide nanoparticles to enhance growth of tomatoes and eggplants in disease infested soil or soilless medium. Environ. Sci. Nano 3 (5), 1072–1079.

Eriksson, S., Nylén, U., Rojas, S., Boutonnet, M., 2004. Preparation of catalysts from microemulsions and their applications in heterogeneous catalysis. Appl. Catal. A Gen. 265 (2), 207–219.

Ershov, B.G., Abkhalimov, E.V., Solovov, R.D., Roldughin, V.I., 2016. Gold nanoparticles in aqueous solutions: influence of size and pH on hydrogen dissociative adsorption and Au (III) ion reduction. Phys. Chem. Chem. Phys. 18 (19), 13459–13466.

Escorihuela, L., Fernández, A., Rallo, R., Martorell, B., 2018a. Molecular dynamics simulations of zinc oxide solubility: from bulk down to nanoparticles. Food Chem. Toxicol. 112, 518–525.

Escorihuela, L., Martorell, B., Rallo, R., Fernández, A., 2018b. Toward computational and experimental characterisation for risk assessment of metal oxide nanoparticles. Environ. Sci. Nano 5 (10), 2241–2251.

Eshed, M., Pol, S., Gedanken, A., Balasubramanian, M., 2011. Zirconium nanoparticles prepared by the reduction of zirconium oxide using the RAPET method. Beilstein J. Nanotechnol. 2 (1), 198–203.

Espitia, P.J.P., Soares, N.D.F.F., dos Reis Coimbra, J.S., de Andrade, N.J., Cruz, R.S., Medeiros, E.A.A., 2012. Zinc oxide nanoparticles: synthesis, antimicrobial activity and food packaging applications. Food Bioprocess Technol. 5 (5), 1447–1464.

Ewing, A.G., Bigelow, J.C., Wightman, R.M., 1983. Direct in vivo monitoring of dopamine released from two striatal compartments in the rat. Science 221 (4606), 169–171.

Fabrega, J., Luoma, S.N., Tyler, C.R., Galloway, T.S., Lead, J.R., 2011. Silver nanoparticles: behaviour and effects in the aquatic environment. Environ. Int. 37 (2), 517–531.

Fahimmunisha, B.A., Ishwarya, R., AlSalhi, M.S., Devanesan, S., Govindarajan, M., Vaseeharan, B., 2020. Green fabrication, characterization and antibacterial potential of zinc oxide nanoparticles using Aloe socotrina leaf extract: a novel drug delivery approach. J. Drug Delivery Sci. Technol. 55, 101465.

Fahmy, B., Cormier, S.A., 2009. Copper oxide nanoparticles induce oxidative stress and cytotoxicity in airway epithelial cells. Toxicol. in Vitro 23 (7), 1365–1371.

Fang, Y., Lu, X., Chen, W., 2017. Application of nano-TiO_2 in cosmetics. Gen. Chem. 3(2).

Fernández-García, M., Rodriguez, J.A., 2011. Metal oxide nanoparticles. In: Encyclopedia of Inorganic and Bioinorganic Chemistry.

Forest, V., Leclerc, L., Hochepied, J.F., Trouvé, A., Sarry, G., Pourchez, J., 2017. Impact of cerium oxide nanoparticles shape on their in vitro cellular toxicity. Toxicol. in Vitro 38, 136–141.

Fraceto, L.F., de Lima, R., Oliveira, H.C., et al., 2018. Future trends in nanotechnology aiming environmental applications. Energ. Ecol. Environ. 3, 69–71. https://doi.org/10.1007/s40974-018-0087-x.
Franke, M.E., Koplin, T.J., Simon, U., 2006. Metal and metal oxide nanoparticles in chemiresistors: does the nanoscale matter? Small 2 (1), 36–50.
Frenk, S., Ben-Moshe, T., Dror, I., Berkowitz, B., Minz, D., 2013. Effect of metal oxide nanoparticles on microbial community structure and function in two different soil types. PLoS ONE. 8(12).
Fröhlich, E., Salar-Behzadi, S., 2014. Toxicological assessment of inhaled nanoparticles: role of in vivo, ex vivo, in vitro, and in silico studies. Int. J. Mol. Sci. 15 (3), 4795–4822.
Fu, L., Liu, Z.H.I.M.I.N., Liu, Y., Han, B., Hu, P., Cao, L., Zhu, D.A.O.B.E.N., 2005. Beaded cobalt oxide nanoparticles along carbon nanotubes: towards more highly integrated electronic devices. Adv. Mater. 17 (2), 217–221.
Gagnon, C., Bruneau, A., Turcotte, P., Pilote, M., Gagné, F., 2018. Fate of cerium oxide nanoparticles in natural waters and immunotoxicity in exposed rainbow trout. J. Nanomed. Nanotechnol. 9(2).
Gan, T., Zhao, A.X., Wang, S.H., Lv, Z., Sun, J.Y., 2016. Hierarchical triple-shelled porous hollow zinc oxide spheres wrapped in graphene oxide as efficient sensor material for simultaneous electrochemical determination of synthetic antioxidants in vegetable oil. Sensors Actuators B Chem. 235, 707–716.
Ganesan, S., Anaimalai Thirumurthi, N., Raghunath, A., Vijayakumar, S., Perumal, E., 2016. Acute and sub-lethal exposure to copper oxide nanoparticles causes oxidative stress and teratogenicity in zebrafish embryos. J. Appl. Toxicol. 36, 554–567.
Ganguli, A.K., Ganguly, A., Vaidya, S., 2010. Microemulsion-based synthesis of nanocrystalline materials. Chem. Soc. Rev. 39 (2), 474–485.
Gao, X., Zhang, T., 2018. An overview: facet-dependent metal oxide semiconductor gas sensors. Sensors Actuators B Chem. 277, 604–633.
Gao, Z., Zhou, J., Gu, Y., Fei, P., Hao, Y., Bao, G., Wang, Z.L., 2009. Effects of piezoelectric potential on the transport characteristics of metal-ZnO nanowire-metal field effect transistor. J. Appl. Phys. 105 (11), 113707.
Gao, S., Wang, X., Wang, S., Zhu, S., Rong, R., Xu, X., 2017a. Complex effect of zinc oxide nanoparticles on cadmium chloride-induced hepatotoxicity in mice: protective role of metallothionein. Metallomics 9 (6), 706–714.
Gao, W., Wang, W., Yao, S., Wu, S., Zhang, H., Zhang, J., Jing, F., Mao, H., Jin, Q., Cong, H., Jia, C., 2017b. Highly sensitive detection of multiple tumor markers for lung cancer using gold nanoparticle probes and microarrays. Anal. Chim. Acta 958, 77–84.
Garcia, C.V., Shin, G.H., Kim, J.T., 2018. Metal oxide-based nanocomposites in food packaging: applications, migration, and regulations. Trends Food Sci. Technol. 82, 21–31.
García-Negrete, C.A., Blasco, J., Volland, M., Rojas, T.C., Hampel, M., Lapresta-Fernández, A., De Haro, M.J., Soto, M., Fernández, A., 2013. Behaviour of Au-citrate nanoparticles in seawater and accumulation in bivalves at environmentally relevant concentrations. Environ. Pollut. 174, 134–141.
Gavrieli, Y., Sherman, Y., Ben-Sasson, S.A., 1992. Identification of programmed cell death in situ via specific labeling of nuclear DNA fragmentation. J. Cell Biol. 119 (3), 493–501.
Gawande, M.B., Goswami, A., Felpin, F.X., Asefa, T., Huang, X., Silva, R., Zou, X., Zboril, R., Varma, R.S., 2016. Cu and Cu-based nanoparticles: synthesis and applications in catalysis. Chem. Rev. 116 (6), 3722–3811.
Ge, Y., Schimel, J.P., Holden, P.A., 2011. Evidence for negative effects of TiO_2 and ZnO nanoparticles on soil bacterial communities. Environ. Sci. Technol. 45 (4), 1659–1664.
Gershon, T.S., Li, N., Sadana, D. and Todorov, T.K., International Business Machines Corp, 2017. Plasmonic Enhancement of Zinc Oxide Light Absorption for Sunscreen Applications. U.S. Patent Application 15/344,871.
Ghasemi, E., Heydari, A., Sillanpää, M., 2017. Superparamagnetic Fe_3O_4@ EDTA nanoparticles as an efficient adsorbent for simultaneous removal of Ag (I), Hg (II), Mn (II), Zn (II), Pb (II) and Cd (II) from water and soil environmental samples. Microchem. J. 131, 51–56.
Ghosh, S., Mashayekhi, H., Pan, B., Bhowmik, P., Xing, B., 2008. Colloidal behavior of aluminum oxide nanoparticles as affected by pH and natural organic matter. Langmuir 24 (21), 12385–12391.

Ghosh, P.R., Fawcett, D., Sharma, S.B., Poinern, G.E., 2017. Production of high-value nanoparticles via biogenic processes using aquacultural and horticultural food waste. Materials 10 (8), 852.

Gliga, A.R., Di Bucchianico, S., Lindvall, J., Fadeel, B., Karlsson, H.L., 2018. RNA-sequencing reveals long-term effects of silver nanoparticles on human lung cells. Sci. Rep. 8 (1), 1–14.

Glinka, Y.D., Lin, S.H., Hwang, L.P., Chen, Y.T., Tolk, N.H., 2001. Size effect in self-trapped exciton photoluminescence from SiO_2-based nanoscale materials. Phys. Rev. B 64 (8), 085421.

Gnanasangeetha, D., Thambavani, D.S., 2013. ZnO nanoparticle entrenched on activated silica as a proficient adsorbent for removal of As^{3+}. Int. J. Res. Pharmaceut. Biomed. Sci. 4 (4), 1295–1304.

Golbamaki, A., Golbamaki, N., Sizochenko, N., Rasulev, B., Leszczynski, J., Benfenati, E., 2018. Genotoxicity induced by metal oxide nanoparticles: a weight of evidence study and effect of particle surface and electronic properties. Nanotoxicology 12 (10), 1113–1129.

Gottschalk, F., Nowack, B., 2011. The release of engineered nanomaterials to the environment. J. Environ. Monit. 13 (5), 1145–1155.

Graves, J.L., 2014. A grain of salt: metallic and metallic oxide nanoparticles as the new antimicrobials. JSM Nanotechnol. Nanomed. 2 (2), 1026–1030.

Grumezescu, A., 2016. Nanobiomaterials in Antimicrobial Therapy: Applications of Nanobiomaterials. William Andrew.

Gu, L., Zhang, M., He, J., Ni, P., 2018. A porous cross-linked gel polymer electrolyte separator for lithium-ion batteries prepared by using zinc oxide nanoparticle as a foaming agent and filler. Electrochim. Acta 292, 769–778.

Gui, X., Deng, Y., Rui, Y., Gao, B., Luo, W., Chen, S., Li, X., Liu, S., Han, Y., Liu, L., Xing, B., 2015. Response difference of transgenic and conventional rice (*Oryza sativa*) to nanoparticles (γFe_2O_3). Environ. Sci. Pollut. Res. 22 (22), 17716–17723.

Guo, N.L., Wan, Y.W., Denvir, J., Porter, D.W., Pacurari, M., Wolfarth, M.G., Castranova, V., Qian, Y., 2012. Multiwalled carbon nanotube-induced gene signatures in the mouse lung: potential predictive value for human lung cancer risk and prognosis. J. Toxic. Environ. Health A 75 (18), 1129–1153.

Guozhong, C., 2004. Nanostructures and Nanomaterials: Synthesis, Properties and Applications. World Scientific.

Gurr, J.R., Wang, A.S., Chen, C.H., Jan, K.Y., 2005. Ultrafine titanium dioxide particles in the absence of photoactivation can induce oxidative damage to human bronchial epithelial cells. Toxicology 213 (1–2), 66–73.

Haider, A.J., Jameel, Z.N., Al-Hussaini, I.H., 2019. Review on: titanium dioxide applications. Energy Procedia 157, 17–29.

Hajibeygi, M., Shabanian, M., Omidi-Ghallemohamadi, M., Khonakdar, H.A., 2017. Optical, thermal and combustion properties of self-colored polyamide nanocomposites reinforced with azo dye surface modified ZnO nanoparticles. Appl. Surf. Sci. 416, 628–638.

Hakuta, Y., Ura, H., Hayashi, H., Arai, K., 2005. Continuous production of $BaTiO_3$ nanoparticles by hydrothermal synthesis. Ind. Eng. Chem. Res. 44 (4), 840–846.

Halliwell, B., Whiteman, M., 2004. Measuring reactive species and oxidative damage in vivo and in cell culture: how should you do it and what do the results mean? Br. J. Pharmacol. 142 (2), 231–255.

Handford, C.E., Dean, M., Henchion, M., Spence, M., Elliott, C.T., Campbell, K., 2014. Implications of nanotechnology for the agri-food industry: opportunities, benefits and risks. Trends Food Sci. Technol. 40 (2), 226–241.

Hanna, S.K., Miller, R.J., Zhou, D., Keller, A.A., Lenihan, H.S., 2013. Accumulation and toxicity of metal oxide nanoparticles in a soft-sediment estuarine amphipod. Aquat. Toxicol. 142, 441–446.

Harano, A., Shimada, K., Okubo, T., Sadakata, M., 2002. Crystal phases of TiO_2 ultrafine particles prepared by laser ablation of solid rods. J. Nanopart. Res. 4 (3), 215–219.

He, S., Feng, Y., Ni, J., Sun, Y., Xue, L., Feng, Y., Yu, Y., Lin, X., Yang, L., 2016. Different responses of soil microbial metabolic activity to silver and iron oxide nanoparticles. Chemosphere 147, 195–202.

Heng, B.C., Zhao, X., Tan, E.C., Khamis, N., Assodani, A., Xiong, S., Ruedl, C., Ng, K.W., Loo, J.S.C., 2011. Evaluation of the cytotoxic and inflammatory potential of differentially shaped zinc oxide nanoparticles. Arch. Toxicol. 85 (12), 1517–1528.

Hernández-Sierra, J.F., Ruiz, F., Pena, D.C.C., Martínez-Gutiérrez, F., Martínez, A.E., Guillén, A.D.J.P., Tapia-Pérez, H., Castañón, G.M., 2008. The antimicrobial sensitivity of *Streptococcus mutans* to nanoparticles of silver, zinc oxide, and gold. Nanomedicine 4 (3), 237–240.

Hillyer, J.F., Albrecht, R.M., 2001. Gastrointestinal persorption and tissue distribution of differently sized colloidal gold nanoparticles. J. Pharm. Sci. 90, 1927–1936.

Hischier, R., 2014. Life cycle assessment of manufactured nanomaterials: inventory modelling rules and application example. Int. J. Life Cycle Assess. 19 (4), 941–943.

Hochella Jr., M.F., Aruguete, D., Kim, B., Madden, A.S., 2012. Naturally occurring inorganic nanoparticles: general assessment and a global budget for one of Earth's last unexplored major geochemical components. In: Nature's Nanostructures.pp. 1–31.

Hong, J., Rico, C.M., Zhao, L., Adeleye, A.S., Keller, A.A., Peralta-Videa, J.R., Gardea-Torresdey, J.L., 2015. Toxic effects of copper-based nanoparticles or compounds to lettuce (*Lactuca sativa*) and alfalfa (*Medicago sativa*). Environ Sci Process Impacts 17 (1), 177–185.

Horie, M., Stowe, M., Tabei, M., Kuroda, E., 2015. Pharyngeal aspiration of metal oxide nanoparticles showed potential of allergy aggravation effect to inhaled ovalbumin. Inhal. Toxicol. 27 (3), 181–190.

Hosny, A.E.D.M., Kashef, M.T., Taher, H.A., El-Bazza, Z.E., 2017. The use of unirradiated and γ-irradiated zinc oxide nanoparticles as a preservative in cosmetic preparations. Int. J. Nanomedicine 12, 6799.

Housaindokht, M.R., Nakhaei Pour, A., 2013. Size control of iron oxide nanoparticles using reverse microemulsion method: morphology, reduction, and catalytic activity in CO hydrogenation. J. Chem. 2013, 781595.

Huang, M., Khor, E., Lim, L.Y., 2004. Uptake and cytotoxicity of chitosan molecules and nanoparticles: effects of molecular weight and degree of deacetylation. Pharm. Res. 21 (2), 344–353.

Huang, Y., Lenaghan, S.C., Xia, L., Burris, J.N., Stewart, C.N.J., Zhang, M., 2013. Characterization of physicochemical properties of ivy nanoparticles for cosmetic application. J. Nanobiotechnology 11 (1), 3.

Huang, K.L., Lee, Y.H., Chen, H.I., Liao, H.S., Chiang, B.L., Cheng, T.J., 2015. Zinc oxide nanoparticles induce eosinophilic airway inflammation in mice. J. Hazard. Mater. 297, 304–312.

Hunt, A., Watkiss, P., 2007. Literature Review on Climate Change Impacts on Urban City Centres: Initial Findings; ENV/EPOC/GSP(2007)10/Final. OECD.

Hussain, T., Batool, R., 2017. Microemulsion route for the synthesis of nano-structured catalytic materials. In: Properties and Uses of Microemulsions.p. 13.

Hussain, S.M., Javorina, A.K., Schrand, A.M., Duhart, H.M., Ali, S.F., Schlager, J.J., 2006. The interaction of manganese nanoparticles with PC-12 cells induces dopamine depletion. Toxicol. Sci. 92 (2), 456–463.

Ibrahem, E.J., Thalij, K.M., Saleh, M.K., Badawy, A.S., 2017. Biosynthesis of zinc oxide nanoparticles and assay of antibacterial activity. Am. J. Biochem. Biotechnol. 13, 63–69.

Ibrahim, R.K., Hayyan, M., AlSaadi, M.A., Hayyan, A., Ibrahim, S., 2016. Environmental application of nanotechnology: air, soil, and water. Environ. Sci. Pollut. Res. 23 (14), 13754–13788.

Immanuel, S., Aparna, T.K., Sivasubramanian, R., 2019. Graphene–metal oxide nanocomposite modified electrochemical sensors. In: Graphene-Based Electrochemical Sensors for Biomolecules. Elsevier, pp. 113–138.

Ingale, A.G., Chaudhari, A.N., 2013. Biogenic synthesis of nanoparticles and potential applications: an eco-friendly approach. J. Nanomed. Nanotechol. 4 (165), 1–7.

ISO (International Organization for Standardization), 2006a. Technical Specification UNI EN14040: Environmental Management-Life Cycle Assessment (LCA)—Principles and Framework. Switzerland: Geneve.

ISO (International Organization for Standardization), 2006b. Technical Specification UNI EN14044: Environmental Management—LCA-Requirement and Guidelines. Switzerland: Geneve.

Itoh, T., Uchida, T., Matsubara, I., Izu, N., Shin, W., Miyazaki, H., Tanjo, H., Kanda, K., 2015. Preparation of γ-alumina large grain particles with large specific surface area via polyol synthesis. Ceram. Int. 41 (3), 3631–3638.

Jamal, S.A., 2013. Application of nanoparticles of ceramics, peptides, silicon, carbon, and diamonds in tissue engineering. Chem. Sci. J. 4 (1), 1.

Janrao, K.K., Gadhave, M.V., Banerjee, S.K., Gaikwad, D.D., 2014. Nanoparticle induced nanotoxicity: an overview. Asian J. Biomed. Pharm. Sci. 4 (32), 1.

Jaworek, A.T.S.A., Sobczyk, A.T., 2008. Electrospraying route to nanotechnology: an overview. J. Electrost. 66 (3–4), 197–219.

Jayaseelan, C., Rahuman, A.A., Kirthi, A.V., Marimuthu, S., Santhoshkumar, T., Bagavan, A., Gaurav, K., Karthik, L., Rao, K.B., 2012. Novel microbial route to synthesize ZnO nanoparticles using *Aeromonas hydrophila* and their activity against pathogenic bacteria and fungi. Spectrochim. Acta A Mol. Biomol. Spectrosc. 90, 78–84.

Jeevanandam, J., Chan, Y.S., Danquah, M.K., 2016. Biosynthesis of metal and metal oxide nanoparticles. ChemBioEng Rev. 3 (2), 55–67.

Jeong, J., Kim, S.H., Lee, S., Lee, D.K., Han, Y., Jeon, S., Cho, W.S., 2018. Differential contribution of constituent metal ions to the cytotoxic effects of fast-dissolving metal-oxide nanoparticles. Front. Pharmacol. 9, 15.

Jevšnik, M., Ovca, A., Bauer, M., Fink, R., Oder, M., Sevšek, F., 2013. Food safety knowledge and practices among elderly in Slovenia. Food Control 31 (2), 284–290.

Jiang, J., Pi, J., Cai, J., 2018. The advancing of zinc oxide nanoparticles for biomedical applications. Bioinorg. Chem. Appl. 2018, 1062562.

Jin, Y., Kannan, S., Wu, M., Zhao, J.X., 2007. Toxicity of luminescent silica nanoparticles to living cells. Chem. Res. Toxicol. 20 (8), 1126–1133.

Jindal, A.K., Pandya, K., Khan, I.D., 2015. Antimicrobial resistance: a public health challenge. Med. J. Armed Forces India 71 (2), 178–181.

Joo, J., Kwon, S.G., Yu, J.H., Hyeon, T., 2005. Synthesis of ZnO nanocrystals with cone, hexagonal cone, and rod shapes via non-hydrolytic ester elimination sol–gel reactions. Adv. Mater. 17 (15), 1873–1877.

Joubert, L.M., 2017. Variable pressure-SEM: a versatile tool for visualization of hydrated and non-conductive specimens. In: Microscopy and Imaging Science: Practical Approaches to Applied Research and Education.pp. 655–662.

Jovanović, B., Palić, D., 2012. Immunotoxicology of non-functionalized engineered nanoparticles in aquatic organisms with special emphasis on fish—review of current knowledge, gap identification, and call for further research. Aquat. Toxicol. 118, 141–151.

Juan, W.A.N.G., Kunhui, S.H.U., Zhang, L.I., Youbin, S.I., 2017. Effects of silver nanoparticles on soil microbial communities and bacterial nitrification in suburban vegetable soils. Pedosphere 27 (3), 482–490.

Julia, X.Y., Li, T.H., 2011. Distinct biological effects of different nanoparticles commonly used in cosmetics and medicine coatings. Cell Biosci. 1 (1), 19.

Jun, Y.W., Seo, J.W., Cheon, J., 2008. Nanoscaling laws of magnetic nanoparticles and their applicabilities in biomedical sciences. Acc. Chem. Res. 41 (2), 179–189.

Ju-Nam, Y., Lead, J., 2016. Properties, sources, pathways, and fate of nanoparticles in the environment. In: Engineered Nanoparticles and the Environment: Biophysicochemical Processes and Toxicity.vol. 4. pp. 95–117.

Kadammattil, A.V., Sajankila, S.P., Prabhu, S., Rao, B.N., Rao, B.S.S., 2018. Systemic toxicity and teratogenicity of copper oxide nanoparticles and copper sulfate. J. Nanosci. Nanotechnol. 18 (4), 2394–2404.

Kahan, D.M., Braman, D., Slovic, P., Gastil, J., Cohen, G., 2009. Cultural cognition of the risks and benefits of nanotechnology. Nat. Nanotechnol. 4 (2), 87–90.

Kahlon, S.K., Sharma, G., Julka, J.M., Kumar, A., Sharma, S., Stadler, F.J., 2018. Impact of heavy metals and nanoparticles on aquatic biota. Environ. Chem. Lett. 16 (3), 919–946.

Kahouli, M., Barhoumi, A., Bouzid, A., Al-Hajry, A., Guermazi, S., 2015. Structural and optical properties of ZnO nanoparticles prepared by direct precipitation method. Superlattice. Microst. 85, 7–23.

Kahru, A., Dubourguier, H.C., 2010. From ecotoxicology to nanoecotoxicology. Toxicology 269 (2–3), 105–119.

Kahru, A., Ivask, A., 2013. Mapping the dawn of nanoecotoxicological research. Acc. Chem. Res. 46 (3), 823–833.

Kanwal, Z., Raza, M.A., Manzoor, F., Riaz, S., Jabeen, G., Fatima, S., Naseem, S., 2019. A comparative assessment of nanotoxicity induced by metal (silver, nickel) and metal oxide (cobalt, chromium) nanoparticles in *Labeo rohita*. Nanomaterials 9 (2), 309.

Kanwar, R., Rathee, J., Patil, M.T., Mehta, S.K., 2018. Microemulsions as nanotemplates: a soft and versatile approach. In: Microemulsion—A Chemical Nanoreactor. IntechOpen.
Karimzadeh, I., Aghazadeh, M., Doroudi, T., Ganjali, M.R., Kolivand, P.H., 2017. Superparamagnetic iron oxide (Fe_3O_4) nanoparticles coated with PEG/PEI for biomedical applications: a facile and scalable preparation route based on the cathodic electrochemical deposition method. Adv. Phys. Chem. 2017, 9437487.
Karmakar, A., Zhang, Q., Zhang, Y., 2014. Neurotoxicity of nanoscale materials. J. Food Drug Anal. 22 (1), 147–160.
Kasana, R.C., Panwar, N.R., Kaul, R.K., Kumar, P., 2017. Biosynthesis and effects of copper nanoparticles on plants. Environ. Chem. Lett. 15 (2), 233–240.
Katz, L.M., Dewan, K., Bronaugh, R.L., 2015. Nanotechnology in cosmetics. Food Chem. Toxicol. 85, 127–137.
Kaur, P., Thakur, R., Chaudhury, A., 2016. Biogenesis of copper nanoparticles using peel extract of *Punica granatum* and their antimicrobial activity against opportunistic pathogens. Green Chem. Lett. Rev. 9 (1), 33–38.
Kaushal, M., Wani, S.P., 2017. Nanosensors: frontiers in precision agriculture. In: Nanotechnology. Springer, Singapore, pp. 279–291.
Kecili, R., Hussain, C.M., 2018. Mechanism of adsorption on nanomaterials. In: Nanomaterials in Chromatography. Elsevier, pp. 89–115.
Keller, A.A., Adeleye, A.S., Conway, J.R., Garner, K.L., Zhao, L., Cherr, G.N., Hong, J., Gardea-Torresdey, J.L., Godwin, H.A., Hanna, S., Ji, Z., 2017. Comparative environmental fate and toxicity of copper nanomaterials. NanoImpact 7, 28–40.
Kendall, K., Dhir, A., Du, S., 2009. A new measure of molecular attractions between nanoparticles near kT adhesion energy. Nanotechnology 20 (27), 275701.
Kermanizadeh, A., Gaiser, B.K., Johnston, H., Brown, D.M., Stone, V., 2014. Toxicological effect of engineered nanomaterials on the liver. Br. J. Pharmacol. 171 (17), 3980–3987.
Kettler, K., Veltman, K., van de Meent, D., van Wezel, A., Hendriks, A.J., 2014. Cellular uptake of nanoparticles as determined by particle properties, experimental conditions, and cell type. Environ. Toxicol. Chem. 33 (3), 481–492.
Khalil, M., Jan, B.M., Tong, C.W., Berawi, M.A., 2017. Advanced nanomaterials in oil and gas industry: design, application and challenges. Appl. Energy 191, 287–310.
Khan, R., Fulekar, M.H., 2016. Biosynthesis of titanium dioxide nanoparticles using *Bacillus amyloliquefaciens* culture and enhancement of its photocatalytic activity for the degradation of a sulfonated textile dye Reactive Red 31. J. Colloid Interface Sci. 475, 184–191.
Khan, M.M., Ansari, S.A., Pradhan, D., Ansari, M.O., Lee, J., Cho, M.H., 2014. Band gap engineered TiO_2 nanoparticles for visible light induced photoelectrochemical and photocatalytic studies. J. Mater. Chem. A 2 (3), 637–644.
Khandel, P., Shahi, S.K., 2018. Mycogenic nanoparticles and their bio-prospective applications: current status and future challenges. J. Nanostructure Chem. 8 (4), 369–391.
Kim, M.H., Lim, B., Lee, E.P., Xia, Y., 2008. Polyol synthesis of Cu_2O nanoparticles: use of chloride to promote the formation of a cubic morphology. J. Mater. Chem. 18 (34), 4069–4073.
Kim, S., Lee, S., Lee, I., 2012. Alteration of phytotoxicity and oxidant stress potential by metal oxide nanoparticles in *Cucumis sativus*. Water Air Soil Pollut. 223 (5), 2799–2806.
Kirthi, A.V., Rahuman, A.A., Rajakumar, G., Marimuthu, S., Santhoshkumar, T., Jayaseelan, C., Elango, G., Zahir, A.A., Kamaraj, C., Bagavan, A., 2011. Biosynthesis of titanium dioxide nanoparticles using bacterium *Bacillus subtilis*. Mater. Lett. 65 (17–18), 2745–2747.
Klaper, R., Arndt, D., Bozich, J., Dominguez, G., 2014. Molecular interactions of nanomaterials and organisms: defining biomarkers for toxicity and high-throughput screening using traditional and next-generation sequencing approaches. Analyst 139 (5), 882–895.
Kononenko, V., Repar, N., Marušič, N., Drašler, B., Romih, T., Hočevar, S., Drobne, D., 2017. Comparative in vitro genotoxicity study of ZnO nanoparticles, ZnO macroparticles and $ZnCl_2$ to MDCK kidney cells: size matters. Toxicol. in Vitro 40, 256–263.

Korotcenkov, G., Cho, B.K., 2017. Metal oxide composites in conductometric gas sensors: achievements and challenges. Sensors Actuators B Chem. 244, 182–210.

Košťáková, E., Zemanová, E.V.A., Mikeš, P., Soukupová, J., Matheisová, H., Klouda, K., 2012. Electrospinning and electrospraying of polymer solutions with spherical fullerenes. In: Proceedings of the NANOCON.

Kronberg, B., Holmberg, K., Lindman, B., 2014. Types of surfactants, their synthesis, and applications. In: Surface Chemistry of Surfactants and Polymers.pp. 1–47.

Kumar, A., Dhawan, A., 2013. Genotoxic and carcinogenic potential of engineered nanoparticles: an update. Arch. Toxicol. 87 (11), 1883–1900.

Kumar, N., Shah, V., Walker, V.K., 2011. Perturbation of an arctic soil microbial community by metal nanoparticles. J. Hazard. Mater. 190 (1–3), 816–822.

Kumar, N., Shah, V., Walker, V.K., 2012. Influence of a nanoparticle mixture on an arctic soil community. Environ. Toxicol. Chem. 31 (1), 131–135.

Kumar, A., Sharma, V., Dhawan, A., 2013. Methods for detection of oxidative stress and genotoxicity of engineered nanoparticles. In: Oxidative Stress and Nanotechnology. Humana Press, Totowa, NJ, pp. 231–246.

Kumar, P., Kumar, A., Fernandes, T., Ayoko, G.A., 2014. Nanomaterials and the environment. Journal of Nanomaterials. 2014. https://doi.org/10.1155/2014/528606.

Kumar, R., Chauhan, M., Sharma, N., Chaudhary, G.R., 2018. Toxic effects of nanomaterials on environment. In: Environmental Toxicity of Nanomaterials. CRC Press, pp. 1–20.

Kumar, G.D., Raja, K., Natarajan, N., Govindaraju, K., Subramanian, K.S., 2020. Invigouration treatment of metal and metal oxide nanoparticles for improving the seed quality of aged chilli seeds (*Capsicum annum* L.). Mater. Chem. Phys. 242, 122492.

Kundu, D., Hazra, C., Chatterjee, A., Chaudhari, A., Mishra, S., 2014. Extracellular biosynthesis of zinc oxide nanoparticles using Rhodococcuspyridinivorans NT2: multifunctional textile finishing, biosafety evaluation and in vitro drug delivery in colon carcinoma. J. Photochem. Photobiol. B Biol. 140, 194–204.

Kwon, J.Y., Koedrith, P., Seo, Y.R., 2014. Current investigations into the genotoxicity of zinc oxide and silica nanoparticles in mammalian models in vitro and in vivo: carcinogenic/genotoxic potential, relevant mechanisms and biomarkers, artifacts, and limitations. Int. J. Nanomedicine 9 (Suppl 2), 271.

Kyzas, G.Z., Matis, K.A., 2015. Nanoadsorbents for pollutants removal: a review. J. Mol. Liq. 203, 159–168.

Lagaron, J.M., Lopez-Rubio, A., 2011. Nanotechnology for bioplastics: opportunities, challenges and strategies. Trends Food Sci. Technol. 22 (11), 611–617.

Lai, R.W., Yeung, K.W., Yung, M.M., Djurišić, A.B., Giesy, J.P., Leung, K.M., 2018a. Regulation of engineered nanomaterials: current challenges, insights and future directions. Environ. Sci. Pollut. Res. 25 (4), 3060–3077.

Lai, X., Zhao, H., Zhang, Y., Guo, K., Xu, Y., Chen, S., Zhang, J., 2018b. Intranasal delivery of copper oxide nanoparticles induces pulmonary toxicity and fibrosis in C57BL/6 mice. Sci. Rep. 8 (1), 1–12.

Laine, R.M., Marchal, J.C., Sun, H.P., Pan, X.Q., 2006. Nano-α-Al_2O_3 by liquid-feed flame spray pyrolysis. Nat. Mater. 5 (9), 710–712.

Laurent, S., Dutz, S., Häfeli, U.O., Mahmoudi, M., 2011. Magnetic fluid hyperthermia: focus on superparamagnetic iron oxide nanoparticles. Adv. Colloid Interf. Sci. 166 (1–2), 8–23.

Lavik, E.B., Kosacki, I., Tuller, H.L., Chiang, Y.M., Ying, J.Y., 1997. Nonstoichiometry and electrical conductivity of nanocrystalline CeO_{2-x}. J. Electroceram. 1 (1), 7–14.

Lazareva, A., Keller, A.A., 2014. Estimating potential life cycle releases of engineered nanomaterials from wastewater treatment plants. ACS Sustain. Chem. Eng. 2 (7), 1656–1665.

Le Van, N., Rui, Y., Cao, W., Shang, J., Liu, S., Nguyen Quang, T., Liu, L., 2016. Toxicity and bio-effects of CuO nanoparticles on transgenic Ipt-cotton. J. Plant Interact. 11 (1), 108–116.

Lee, K.J., Nallathamby, P.D., Browning, L.M., Osgood, C.J., Xu, X.H.N., 2007. In vivo imaging of transport and biocompatibility of single silver nanoparticles in early development of zebrafish embryos. ACS Nano 1 (2), 133–143.

Lee, W.M., An, Y.J., Yoon, H., Kweon, H.S., 2008a. Toxicity and bioavailability of copper nanoparticles to the terrestrial plants mung bean (*Phaseolus radiatus*) and wheat (*Triticum aestivum*): plant agar test for water-insoluble nanoparticles. Environ. Toxicol. Chem. 27 (9), 1915–1921.

Lee, S., Jeong, S., Kim, D., Hwang, S., Jeon, M., Moon, J., 2008b. ZnO nanoparticles with controlled shapes and sizes prepared using a simple polyol synthesis. Superlattice. Microst. 43 (4), 330–339.
Lee, J., Mahendra, S., Alvarez, P.J., 2010a. Nanomaterials in the construction industry: a review of their applications and environmental health and safety considerations. ACS Nano 4 (7), 3580–3590.
Lee, C.W., Mahendra, S., Zodrow, K., Li, D., Tsai, Y.C., Braam, J., Alvarez, P.J., 2010b. Developmental phytotoxicity of metal oxide nanoparticles to *Arabidopsis thaliana*. Environ. Toxicol. Chem. 29 (3), 669–675.
Lee, K.M., Lai, C.W., Ngai, K.S., Juan, J.C., 2016. Recent developments of zinc oxide based photocatalyst in water treatment technology: a review. Water Res. 88, 428–448.
Lei, R., Wu, C., Yang, B., Ma, H., Shi, C., Wang, Q., Wang, Q., Yuan, Y., Liao, M., 2008. Integrated metabolomic analysis of the nano-sized copper particle-induced hepatotoxicity and nephrotoxicity in rats: a rapid in vivo screening method for nanotoxicity. Toxicol. Appl. Pharmacol. 232 (2), 292–301.
Lei, Y., Chen, F., Luo, Y., Zhang, L., 2014. Three-dimensional magnetic graphene oxide foam/Fe_3O_4 nanocomposite as an efficient absorbent for Cr (VI) removal. J. Mater. Sci. 49 (12), 4236–4245.
Lekamge, S., Ball, A.S., Shukla, R., Nugegoda, D., 2020. The toxicity of nanoparticles to organisms in freshwater. Rev. Environ. Contam. Toxicol. 248, 1–80.
Lepeshev, A.A., Karpov, I.V., Ushakov, A.V., Fedorov, L.Y., Shaihadinov, A.A., 2016. Synthesis of nanosized titanium oxide and nitride through vacuum arc plasma expansion technique. Int. J. Nanosci. 15 (01n02), 1550027.
Lewicka, Z.A., Benedetto, A.F., Benoit, D.N., William, W.Y., Fortner, J.D., Colvin, V.L., 2011. The structure, composition, and dimensions of TiO_2 and ZnO nanomaterials in commercial sunscreens. J. Nanopart. Res. 13 (9), 3607.
Lewinski, N.A., Zhu, H., Jo, H.J., Pham, D., Kamath, R.R., Ouyang, C.R., Vulpe, C.D., Colvin, V.L., Drezek, R.A., 2010. Quantification of water solubilized CdSe/ZnS quantum dots in Daphnia magna. Environ. Sci. Technol. 44 (5), 1841–1846.
Li, G.Y., Osborne, N.N., 2008. Oxidative-induced apoptosis to an immortalized ganglion cell line is caspase independent but involves the activation of poly (ADP-ribose) polymerase and apoptosis-inducing factor. Brain Res. 1188, 35–43.
Li, W.M., Wang, W.X., 2013. Distinct biokinetic behavior of ZnO nanoparticles in *Daphnia magna* quantified by synthesizing 65Zn tracer. Water Res. 47 (2), 895–902.
Li, J.W., Liu, X.J., Yang, L.W., Zhou, Z.F., Xie, G.F., Pan, Y., Wang, X.H., Zhou, J., Li, L.T., Pan, L., Sun, Z., 2009. Photoluminescence and photoabsorption blueshift of nanostructured ZnO: skin-depth quantum trapping and electron-phonon coupling. Appl. Phys. Lett. 95 (3), 031906.
Li, L., Jiang, W., Luo, K., Song, H., Lan, F., Wu, Y., Gu, Z., 2013. Superparamagnetic iron oxide nanoparticles as MRI contrast agents for non-invasive stem cell labeling and tracking. Theranostics 3 (8), 595.
Liang, D., Lu, Z., Yang, H., Gao, J., Chen, R., 2016. Novel asymmetric wettable AgNPs/chitosan wound dressing: in vitro and in vivo evaluation. ACS Appl. Mater. Interfaces 8 (6), 3958–3968.
Lin, D., Xing, B., 2007. Phytotoxicity of nanoparticles: inhibition of seed germination and root growth. Environ. Pollut. 150 (2), 243–250.
Liu, W.T., 2006. Nanoparticles and their biological and environmental applications. J. Biosci. Bioeng. 102 (1), 1–7.
Liu, H.H., Cohen, Y., 2014. Multimedia environmental distribution of engineered nanomaterials. Environmental science & technology 48 (6), 3281–3292.
Liu, H.H., Cohen, Y., 2014. Multimedia environmental distribution of engineered nanomaterials. Environ. Sci. Technol. 48 (6), 3281–3292.
Liu, J., Hurt, R.H., 2010. Ion release kinetics and particle persistence in aqueous nano-silver colloids. Environ. Sci. Technol. 44 (6), 2169–2175.
Liu, Y., Li, W., Lao, F., Liu, Y., Wang, L., Bai, R., Zhao, Y., Chen, C., 2011a. Intracellular dynamics of cationic and anionic polystyrene nanoparticles without direct interaction with mitotic spindle and chromosomes. Biomaterials 32 (32), 8291–8303.
Liu, R., Rallo, R., George, S., Ji, Z., Nair, S., Nel, A.E., Cohen, Y., 2011b. Classification NanoSAR development for cytotoxicity of metal oxide nanoparticles. Small 7 (8), 1118–1126.

Liu, X.D., Chen, H., Liu, S.S., Ye, L.Q., Li, Y.P., 2015. Hydrothermal synthesis of superparamagnetic Fe_3O_4 nanoparticles with ionic liquids as stabilizer. Mater. Res. Bull. 62, 217–221.

Liu, R., Zhang, H., Lal, R., 2016. Effects of stabilized nanoparticles of copper, zinc, manganese, and iron oxides in low concentrations on lettuce (*Lactuca sativa*) seed germination: nanotoxicants or nanonutrients? Water Air Soil Pollut. 227 (1), 42.

Liu, X., Sui, B., Sun, J., 2017. Size-and shape-dependent effects of titanium dioxide nanoparticles on the permeabilization of the blood–brain barrier. J. Mater. Chem. B 5 (48), 9558–9570.

Llorens, A., Lloret, E., Picouet, P.A., Trbojevich, R., Fernandez, A., 2012. Metallic-based micro and nano-composites in food contact materials and active food packaging. Trends Food Sci. Technol. 24 (1), 19–29.

Lodhia, J., Mandarano, G., Ferris, N.J., Eu, P., Cowell, S.F., 2010. Development and use of iron oxide nanoparticles (part 1): synthesis of iron oxide nanoparticles for MRI. Biomed. Imaging Intervention J. 6 (2), e12.

López-Muñoz, D., Ochoa-Zapater, M.A., Torreblanca, A., Garcerá, M.D., 2019. Evaluation of the effects of titanium dioxide and aluminum oxide nanoparticles through tarsal contact exposure in the model insect *Oncopeltus fasciatus*. Sci. Total Environ. 666, 759–765.

Lozano, T., Rey, M., Rojas, E., Moya, S., Fleddermann, J., Estrela-Lopis, I., Donath, E., Wang, B., Mao, Z., Gao, C., González-Fernández, Á., 2011. Cytotoxicity effects of metal oxide nanoparticles in human tumor cell lines. J. Phys. Conf. Ser. 304 (1), 012046 IOP Publishing.

Lu, P.J., Huang, S.C., Chen, Y.P., Chiueh, L.C., Shih, D.Y.C., 2015. Analysis of titanium dioxide and zinc oxide nanoparticles in cosmetics. J. Food Drug Anal. 23 (3), 587–594.

Lu, A.X., McEntee, M., Browe, M.A., Hall, M.G., DeCoste, J.B., Peterson, G.W., 2017. MOFabric: electrospun nanofiber mats from PVDF/UiO-66-NH2 for chemical protection and decontamination. ACS Appl. Mater. Interfaces 9 (15), 13632–13636.

Luoma, S.N., Khan, F.R., Croteau, M.N., 2014. Bioavailability and bioaccumulation of metal-based engineered nanomaterials in aquatic environments: concepts and processes. Front. Nanosci. 7, 157–193 Elsevier.

Ma, F.X., Wang, P.P., Xu, C.Y., Yu, J., Fang, H.T., Zhen, L., 2014. Synthesis of self-stacked $CuFe_2O_4$–Fe_2O_3 porous nanosheets as a high performance Li-ion battery anode. J. Mater. Chem. A 2 (45), 19330–19337.

Madan, H.R., Sharma, S.C., Suresh, D., Vidya, Y.S., Nagabhushana, H., Rajanaik, H., Anantharaju, K.S., Prashantha, S.C., Maiya, P.S., 2016. Facile green fabrication of nanostructure ZnO plates, bullets, flower, prismatic tip, closed pine cone: their antibacterial, antioxidant, photoluminescent and photocatalytic properties. Spectrochim. Acta A Mol. Biomol. Spectrosc. 152, 404–416.

Magaye, R., Zhao, J., 2012. Recent progress in studies of metallic nickel and nickel-based nanoparticles' genotoxicity and carcinogenicity. Environ. Toxicol. Pharmacol. 34 (3), 644–650.

Magaye, R., Zhao, J., Bowman, L., Ding, M.I.N., 2012. Genotoxicity and carcinogenicity of cobalt-, nickel and copper based nanoparticles. Exp. Ther. Med. 4 (4), 551–561.

Magder, S., 2006. Reactive oxygen species: toxic molecules or spark of life? Crit. Care 10 (1), 208.

Mahdavi, S., Afkhami, A., Jalali, M., 2015. Reducing leachability and bioavailability of soil heavy metals using modified and bare Al_2O_3 and ZnO nanoparticles. Environ. Earth Sci. 73 (8), 4347–4371.

Mahmoudi, M., Sant, S., Wang, B., Laurent, S., Sen, T., 2011. Superparamagnetic iron oxide nanoparticles (SPIONs): development, surface modification and applications in chemotherapy. Adv. Drug Deliv. Rev. 63 (1–2), 24–46.

Malaikozhundan, B., Vinodhini, J., 2018. Nanopesticidal effects of *Pongamia pinnata* leaf extract coated zinc oxide nanoparticle against the Pulse beetle, *Callosobruchus maculatus*. Mater. Today Commun. 14, 106–115.

Malik, M.A., Wani, M.Y., Hashim, M.A., 2012. Microemulsion method: a novel route to synthesize organic and inorganic nanomaterials: 1st nano update. Arab. J. Chem. 5 (4), 397–417.

Marchand, K.E., Tarret, M., Lechaire, J.P., Normand, L., Kasztelan, S., Cseri, T., 2003. Investigation of AOT-based microemulsions for the controlled synthesis of MoS_x nanoparticles: an electron microscopy study. Colloids Surf. A Physicochem. Eng. Asp. 214 (1–3), 239–248.

Markus, A.A., Parsons, J.R., Roex, E.W., de Voogt, P., Laane, R.W., 2016. Modelling the release, transport and fate of engineered nanoparticles in the aquatic environment–a review. In: Reviews of Environmental Contamination and Toxicology. vol. 243. Springer, Cham, pp. 53–87.

Martínez-Fernández, D., Barroso, D., Komárek, M., 2016. Root water transport of *Helianthus annuus* L. under iron oxide nanoparticle exposure. Environ. Sci. Pollut. Res. 23 (2), 1732–1741.
Maryanti, E., Damayanti, D., Gustian, I., 2014. Synthesis of ZnO nanoparticles by hydrothermal method in aqueous rinds extracts of Sapindusrarak DC. Mater. Lett. 118, 96–98.
McGillicuddy, E., Murray, I., Kavanagh, S., Morrison, L., Fogarty, A., Cormican, M., Dockery, P., Prendergast, M., Rowan, N., Morris, D., 2017. Silver nanoparticles in the environment: sources, detection and ecotoxicology. Sci. Total Environ. 575, 231–246.
McNeil, S.E., 2005. Nanotechnology for the biologist. J. Leukoc. Biol. 78 (3), 585–594.
Mehta, C.M., Srivastava, R., Arora, S., Sharma, A.K., 2016. Impact assessment of silver nanoparticles on plant growth and soil bacterial diversity. 3 Biotech 6 (2), 254.
Mihranyan, A., Ferraz, N., Strømme, M., 2012. Current status and future prospects of nanotechnology in cosmetics. Prog. Mater. Sci. 57 (5), 875–910.
Miranda, R.R., Bezerra Jr., A.G., Ribeiro, C.A.O., Randi, M.A.F., Voigt, C.L., Skytte, L., Rasmussen, K.L., Kjeldsen, F., Neto, F.F., 2017. Toxicological interactions of silver nanoparticles and non-essential metals in human hepatocarcinoma cell line. Toxicol. in Vitro 40, 134–143.
Mirza, S., Ahmad, M.S., Shah, M.I.A., Ateeq, M., 2020. Magnetic nanoparticles: drug delivery and bioimaging applications. In: Metal Nanoparticles for Drug Delivery and Diagnostic Applications. Elsevier, pp. 189–213.
Mirzaei, H., Darroudi, M., 2017. Zinc oxide nanoparticles: biological synthesis and biomedical applications. Ceram. Int. 43 (1), 907–914.
Mirzaei, A., Leonardi, S.G., Neri, G., 2016. Detection of hazardous volatile organic compounds (VOCs) by metal oxide nanostructures-based gas sensors: a review. Ceram. Int. 42 (14), 15119–15141.
Miseljic, M., Olsen, S.I., 2014. Life-cycle assessment of engineered nanomaterials: a literature review of assessment status. J. Nanopart. Res. 16 (6), 2427.
Mishra, A., Fischer, M.K., Bäuerle, P., 2009. Metal-free organic dyes for dye-sensitized solar cells: from structure: property relationships to design rules. Angew. Chem. Int. Ed. 48 (14), 2474–2499.
Mishra, P.K., Mishra, H., Ekielski, A., Talegaonkar, S., Vaidya, B., 2017. Zinc oxide nanoparticles: a promising nanomaterial for biomedical applications. Drug Discov. Today 22 (12), 1825–1834.
Misra, S.K., Nuseibeh, S., Dybowska, A., Berhanu, D., Tetley, T.D., Valsami-Jones, E., 2014. Comparative study using spheres, rods and spindle-shaped nanoplatelets on dispersion stability, dissolution and toxicity of CuO nanomaterials. Nanotoxicology 8 (4), 422–432.
Mistrik, J., Kasap, S., Ruda, H.E., Koughia, C., Singh, J., 2017. Optical properties of electronic materials: fundamentals and characterization. In: Springer Handbook of Electronic and Photonic Materials. Springer, Cham, p. 1.
Mittal, A.K., Chisti, Y., Banerjee, U.C., 2013. Synthesis of metallic nanoparticles using plant extracts. Biotechnol. Adv. 31 (2), 346–356.
Mo, Y., Lim, L.Y., 2005. Paclitaxel-loaded PLGA nanoparticles: potentiation of anticancer activity by surface conjugation with wheat germ agglutinin. J. Control. Release 108 (2–3), 244–262.
Molinari, B.L., Tasat, D.R., Palmieri, M.A., O'Connor, S.E., Cabrini, R.L., 2003. Cell-based quantitative evaluation of the MTT assay. Anal. Quant. Cytol. Histol. 25 (5), 254–262.
Monteiro-Riviere, N.A., Nemanich, R.J., Inman, A.O., Wang, Y.Y., Riviere, J.E., 2005. Multi-walled carbon nanotube interactions with human epidermal keratinocytes. Toxicol. Lett. 155 (3), 377–384.
Montes, A., Bisson, M.A., Gardella Jr., J.A., Aga, D.S., 2017. Uptake and transformations of engineered nanomaterials: critical responses observed in terrestrial plants and the model plant *Arabidopsis thaliana*. Sci. Total Environ. 607, 1497–1516.
Montini, T., Melchionna, M., Monai, M., Fornasiero, P., 2016. Fundamentals and catalytic applications of CeO_2-based materials. Chem. Rev. 116 (10), 5987–6041.
Moon, Y.S., Park, E.S., Kim, T.O., Lee, H.S., Lee, S.E., 2014. SELDI-TOF MS-based discovery of a biomarker in *Cucumis sativus* seeds exposed to CuO nanoparticles. Environ. Toxicol. Pharmacol. 38 (3), 922–931.
Morganti, P., 2010. Use and potential of nanotechnology in cosmetic dermatology. Clin. Cosmet. Investig. Dermatol. 3, 5.

Mourdikoudis, S., Pallares, R.M., Thanh, N.T., 2018. Characterization techniques for nanoparticles: comparison and complementarity upon studying nanoparticle properties. Nanoscale 10 (27), 12871–12934.
Mousavi, S.R., Rezaei, M., 2011. Nanotechnology in agriculture and food production. J. Appl. Environ. Biol. Sci. 1 (10), 414–419.
Mu, L., Sprando, R.L., 2010. Application of nanotechnology in cosmetics. Pharm. Res. 27 (8), 1746–1749.
Mudshinge, S.R., Deore, A.B., Patil, S., Bhalgat, C.M., 2011. Nanoparticles: emerging carriers for drug delivery. Saudi Pharm. J. 19 (3), 129–141.
Mukherjee, A., Peralta-Videa, J.R., Bandyopadhyay, S., Rico, C.M., Zhao, L., Gardea-Mukherjee, A., Peralta-Videa, J.R., Bandyopadhyay, S., Rico, C.M., Zhao, L., Gardea-Torresdey, J.L., 2014. Physiological effects of nanoparticulate ZnO in green peas (*Pisum sativum* L.) cultivated in soil. Metallomics 6 (1), 132–138.
Mukhopadhyay, S.S., 2014. Nanotechnology in agriculture: prospects and constraints. Nanotechnol. Sci. Appl. 7, 63–71.
Nair, P.M.G., Chung, I.M., 2014. A mechanistic study on the toxic effect of copper oxide nanoparticles in soybean (*Glycine max* L.) root development and lignification of root cells. Biol. Trace Elem. Res. 162 (1–3), 342–352.
Nair, P.M.G., Chung, I.M., 2015. Study on the correlation between copper oxide nanoparticles induced growth suppression and enhanced lignification in Indian mustard (*Brassica juncea* L.). Ecotoxicol. Environ. Saf. 113, 302–313.
Nair, P.M.G., Kim, S.H., Chung, I.M., 2014. Copper oxide nanoparticle toxicity in mung bean (*Vigna radiata* L.) seedlings: physiological and molecular level responses of in vitro grown plants. Acta Physiol. Plant. 36 (11), 2947–2958.
Nasrollahzadeh, M., Sajadi, S.M., 2015. Synthesis and characterization of titanium dioxide nanoparticles using *Euphorbia heteradena Jaub* root extract and evaluation of their stability. Ceram. Int. 41 (10), 14435–14439.
Natarajan, M., Mohan, S., Martinez, B.R., Meltz, M.L., Herman, T.S., 2000. Antioxidant compounds interfere with the 3. Cancer Detect. Prev. 24 (5), 405–414.
Noubactep, C., Caré, S., Crane, R., 2012. Nanoscale metallic iron for environmental remediation: prospects and limitations. Water Air Soil Pollut. 223 (3), 1363–1382.
Oberdörster, G., Oberdörster, E., Oberdörster, J., 2005. Nanotoxicology: an emerging discipline evolving from studies of ultrafine particles. Environ. Health Perspect. 113 (7), 823–839.
Oberdörster, G., Stone, V., Donaldson, K., 2007. Toxicology of nanoparticles: a historical perspective. Nanotoxicology 1 (1), 2–25.
OECD, 2016. Physical-Chemical Parameters: Measurements and Methods Relevant for the Regulation of Nanomaterials. p. JT03389225.
Olchowik, J., Bzdyk, R.M., Studnicki, M., Bederska-Błaszczyk, M., Urban, A., Aleksandrowicz Trzcińska, M., 2017. The effect of silver and copper nanoparticles on the condition of english oak (*Quercus robur* L.) seedlings in a container nursery experiment. Forests 8 (9), 310.
Palchoudhury, S., Jungjohann, K.L., Weerasena, L., Arabshahi, A., Gharge, U., Albattah, A., Miller, J., Patel, K., Holler, R.A., 2018. Enhanced legume root growth with pre-soaking in α-Fe_2O_3 nanoparticle fertilizer. RSC Adv. 8 (43), 24075–24083.
Pallares, R.M., Su, X., Lim, S.H., Thanh, N.T., 2016. Fine-tuning of gold nanorod dimensions and plasmonic properties using the Hofmeister effects. J. Mater. Chem. C 4 (1), 53–61.
Pan, Y., Neuss, S., Leifert, A., Fischler, M., Wen, F., Simon, U., Schmid, G., Brandau, W., Jahnen-Dechent, W., 2007. Size-dependent cytotoxicity of gold nanoparticles. Small 3 (11), 1941–1949.
Pan, Y., Li, T., Cheng, J., Telesca, D., Zink, J.I., Jiang, J., 2016. Nano-QSAR modeling for predicting the cytotoxicity of metal oxide nanoparticles using novel descriptors. RSC Adv. 6 (31), 25766–25775.
Pandey, S., Giri, K., Kumar, R., Mishra, G., Rishi, R.R., 2018. Nanopesticides: opportunities in crop protection and associated environmental risks. Proc. Natl. Acad. Sci. India Sect. B. Biol. Sci. 88 (4), 1287–1308.
Peijnenburg, W., Praetorius, A., Scott-Fordsmand, J., Cornelis, G., 2016. Fate assessment of engineered nanoparticles in solids dominated media—current insights and the way forward. Environ. Pollut. 218, 1365–1369.

Pemartin-Biernath, K., Vela-González, A.V., Moreno-Trejo, M.B., Leyva-Porras, C., Castañeda-Reyna, I.E., Juárez-Ramírez, I., Solans, C., Sánchez-Domínguez, M., 2016. Synthesis of mixed Cu/Ce oxide nanoparticles by the oil-in-water microemulsion reaction method. Materials 9 (6), 480.

Peng, C., Duan, D., Xu, C., Chen, Y., Sun, L., Zhang, H., Yuan, X., Zheng, L., Yang, Y., Yang, J., Zhen, X., 2015. Translocation and biotransformation of CuO nanoparticles in rice (*Oryza sativa* L.) plants. Environ. Pollut. 197, 99–107.

Peters, A., Veronesi, B., Calderón-Garcidueñas, L., Gehr, P., Chen, L.C., Geiser, M., Reed, W., Rothen-Rutishauser, B., Schürch, S., Schulz, H., 2006. Translocation and potential neurological effects of fine and ultrafine particles a critical update. Part. Fibre Toxicol. 3 (1), 13.

Petty, M.C., 2019. Organic and Molecular Electronics: From Principles to Practice. John Wiley & Sons.

Piccinno, F., Gottschalk, F., Seeger, S., Nowack, B., 2012. Industrial production quantities and uses of ten engineered nanomaterials in Europe and the world. J. Nanopart. Res. 14 (9), 1109.

Pietroiusti, A., Magrini, A., 2015. Engineered nanoparticles at the workplace: current knowledge about workers' risk. Occup. Med. 65 (2), 171–173.

Pokhrel, S., Birkenstock, J., Schowalter, M., Rosenauer, A., Mädler, L., 2010. Growth of ultrafine single crystalline WO3 nanoparticles using flame spray pyrolysis. Cryst. Growth Des. 10 (2), 632–639.

Polarz, S., Roy, A., Merz, M., Halm, S., Schröder, D., Schneider, L., Bacher, G., Kruis, F.E., Driess, M., 2005. Chemical vapor synthesis of size-selected zinc oxide nanoparticles. Small 1 (5), 540–552.

Popovici, E., Dumitrache, F., Morjan, I., Alexandrescu, R., Ciupina, V., Prodan, G., Vekas, L., Bica, D., Marinica, O., Vasile, E., 2007. Iron/iron oxides core–shell nanoparticles by laser pyrolysis: structural characterization and enhanced particle dispersion. Appl. Surf. Sci. 254 (4), 1048–1052.

Prasad, R., Bhattacharyya, A., Nguyen, Q.D., 2017. Nanotechnology in sustainable agriculture: recent developments, challenges, and perspectives. Front. Microbiol. 8, 1014.

Priester, J.H., Moritz, S.C., Espinosa, K., Ge, Y., Wang, Y., Nisbet, R.M., Schimel, J.P., Goggi, A.S., Gardea-Torresdey, J.L., Holden, P.A., 2017. Damage assessment for soybean cultivated in soil with either CeO_2 or ZnO manufactured nanomaterials. Sci. Total Environ. 579, 1756–1768.

Pruna, A., 2019. Nanocoatings for protection against steel corrosion. In: Nanotechnology in Eco-efficient Construction. Woodhead Publishing, pp. 337–359.

Punshon, G., Vara, D.S., Sales, K.M., Kidane, A.G., Salacinski, H.J., Seifalian, A.M., 2005. Interactions between endothelial cells and a poly (carbonate-silsesquioxane-bridge-urea) urethane. Biomaterials 26 (32), 6271–6279.

Qiao, R., Yang, C., Gao, M., 2009. Superparamagnetic iron oxide nanoparticles: from preparations to in vivo MRI applications. J. Mater. Chem. 19 (35), 6274–6293.

Raghunath, A., Perumal, E., 2017. Metal oxide nanoparticles as antimicrobial agents: a promise for the future. Int. J. Antimicrob. Agents 49 (2), 137–152.

Rahman, M.M., Alam, M.M., Asiri, A.M., Islam, M.A., 2017. Fabrication of selective chemical sensor with ternary ZnO/SnO2/Yb_2O_3 nanoparticles. Talanta 170, 215–223.

Rai, M., Kon, K., Ingle, A., Duran, N., Galdiero, S., Galdiero, M., 2014. Broad-spectrum bioactivities of silver nanoparticles: the emerging trends and future prospects. Appl. Microbiol. Biotechnol. 98 (5), 1951–1961.

Rajakumar, G., Rahuman, A.A., Roopan, S.M., Khanna, V.G., Elango, G., Kamaraj, C., Zahir, A.A., Velayutham, K., 2012. Fungus-mediated biosynthesis and characterization of TiO_2 nanoparticles and their activity against pathogenic bacteria. Spectrochim. Acta A Mol. Biomol. Spectrosc. 91, 23–29.

Rajput, V.D., Chen, Y., Ayup, M., 2015. Effects of high salinity on physiological and anatomical indices in the early stages of *Populus euphratica* growth. Russ. J. Plant Physiol. 62 (2), 229–236.

Rajput, V.D., Tstitsuashvili, V.S., Sushkova, S.N., Nevidomskaya, D.G., 2017, March. Effects of ZnO and CuO nanoparticles on soil, plant and microbial community. In: Материалы Международной научной конференции XX Докучаев-ские молодежные чтения «Почва и устойчивое развитие государства».p. 12.

Rajput, V.D., Minkina, T.M., Behal, A., Sushkova, S.N., Mandzhieva, S., Singh, R., Gorovtsov, A., Tsitsuashvili, V.S., Purvis, W.O., Ghazaryan, K.A., Movsesyan, H.S., 2018a. Effects of zinc-oxide nanoparticles on soil, plants, animals and soil organisms: a review. Environ. Nanotechnol. Monit. Manag. 9, 76–84.

Rajput, V., Minkina, T., Fedorenko, A., Sushkova, S., Mandzhieva, S., Lysenko, V., Duplii, N., Fedorenko, G., Dvadnenko, K., Ghazaryan, K., 2018b. Toxicity of copper oxide nanoparticles on spring barley (*Hordeum sativum* distichum). Sci. Total Environ. 645, 1103–1113.

Rajput, V.D., Minkina, T., Sushkova, S., Tsitsuashvili, V., Mandzhieva, S., Gorovtsov, A., Nevidomskyaya, D., Gromakova, N., 2018c. Effect of nanoparticles on crops and soil microbial communities. J. Soils Sediments 18 (6), 2179–2187.

Rajput, V.D., Minkina, T., Suskova, S., Mandzhieva, S., Tsitsuashvili, V., Chapligin, V., Fedorenko, A., 2018d. Effects of copper nanoparticles (CuO NPs) on crop plants: a mini review. BioNanoScience 8 (1), 36–42.

Rajput, V., Minkina, T., Sushkova, S., Behal, A., Maksimov, A., Blicharska, E., Ghazaryan, K., Movsesyan, H., Barsova, N., 2020. ZnO and CuO nanoparticles: a threat to soil organisms, plants, and human health. Environ. Geochem. Health 42 (1), 147–158.

Raliya, R., Tarafdar, J.C., 2013. ZnO nanoparticle biosynthesis and its effect on phosphorous-mobilizing enzyme secretion and gum contents in Clusterbean (*Cyamopsis tetragonoloba* L.). Agric. Res. 2 (1), 48–57.

Ramallo-Lopez, J.M., Giovanetti, L., Craievich, A.F., Vicentin, F.C., Marín-Almazo, M., José-Yacaman, M., Requejo, F.G., 2007. XAFS, SAXS and HREM characterization of Pd nanoparticles capped with n-alkyl thiol molecules. Phys. B Condens. Matter 389 (1), 150–154.

Ramskov, T., Croteau, M.N., Forbes, V.E., Selck, H., 2015. Biokinetics of different-shaped copper oxide nanoparticles in the freshwater gastropod, *Potamopyrgus antipodarum*. Aquat. Toxicol. 163, 71–80.

Rao, S., Shekhawat, G.S., 2016. Phytotoxicity and oxidative stress perspective of two selected nanoparticles in *Brassica juncea*. 3 Biotech 6 (2), 244.

Rasmussen, J.W., Martinez, E., Louka, P., Wingett, D.G., 2010. Zinc oxide nanoparticles for selective destruction of tumor cells and potential for drug delivery applications. Expert Opin. Drug Deliv. 7 (9), 1063–1077.

Reddy, B.M., Kumar, T.V., Durgasri, N., 2013. New developments in ceria-based mixed oxide synthesis and reactivity in combustion and oxidation reactions. In: Catalysis by Ceria and Related Materials.vol. 12. p. 397.

Rico, C.M., Majumdar, S., Duarte-Gardea, M., Peralta-Videa, J.R., Gardea-Torresdey, J.L., 2011. Interaction of nanoparticles with edible plants and their possible implications in the food chain. J. Agric. Food Chem. 59 (8), 3485–3498.

Rico, C.M., Hong, J., Morales, M.I., Zhao, L., Barrios, A.C., Zhang, J.Y., Peralta-Videa, J.R., Gardea-Torresdey, J.L., 2013. Effect of cerium oxide nanoparticles on rice: a study involving the antioxidant defense system and in vivo fluorescence imaging. Environ. Sci. Technol. 47 (11), 5635–5642.

Rico, C.M., Lee, S.C., Rubenecia, R., Mukherjee, A., Hong, J., Peralta-Videa, J.R., Gardea-Torresdey, J.L., 2014. Cerium oxide nanoparticles impact yield and modify nutritional parameters in wheat (*Triticum aestivum* L.). J. Agric. Food Chem. 62 (40), 9669–9675.

Rizwan, M., Ali, S., Qayyum, M.F., Ok, Y.S., Adrees, M., Ibrahim, M., Zia-ur-Rehman, M., Farid, M., Abbas, F., 2017. Effect of metal and metal oxide nanoparticles on growth and physiology of globally important food crops: a critical review. J. Hazard. Mater. 322, 2–16.

Rizwan, M., Ali, S., Ali, B., Adrees, M., Arshad, M., Hussain, A., Zia ur Rehman, M., Waris, A.A., 2019. Zinc and iron oxide nanoparticles improved the plant growth and reduced the oxidative stress and cadmium concentration in wheat. Chemosphere 214, 269–277.

Robinson, R.L.M., Cronin, M.T., Richarz, A.N., Rallo, R., 2015. An ISA-TAB-nano based data collection framework to support data-driven modelling of nanotoxicology. Beilstein J. Nanotechnol. 6 (1), 1978–1999.

Rocha, T.L., Gomes, T., Durigon, E.G., Bebianno, M.J., 2016. Subcellular partitioning kinetics, metallothionein response and oxidative damage in the marine mussel *Mytilus galloprovincialis* exposed to cadmium-based quantum dots. Sci. Total Environ. 554, 130–141.

Rodríguez, J.A., Fernández-García, M. (Eds.), 2007. Synthesis, Properties, and Applications of Oxide Nanomaterials. John Wiley & Sons.

Rönkkö, T., Timonen, H., 2019. Overview of sources and characteristics of nanoparticles in Urban traffic-influenced areas. J. Alzheimers Dis. 1–14 (Preprint).

Rosenberg, J.T., Yuan, X., Helsper, S.N., Bagdasarian, F.A., Ma, T., Grant, S.C., 2018. Effects of labeling human mesenchymal stem cells with superparamagnetic iron oxides on cellular functions and magnetic resonance contrast in hypoxic environments and long-term monitoring. Brain Circ. 4 (3), 133.
Rufus, A., Sreeju, N., Vilas, V., Philip, D., 2017. Biosynthesis of hematite (α-Fe_2O_3) nanostructures: size effects on applications in thermal conductivity, catalysis, and antibacterial activity. J. Mol. Liq. 242, 537–549.
Rydosz, A., Brudnik, A., Staszek, K., 2019. Metal oxide thin films prepared by magnetron sputtering technology for volatile organic compound detection in the microwave frequency range. Materials 12 (6), 877.
Ryter, S.W., Kim, H.P., Hoetzel, A., Park, J.W., Nakahira, K., Wang, X., Choi, A.M., 2007. Mechanisms of cell death in oxidative stress. Antioxid. Redox Signal. 9 (1), 49–89.
Saba, M.K., Amini, R., 2017. Nano-ZnO/carboxymethyl cellulose-based active coating impact on ready-to-use pomegranate during cold storage. Food Chem. 232, 721–726.
Sakiyama, K., Koga, K., Seto, T., Hirasawa, M., Orii, T., 2004. Formation of size-selected Ni/NiO core-shell particles by pulsed laser ablation. J. Phys. Chem. B 108 (2), 523–529.
Salah, N., Habib, S.S., Khan, Z.H., Memic, A., Azam, A., Alarfaj, E., Zahed, N., Al-Hamedi, S., 2011. High-energy ball milling technique for ZnO nanoparticles as antibacterial material. Int. J. Nanomedicine 6, 863.
Salaheldin, T.A., Husseiny, S.M., Al-Enizi, A.M., Elzatahry, A., Cowley, A.H., 2016. Evaluation of the cytotoxic behavior of fungal extracellular synthesized Ag nanoparticles using confocal laser scanning microscope. Int. J. Mol. Sci. 17 (3), 329.
Salata, O., 2004. Applications of nanoparticles in biology and medicine. J. Nanobiotechnol. 2.
Salieri, B., Righi, S., Pasteris, A., Olsen, S.I., 2015. Freshwater ecotoxicity characterisation factor for metal oxide nanoparticles: a case study on titanium dioxide nanoparticle. Sci. Total Environ. 505, 494–502.
Saliner, A.G., Burello, E., Worth, A., 2008. Review of computational approaches for predicting the physicochemical and biological properties of nanoparticles. JRC Sci. Tech. Rep. 23974-2009.
Salmani, M.H., Zarei, S., Ehrampoush, M.H., Danaie, S., 2014. Evaluations of pH and high ionic strength solution effect in cadmium removal by zinc oxide nanoparticles. J. Appl. Sci. Environ. Manag. 17 (4), 583–593.
Sanders, K., Degn, L.L., Mundy, W.R., Zucker, R.M., Dreher, K., Zhao, B., Roberts, J.E., Boyes, W.K., 2012. In vitro phototoxicity and hazard identification of nano-scale titanium dioxide. Toxicol. Appl. Pharmacol. 258 (2), 226–236.
Santos, C.S., Gabriel, B., Blanchy, M., Menes, O., García, D., Blanco, M., Arconada, N., Neto, V., 2015. Industrial applications of nanoparticles–a prospective overview. Mater. Today Proc. 2 (1), 456–465.
Sayes, C.M., Reed, K.L., Warheit, D.B., 2007. Assessing toxicity of fine and nanoparticles: comparing in vitro measurements to in vivo pulmonary toxicity profiles. Toxicol. Sci. 97 (1), 163–180.
Schaeublin, N.M., Braydich-Stolle, L.K., Schrand, A.M., Miller, J.M., Hutchison, J., Schlager, J.J., Hussain, S.M., 2011. Surface charge of gold nanoparticles mediates mechanism of toxicity. Nanoscale 3 (2), 410–420.
Schneider, S.L., Lim, H.W., 2019. A review of inorganic UV filters zinc oxide and titanium dioxide. Photodermatol. Photoimmunol. Photomed. 35 (6), 442–446.
Schwabe, F., Tanner, S., Schulin, R., Rotzetter, A., Stark, W., Von Quadt, A., Nowack, B., 2015. Dissolved cerium contributes to uptake of Ce in the presence of differently sized CeO_2-nanoparticles by three crop plants. Metallomics 7 (3), 466–477.
Schwarzenbach, R.P., Egli, T., Hofstetter, T.B., Von Gunten, U., Wehrli, B., 2010. Global water pollution and human health. Annu. Rev. Environ. Resour. 35, 109–136.
Sel, S., Duygulu, O., Kadiroglu, U., Machin, N.E., 2014. Synthesis and characterization of nano-V_2O_5 by flame spray pyrolysis, and its cathodic performance in Li-ion rechargeable batteries. Appl. Surf. Sci. 318, 150–156.
Selck, H., Handy, R.D., Fernandes, T.F., Klaine, S.J., Petersen, E.J., 2016. Nanomaterials in the aquatic environment: a European Union–United States perspective on the status of ecotoxicity testing, research priorities, and challenges ahead. Environ. Toxicol. Chem. 35 (5), 1055–1067.
Shah, V., Collins, D., Walker, V.K., Shah, S., 2014. The impact of engineered cobalt, iron, nickel and silver nanoparticles on soil bacterial diversity under field conditions. Environ. Res. Lett. 9 (2), 024001.

Shah, M., Fawcett, D., Sharma, S., Tripathy, S.K., Poinern, G.E.J., 2015. Green synthesis of metallic nanoparticles via biological entities. Materials 8 (11), 7278–7308.

Shams, M., Yildirim, E., Guleray, A.G.A.R., Ercisli, S., Dursun, A., Ekinci, M., Raziye, K.U.L., 2018. Nitric oxide alleviates copper toxicity in germinating seed and seedling growth of *Lactuca sativa* L. Not. Bot. Horti. Agrobot. Cluj Napoca 46 (1), 167–172.

Shanker Sharma, H., Sharma, A., 2012. Neurotoxicity of engineered nanoparticles from metals. CNS Neurol. Disord. Drug Targets 11 (1), 65–80.

Sharifi, S., Behzadi, S., Laurent, S., Forrest, M.L., Stroeve, P., Mahmoudi, M., 2012. Toxicity of nanomaterials. Chem. Soc. Rev. 41 (6), 2323–2343.

Sharma, R.C., Rawat, J.S., 2009. Monitoring of aquatic macroinvertebrates as bioindicator for assessing the health of wetlands: a case study in the Central Himalayas, India. Ecol. Indic. 9 (1), 118–128.

Sharma, M.K., Shah, D.O., 1985. Introduction to Macro-and Microemulsions.

Sharma, A., Muresanu, D.F., Patnaik, R., Sharma, H.S., 2013. Size-and age-dependent neurotoxicity of engineered metal nanoparticles in rats. Mol. Neurobiol. 48 (2), 386–396.

Sharma, D., Sabela, M.I., Kanchi, S., Mdluli, P.S., Singh, G., Stenström, T.A., Bisetty, K., 2016. Biosynthesis of ZnO nanoparticles using *Jacaranda mimosifolia* flowers extract: synergistic antibacterial activity and molecular simulated facet specific adsorption studies. J. Photochem. Photobiol. B Biol. 162, 199–207.

Sharon, M., Choudhary, A.K., Kumar, R., 2010. Nanotechnology in agricultural diseases and food safety. J. Phytol. 2 (4), 83–92.

Shaw, A.K., Hossain, Z., 2013. Impact of nano-CuO stress on rice (Oryza sativa L.) seedlings. Chemosphere 93 (6), 906–915.

Shaw, A.K., Ghosh, S., Kalaji, H.M., Bosa, K., Brestic, M., Zivcak, M., Hossain, Z., 2014. Nano-CuO stress induced modulation of antioxidative defense and photosynthetic performance of Syrian barley (*Hordeum vulgare* L.). Environ. Exp. Bot. 102, 37–47.

Sheykhbaglou, R., Sedghi, M., Shishevan, M.T., Sharifi, R.S., 2010. Effects of nano-iron oxide particles on agronomic traits of soybean. Not. Sci. Biol. 2 (2), 112–113.

Shi, H., Magaye, R., Castranova, V., Zhao, J., 2013a. Titanium dioxide nanoparticles: a review of current toxicological data. Part. Fibre Toxicol. 10 (1), 15.

Shi, W., Song, S., Zhang, H., 2013b. Hydrothermal synthetic strategies of inorganic semiconducting nanostructures. Chem. Soc. Rev. 42 (13), 5714–5743.

Siddiqi, K.S., Husen, A., 2017. Plant response to engineered metal oxide nanoparticles. Nanoscale Res. Lett. 12 (1), 92.

Siegrist, M., Stampfli, N., Kastenholz, H., Keller, C., 2008. Perceived risks and perceived benefits of different nanotechnology foods and nanotechnology food packaging. Appetite 51 (2), 283–290.

Silva, M.T., 2010. Secondary necrosis: the natural outcome of the complete apoptotic program. FEBS Lett. 584 (22), 4491–4499.

Silveira, C., Shimabuku, Q.L., Fernandes Silva, M., Bergamasco, R., 2018. Iron-oxide nanoparticles by the green synthesis method using *Moringa oleifera* leaf extract for fluoride removal. Environ. Technol. 39 (22), 2926–2936.

Simeone, F.C., Costa, A.L., 2019. Assessment of cytotoxicity of metal oxide nanoparticles on the basis of fundamental physical–chemical parameters: a robust approach to grouping. Environ. Sci. Nano 6 (10), 3102–3112.

Simonin, M., Richaume, A., Guyonnet, J.P., Dubost, A., Martins, J.M., Pommier, T., 2016. Titanium dioxide nanoparticles strongly impact soil microbial function by affecting archaeal nitrifiers. Sci. Rep. 6 (1), 1–10.

Singh, S., 2019. Zinc oxide nanoparticles impacts: cytotoxicity, genotoxicity, developmental toxicity, and neurotoxicity. Toxicol. Mech. Methods 29 (4), 300–311.

Singh, S.C., Gopal, R., 2007. Zinc nanoparticles in solution by laser ablation technique. Bull. Mater. Sci. 30 (3), 291–293.

Singh, D., Kumar, A., 2016. Impact of irrigation using water containing CuO and ZnO nanoparticles on *Spinach oleracea* grown in soil media. Bull. Environ. Contam. Toxicol. 97 (4), 548–553.

Singh, J., Mukherjee, A., Sengupta, S.K., Im, J., Peterson, G.W., Whitten, J.E., 2012. Sulfur dioxide and nitrogen dioxide adsorption on zinc oxide and zirconium hydroxide nanoparticles and the effect on photoluminescence. Appl. Surf. Sci. 258 (15), 5778–5785.

Singh, J., Dutta, T., Kim, K.H., Rawat, M., Samddar, P., Kumar, P., 2018. 'Green'synthesis of metals and their oxide nanoparticles: applications for environmental remediation. J. Nanobiotechnology 16 (1), 84.
Singhal, P., Jha, S.K., Pandey, S.P., Neogy, S., 2017. Rapid extraction of uranium from sea water using Fe_3O_4 and humic acid coated Fe_3O_4 nanoparticles. J. Hazard. Mater. 335, 152–161.
Smijs, T.G., Pavel, S., 2011. Titanium dioxide and zinc oxide nanoparticles in sunscreens: focus on their safety and effectiveness. Nanotechnol. Sci. Appl. 4, 95.
Snell, T.W., Hicks, D.G., 2011. Assessing toxicity of nanoparticles using *Brachionus manjavacas* (Rotifera). Environ. Toxicol. 26 (2), 146–152.
Sohail, M.I., Waris, A.A., Ayub, M.A., Usman, M., ur Rehman, M.Z., Sabir, M., Faiz, T., 2019. Environmental application of nanomaterials: A promise to sustainable future. In: *Comprehensive Analytical Chemistry*. vol. 87. Elsevier, pp. 1–54.
Sokolowski, M.A.A.B., Sokolowska, A., Michalski, A., Gokieli, B., 1977. The "in-flame-reaction" method for Al_2O_3 aerosol formation. J. Aerosol Sci. 8 (4), 219–230.
Solaiman, S.M., Algie, J., Bakand, S., Sluyter, R., Sencadas, V., Lerch, M., Huang, X.F., Konstantinov, K., Barker, P.J., 2019. Nano-sunscreens–a double-edged sword in protecting consumers from harm: viewing Australian regulatory policies through the lenses of the European Union. Crit. Rev. Toxicol. 49 (2), 122–139.
Song, Z., Cai, T., Chang, Z., Liu, G., Rodriguez, J.A., Hrbek, J., 2003. Molecular level study of the formation and the spread of MoO_3 on Au (111) by scanning tunneling microscopy and X-ray photoelectron spectroscopy. J. Am. Chem. Soc. 125 (26), 8059–8066.
Song, Y., Li, X., Du, X., 2009. Exposure to nanoparticles is related to pleural effusion, pulmonary fibrosis and granuloma. Eur. Respir. J. 34 (3), 559–567.
Soytaş, S.H., Oğuz, O., Menceloğlu, Y.Z., 2019. Polymer nanocomposites with decorated metal oxides. In: Polymer Composites with Functionalized Nanoparticles. Elsevier, pp. 287–323.
Srabionyan, V.V., Pryadchenko, V.V., Kurzin, A.A., Belenov, S.V., Avakyan, L.A., Guterman, V.E., Bugaev, L.A., 2016. Atomic structure of PtCu nanoparticles in PtCu/C catalysts from EXAFS spectroscopy data. Phys. Solid State 58 (4), 752–762.
Srdić, V.V., Winterer, M., Hahn, H., 2000. Sintering behavior of nanocrystalline zirconia doped with alumina prepared by chemical vapor synthesis. J. Am. Ceram. Soc. 83 (8), 1853–1860.
Srisitthiratkul, C., Pongsorrarith, V., Intasanta, N., 2011. The potential use of nanosilver-decorated titanium dioxide nanofibers for toxin decomposition with antimicrobial and self-cleaning properties. Appl. Surf. Sci. 257 (21), 8850–8856.
Srivastav, A.K., Kumar, M., Ansari, N.G., Jain, A.K., Shankar, J., Arjaria, N., Jagdale, P., Singh, D., 2016. A comprehensive toxicity study of zinc oxide nanoparticles versus their bulk in Wistar rats: toxicity study of zinc oxide nanoparticles. Hum. Exp. Toxicol. 35 (12), 1286–1304.
Srivastava, A.K., Dev, A., Karmakar, S., 2018. Nanosensors and nanobiosensors in food and agriculture. Environ. Chem. Lett. 16 (1), 161–182.
Stampoulis, D., Sinha, S.K., White, J.C., 2009. Assay-dependent phytotoxicity of nanoparticles to plants. Environ. Sci. Technol. 43 (24), 9473–9479.
Stankic, S., Suman, S., Haque, F., Vidic, J., 2016. Pure and multi metal oxide nanoparticles: synthesis, antibacterial and cytotoxic properties. J. Nanobiotechnology 14 (1), 73.
Steckiewicz, K.P., Zwara, J., Jaskiewicz, M., Kowalski, S., Kamysz, W., Zaleska-Medynska, A., Inkielewicz-Stepniak, I., 2019. Shape-depended biological properties of Ag_3PO_4 microparticles: evaluation of antimicrobial properties and cytotoxicity in in vitro model—safety assessment of potential clinical usage. Oxidative Med. Cell. Longev. 2019, 6740325.
Stelmakh, S., Skrobas, K., Gierlotka, S., Palosz, B., 2017. Effect of interaction of external surfaces on the symmetry and lattice distortion of CdSe nanocrystals by molecular dynamics simulations. J. Nanopart. Res. 19 (12), 391.
Stijepovic, I., Djenadic, R., Srdic, V.V., Winterer, M., 2015. Chemical vapour synthesis of lanthanum gallium oxide nanoparticles. J. Eur. Ceram. Soc. 35 (13), 3545–3552.
Stoimenov, P.K., Klinger, R.L., Marchin, G.L., Klabunde, K.J., 2002. Metal oxide nanoparticles as bactericidal agents. Langmuir 18 (17), 6679–6686.

Strambeanu, N., Demetrovici, L., Dragos, D., 2015. Natural sources of nanoparticles. In: Lungu, M. et al., (Ed.), Nanoparticles' Promises and Risks. Springer International Publishing, Switzerland, pp. 9–19.
Strober, W., 2015. Trypan blue exclusion test of cell viability. Curr. Protoc. Immunol. 111 (1), A3.B.1–A3. B.3.
Subramaniam, V.D., Prasad, S.V., Banerjee, A., Gopinath, M., Murugesan, R., Marotta, F., Sun, X.F., Pathak, S., 2019. Health hazards of nanoparticles: understanding the toxicity mechanism of nanosized ZnO in cosmetic products. Drug Chem. Toxicol. 42 (1), 84–93.
Sue, K., Kawasaki, S.I., Suzuki, M., Hakuta, Y., Hayashi, H., Arai, K., Takebayashi, Y., Yoda, S., Furuya, T., 2011. Continuous hydrothermal synthesis of Fe_2O_3, NiO, and CuO nanoparticles by super-rapid heating using a T-type micro mixer at 673 K and 30 MPa. Chem. Eng. J. 166 (3), 947–953.
Sukhanova, A., Bozrova, S., Sokolov, P., Berestovoy, M., Karaulov, A., Nabiev, I., 2018. Dependence of nanoparticle toxicity on their physical and chemical properties. Nanoscale Res. Lett. 13 (1), 44.
Sun, C., Veiseh, O., Gunn, J., Fang, C., Hansen, S., Lee, D., Sze, R., Ellenbogen, R.G., Olson, J., Zhang, M., 2008. In vivo MRI detection of gliomas by chlorotoxin-conjugated superparamagnetic nanoprobes. Small 4 (3), 372–379.
Sun, J., Wang, S., Zhao, D., Hun, F.H., Weng, L., Liu, H., 2011. Cytotoxicity, permeability, and inflammation of metal oxide nanoparticles in human cardiac microvascular endothelial cells. Cell Biol. Toxicol. 27 (5), 333–342.
Sun, Y., Zhang, G., He, Z., Wang, Y., Cui, J., Li, Y., 2016. Effects of copper oxide nanoparticles on developing zebrafish embryos and larvae. Int. J. Nanomedicine 11, 905.
Sundrarajan, M., Bama, K., Bhavani, M., Jegatheeswaran, S., Ambika, S., Sangili, A., Nithya, P., Sumathi, R., 2017. Obtaining titanium dioxide nanoparticles with spherical shape and antimicrobial properties using *M. citrifolia* leaves extract by hydrothermal method. J. Photochem. Photobiol. B Biol. 171, 117–124.
Suppakul, P., Miltz, J., Sonneveld, K., Bigger, S.W., 2003. Active packaging technologies with an emphasis on antimicrobial packaging and its applications. J. Food Sci. 68 (2), 408–420.
Surendiran, A., Sandhiya, S., Pradhan, S.C., Adithan, C., 2009. Novel applications of nanotechnology in medicine. Indian J. Med. Res. 130(6).
Swain, P.S., Rao, S.B., Rajendran, D., Dominic, G., Selvaraju, S., 2016. Nano zinc, an alternative to conventional zinc as animal feed supplement: a review. Anim. Nutr. 2 (3), 134–141.
Tadros, T.F., 2009. Emulsion Science and Technology: A General Introduction. Wiley-VCH, Weinheim, pp. 1–56.
Tang, L., Cheng, J., 2013. Nonporous silica nanoparticles for nanomedicine application. Nano Today 8 (3), 290–312.
Taze, C., Panetas, I., Kalogiannis, S., Feidantsis, K., Gallios, G.P., Kastrinaki, G., Konstandopoulos, A.G., Václavíková, M., Ivanicova, L., Kaloyianni, M., 2016. Toxicity assessment and comparison between two types of iron oxide nanoparticles in *Mytilus galloprovincialis*. Aquat. Toxicol. 172, 9–20.
Teoh, W.Y., Amal, R., Mädler, L., 2010. Flame spray pyrolysis: an enabling technology for nanoparticles design and fabrication. Nanoscale 2 (8), 1324–1347.
Thakkar, K.N., Mhatre, S.S., Parikh, R.Y., 2010. Biological synthesis of metallic nanoparticles. Nanomedicine 6 (2), 257–262.
Tharchanaa, S.B., Priyanka, K., Preethi, K., Shanmugavelayutham, G., 2020. Facile synthesis of Cu and CuO nanoparticles from copper scrap using plasma arc discharge method and evaluation of antibacterial activity. Mater. Technol. 1–8.
Thiel, B., 2019. Variable pressure scanning electron microscopy. In: Springer Handbook of Microscopy. Springer, Cham, p. 2.
Thiéry, A., Jong, L.D., Issartel, J., Moreau, X., Saez, G., Barthé, L.P., Bestel, I., Santaella, C., Achouak, W., 2012. Effects of metallic and metal oxide nanoparticles in aquatic and terrestrial food chains. Biomarkers responses in invertebrates and bacteria. Int. J. Nanotechnol. 9 (3), 181.
Thorat, N.D., Bauer, J., 2020. Nanomedicine: next generation modality of breast cancer therapeutics. In: Nanomedicines for Breast Cancer Theranostics. Elsevier, pp. 3–16.
Tian, L., Lin, B., Wu, L., Li, K., Liu, H., Yan, J., Liu, X., Xi, Z., 2015. Neurotoxicity induced by zinc oxide nanoparticles: age-related differences and interaction. Sci. Rep. 5, 16117.

Tiede, K., Boxall, A.B., Wang, X., Gore, D., Tiede, D., Baxter, M., David, H., Tear, S.P., Lewis, J., 2010. Application of hydrodynamic chromatography-ICP-MS to investigate the fate of silver nanoparticles in activated sludge. J. Anal. At. Spectrom. 25 (7), 1149–1154.

Tiwari, E., Mondal, M., Singh, N., Khandelwal, N., Monikh, F.A., Darbha, G.K., 2020. Effect of the irrigation water type and other environmental parameters on CeO_2 nanopesticide–clay colloid interactions. Environ Sci Process Impacts 22 (1), 84–94.

Tolaymat, T., El Badawy, A., Genaidy, A., Abdelraheem, W., Sequeira, R., 2017. Analysis of metallic and metal oxide nanomaterial environmental emissions. J. Clean. Prod. 143, 401–412.

Tricoli, A., Graf, M., Pratsinis, S.E., 2008. Optimal doping for enhanced SnO_2 sensitivity and thermal stability. Adv. Funct. Mater. 18 (13), 1969–1976.

Tricoli, A., Nasiri, N., Chen, H., Wallerand, A.S., Righettoni, M., 2016. Ultra-rapid synthesis of highly porous and robust hierarchical ZnO films for dye sensitized solar cells. Sol. Energy 136, 553–559.

Trivedi, M., Murase, J., 2017. Titanium dioxide in sunscreen. In: Application of Titanium Dioxide. IntechOpen, London, pp. 61–71.

Trujillo-Reyes, J., Majumdar, S., Botez, C.E., Peralta-Videa, J.R., Gardea-Torresdey, J.L., 2014a. Exposure studies of core–shell Fe/Fe_3O_4 and Cu/CuO NPs to lettuce (*Lactuca sativa*) plants: are they a potential physiological and nutritional hazard? J. Hazard. Mater. 267, 255–263.

Trujillo-Reyes, J., Peralta-Videa, J.R., Gardea-Torresdey, J.L., 2014b. Supported and unsupported nanomaterials for water and soil remediation: are they a useful solution for worldwide pollution? J. Hazard. Mater. 280, 487–503.

Tsui, M.T.K., Wang, W.X., 2007. Biokinetics and tolerance development of toxic metals in *Daphnia magna*. Environ. Toxicol. Chem. 26 (5), 1023–1032.

Tuli, H.S., Kashyap, D., Bedi, S.K., Kumar, P., Kumar, G., Sandhu, S.S., 2015. Molecular aspects of metal oxide nanoparticle (MO-NPs) mediated pharmacological effects. Life Sci. 143, 71–79.

Turaga, U., Kendall, R.J., Singh, V., Lalagiri, M., Ramkumar, S.S., 2012. Advances in materials for chemical, biological, radiological and nuclear (CBRN) protective clothing. In: Advances in Military Textiles and Personal Equipment. Woodhead Publishing, pp. 260–287.

Ul-Islam, S., Butola, B.S. (Eds.), 2018. Nanomaterials in the Wet Processing of Textiles. John Wiley & Sons.

UNEP (United Nations Environment Programme), 2011. Global guidance principles for life cycle assessment databases. A basis for greener processes and products. Report available atlcinitiative.unep.fr.

Vale, G., Mehennaoui, K., Cambier, S., Libralato, G., Jomini, S., Domingos, R.F., 2016. Manufactured nanoparticles in the aquatic environment-biochemical responses on freshwater organisms: a critical overview. Aquat. Toxicol. 170, 162–174.

van der Zande, M., Vandebriel, R.J., Van Doren, E., Kramer, E., Herrera Rivera, Z., Serrano-Rojero, C.S., Gremmer, E.R., Mast, J., Peters, R.J., Hollman, P.C., Hendriksen, P.J., 2012. Distribution, elimination, and toxicity of silver nanoparticles and silver ions in rats after 28-day oral exposure. ACS Nano 6 (8), 7427–7442.

Van Devener, B., Anderson, S.L., 2006. Breakdown and combustion of JP-10 fuel catalyzed by nanoparticulate CeO_2 and Fe_2O_3. Energy Fuel 20 (5), 1886–1894.

Van Nhan, L., Ma, C., Rui, Y., Cao, W., Deng, Y., Liu, L., Xing, B., 2016. The effects of Fe_2O_3 nanoparticles on physiology and insecticide activity in non-transgenic and Bt-transgenic cotton. Front. Plant Sci. 6, 1263.

Vance, M.E., Kuiken, T., Vejerano, E.P., McGinnis, S.P., Hochella Jr., M.F., Rejeski, D., Hull, M.S., 2015. Nanotechnology in the real world: redeveloping the nanomaterial consumer products inventory. Beilstein J. Nanotechnol. 6 (1), 1769–1780.

Varaprasad, K., Pariguana, M., Raghavendra, G.M., Jayaramudu, T., Sadiku, E.R., 2017. Development of biodegradable metaloxide/polymer nanocomposite films based on poly-ε-caprolactone and terephthalic acid. Mater. Sci. Eng. C 70, 85–93.

Varma, A., James, A.R., Daniel, S.A., 2019, May. A review on nano TiO_2—a repellent in paint. In: National Conference on Structural Engineering and Construction Management. Springer, Cham, pp. 909–918.

Veintemillas-Verdaguer, S., del Puerto Morales, M., Bomati-Miguel, O., Bautista, C., Zhao, X., Bonville, P., de Alejo, R.P., Ruiz-Cabello, J., Santos, M., Tendillo-Cortijo, F.J., Ferreirós, J., 2004. Colloidal dispersions of maghemite nanoparticles produced by laser pyrolysis with application as NMR contrast agents. J. Phys. D. Appl. Phys. 37 (15), 2054.

Velusamy, P., Kumar, G.V., Jeyanthi, V., Das, J., Pachaiappan, R., 2016. Bio-inspired green nanoparticles: synthesis, mechanism, and antibacterial application. Toxicol. Res. 32 (2), 95–102.
Vennemann, A., Alessandrini, F., Wiemann, M., 2017. Differential effects of surface-functionalized zirconium oxide nanoparticles on alveolar macrophages, rat lung, and a mouse allergy model. Nanomaterials 7 (9), 280.
Vijayalakshmi, R., Rajendran, V., 2012. Synthesis and characterization of nano-TiO_2 via different methods. Arch. Appl. Sci. Res. 4 (2), 1183–1190.
Volkovova, K., Handy, R.D., Staruchova, M., Tulinska, J., Kebis, A., Pribojova, J., Ulicna, O., Kucharská, J., Dusinska, M., 2015. Health effects of selected nanoparticles in vivo: liver function and hepatotoxicity following intravenous injection of titanium dioxide and Na-oleate-coated iron oxide nanoparticles in rodents. Nanotoxicology 9 (sup1), 95–105.
Vuong, N.M., Chinh, N.D., Huy, B.T., Lee, Y.I., 2016. CuO-decorated ZnO hierarchical nanostructures as efficient and established sensing materials for H_2S gas sensors. Sci. Rep. 6 (1), 1–13.
Wallace Hayes, A., Kruger, C.L., 2014. Hayes' Principles and Methods of Toxicology.
Wallace, R., Brown, A.P., Brydson, R., Wegner, K., Milne, S.J., 2013. Synthesis of ZnO nanoparticles by flame spray pyrolysis and characterisation protocol. J. Mater. Sci. 48 (18), 6393–6403.
Wang, S., Gao, L., 2019. Laser-driven nanomaterials and laser enabled nanofabrication for industrial applications. In: Industrial Applications of Nanomaterials. Elsevier, pp. 181–203.
Wang, J.J., Sanderson, B.J., Wang, H., 2007a. Cyto-and genotoxicity of ultrafine TiO_2 particles in cultured human lymphoblastoid cells. Mutat. Res. Genet. Toxicol. Environ. Mutagen. 628 (2), 99–106.
Wang, J., Zhou, G., Chen, C., Yu, H., Wang, T., Ma, Y., Jia, G., Gao, Y., Li, B., Sun, J., Li, Y., 2007b. Acute toxicity and biodistribution of different sized titanium dioxide particles in mice after oral administration. Toxicol. Lett. 168 (2), 176–185.
Wang, Y., Aker, W.G., Hwang, H.M., Yedjou, C.G., Yu, H., Tchounwou, P.B., 2011a. A study of the mechanism of in vitro cytotoxicity of metal oxide nanoparticles using catfish primary hepatocytes and human HepG2 cells. Sci. Total Environ. 409 (22), 4753–4762.
Wang, H., Kou, X., Pei, Z., Xiao, J.Q., Shan, X., Xing, B., 2011b. Physiological effects of magnetite (Fe_3O_4) nanoparticles on perennial ryegrass (*Lolium perenne* L.) and pumpkin (*Cucurbita mixta*) plants. Nanotoxicology 5 (1), 30–42.
Wang, Z., Xie, X., Zhao, J., Liu, X., Feng, W., White, J.C., Xing, B., 2012. Xylem-and phloem-based transport of CuO nanoparticles in maize (*Zea mays* L.). Environ. Sci. Technol. 46 (8), 4434–4441.
Wang, B., He, X., Zhang, Z., Zhao, Y., Feng, W., 2013a. Metabolism of nanomaterials in vivo: blood circulation and organ clearance. Acc. Chem. Res. 46 (3), 761–769.
Wang, N., Hsu, C., Zhu, L., Tseng, S., Hsu, J.P., 2013b. Influence of metal oxide nanoparticles concentration on their zeta potential. J. Colloid Interface Sci. 407, 22–28.
Wang, D., Lin, Z., Wang, T., Yao, Z., Qin, M., Zheng, S., Lu, W., 2016a. Where does the toxicity of metal oxide nanoparticles come from: the nanoparticles, the ions, or a combination of both? J. Hazard. Mater. 308, 328–334.
Wang, S., Wang, F., Gao, S., Wang, X., 2016b. Heavy metal accumulation in different rice cultivars as influenced by foliar application of nano-silicon. Water Air Soil Pollut. 227 (7), 228.
Wang, Z., Xu, L., Zhao, J., Wang, X., White, J.C., Xing, B., 2016c. CuO nanoparticle interaction with *Arabidopsis thaliana*: toxicity, parent-progeny transfer, and gene expression. Environ. Sci. Technol. 50 (11), 6008–6016.
Wang, Y., Chen, B.Z., Liu, Y.J., Wu, Z.M., Guo, X.D., 2017a. Application of mesoscale simulation to explore the aggregate morphology of pH-sensitive nanoparticles used as the oral drug delivery carriers under different conditions. Colloids Surf. B: Biointerfaces 151, 280–286.
Wang, Y., Ding, L., Yao, C., Li, C., Xing, X., Huang, Y., Gu, T., Wu, M., 2017b. Toxic effects of metal oxide nanoparticles and their underlying mechanisms. Sci. China Mater. 60 (2), 93–108.
Wang, Y., Lin, Y., Xu, Y., Yin, Y., Guo, H., Du, W., 2019. Divergence in response of lettuce (var. ramosa Hort.) to copper oxide nanoparticles/microparticles as potential agricultural fertilizer. Environ. Pollut. Bioavail. 31 (1), 80–84.
Wei, Y., Li, Y., Jia, J., Jiang, Y., Zhao, B., Zhang, Q., Yan, B., 2016. Aggravated hepatotoxicity occurs in aged mice but not in young mice after oral exposure to zinc oxide nanoparticles. NanoImpact 3, 1–11.

West, G.H., Cooper, M.R., Burrelli, L.G., Dresser, D., Lippy, B.E., 2019. Exposure to airborne nano-titanium dioxide during airless spray painting and sanding. J. Occup. Environ. Hyg. 16 (3), 218–228.

Wierzbinski, K.R., Szymanski, T., Rozwadowska, N., Rybka, J.D., Zimna, A., Zalewski, T., Nowicka-Bauer, K., Malcher, A., Nowaczyk, M., Krupinski, M., Fiedorowicz, M., 2018. Potential use of superparamagnetic iron oxide nanoparticles for in vitro and in vivo bioimaging of human myoblasts. Sci. Rep. 8 (1), 1–17.

Witharana, S., Hodges, C., Xu, D., Lai, X., Ding, Y., 2012. Aggregation and settling in aqueous polydisperse alumina nanoparticle suspensions. J. Nanopart. Res. 14 (5), 851.

Wong, H.W., Choi, S.M., Phillips, D.L., Ma, C.Y., 2009. Raman spectroscopic study of deamidated food proteins. Food Chem. 113 (2), 363–370.

Wongrakpanich, A., Mudunkotuwa, I.A., Geary, S.M., Morris, A.S., Mapuskar, K.A., Spitz, D.R., Grassian, V.H., Salem, A.K., 2016. Size-dependent cytotoxicity of copper oxide nanoparticles in lung epithelial cells. Environ. Sci. Nano 3 (2), 365–374.

World Health Organization, 2018. Antimicrobial Resistance and Primary Health Care: Brief (No. WHO/HIS/SDS/2018.57). World Health Organization.

Wu, M., Lin, G., Chen, D., Wang, G., He, D., Feng, S., Xu, R., 2002. Sol-hydrothermal synthesis and hydrothermally structural evolution of nanocrystal titanium dioxide. Chem. Mater. 14 (5), 1974–1980.

Wu, Q., Nouara, A., Li, Y., Zhang, M., Wang, W., Tang, M., Ye, B., Ding, J., Wang, D., 2013. Comparison of toxicities from three metal oxide nanoparticles at environmental relevant concentrations in nematode *Caenorhabditis elegans*. Chemosphere 90 (3), 1123–1131.

Wu, B., Zhu, L., Le, X.C., 2017. Metabolomics analysis of TiO_2 nanoparticles induced toxicological effects on rice (*Oryza sativa* L.). Environ. Pollut. 230, 302–310.

Xia, T., Kovochich, M., Brant, J., Hotze, M., Sempf, J., Oberley, T., Sioutas, C., Yeh, J.I., Wiesner, M.R., Nel, A.E., 2006. Comparison of the abilities of ambient and manufactured nanoparticles to induce cellular toxicity according to an oxidative stress paradigm. Nano Lett. 6 (8), 1794–1807.

Xie, J., Chen, K., Huang, J., Lee, S., Wang, J., Gao, J., Li, X., Chen, X., 2010. PET/NIRF/MRI triple functional iron oxide nanoparticles. Biomaterials 31 (11), 3016–3022.

Xing, T., Sunarso, J., Yang, W., Yin, Y., Glushenkov, A.M., Li, L.H., Howlett, P.C., Chen, Y., 2013. Ball milling: a green mechanochemical approach for synthesis of nitrogen doped carbon nanoparticles. Nanoscale 5 (17), 7970–7976.

Xu, S., Wang, Z.L., 2011. One-dimensional ZnO nanostructures: solution growth and functional properties. Nano Res. 4 (11), 1013–1098.

Xu, C., Peng, C., Sun, L., Zhang, S., Huang, H., Chen, Y., Shi, J., 2015. Distinctive effects of TiO_2 and CuO nanoparticles on soil microbes and their community structures in flooded paddy soil. Soil Biol. Biochem. 86, 24–33.

Yameen, B., Choi, W.I., Vilos, C., Swami, A., Shi, J., Farokhzad, O.C., 2014. Insight into nanoparticle cellular uptake and intracellular targeting. J. Control. Release 190, 485–499.

Yan, A., Chen, Z., 2018. Detection methods of nanoparticles in plant tissues. In: New Visions in Plant Science.p. 99.

Yan, L., Yu, R., Chen, J., Xing, X., 2008. Template-free hydrothermal synthesis of CeO_2 nano-octahedrons and nanorods: investigation of the morphology evolution. Cryst. Growth Des. 8 (5), 1474–1477.

Yang, K., Ma, Y.Q., 2010. Computer simulation of the translocation of nanoparticles with different shapes across a lipid bilayer. Nat. Nanotechnol. 5 (8), 579.

Yang, H., Liu, C., Yang, D., Zhang, H., Xi, Z., 2009. Comparative study of cytotoxicity, oxidative stress and genotoxicity induced by four typical nanomaterials: the role of particle size, shape and composition. J. Appl. Toxicol. 29 (1), 69–78.

Yang, Z., Chen, J., Dou, R., Gao, X., Mao, C., Wang, L., 2015. Assessment of the phytotoxicity of metal oxide nanoparticles on two crop plants, maize (*Zea mays* L.) and rice (*Oryza sativa* L.). Int. J. Environ. Res. Public Health 12 (12), 15100–15109.

Yang, L., Kuang, H., Liu, Y., Xu, H., Aguilar, Z.P., Xiong, Y., Wei, H., 2016a. Mechanism of enhanced antibacterial activity of ultra-fine ZnO in phosphate buffer solution with various organic acids. Environ. Pollut. 218, 863–869.

Yang, C., Tian, A., Li, Z., 2016b. Reversible cardiac hypertrophy induced by PEG-coated gold nanoparticles in mice. Sci. Rep. 6 (1), 1–12.

Yao, W.T., Yu, S.H., Zhou, Y., Jiang, J., Wu, Q.S., Zhang, L., Jiang, J., 2005. Formation of uniform CuO nanorods by spontaneous aggregation: selective synthesis of CuO, Cu_2O, and Cu nanoparticles by a solid-liquid phase arc discharge process. J. Phys. Chem. B 109 (29), 14011–14016.

Ying, J., Zhang, T., Tang, M., 2015. Metal oxide nanomaterial QNAR models: available structural descriptors and understanding of toxicity mechanisms. Nano 5 (4), 1620–1637.

Yoon, S.J., Kwak, J.I., Lee, W.M., Holden, P.A., An, Y.J., 2014. Zinc oxide nanoparticles delay soybean development: a standard soil microcosm study. Ecotoxicol. Environ. Saf. 100, 131–137.

You, T., Liu, D., Chen, J., Yang, Z., Dou, R., Gao, X., Wang, L., 2018. Effects of metal oxide nanoparticles on soil enzyme activities and bacterial communities in two different soil types. J. Soils Sediments 18 (1), 211–221.

Younes, N.A., Hassan, H.S., Elkady, M.F., Hamed, A.M., Dawood, M.F., 2020. Impact of synthesized metal oxide nanomaterials on seedlings production of three Solanaceae crops. Heliyon 6 (1), e03188.

Yousefi, R., Mahmoudian, M.R., 2019. The use of nanotechnology in preventing corrosion of metal pipe and equipment of Shahid Abbaspour Dam. J. Dam Hydroelectr. Powerplant 6 (20), 31–37.

Yuan, H., Chen, C.Y., Chai, G.H., Du, Y.Z., Hu, F.Q., 2013. Improved transport and absorption through gastrointestinal tract by PEGylated solid lipid nanoparticles. Mol. Pharm. 10 (5), 1865–1873.

Zafar, H., Ali, A., Zia, M., 2017. CuO nanoparticles inhibited root growth from *Brassica nigra* seedlings but induced root from stem and leaf explants. Appl. Biochem. Biotechnol. 181 (1), 365–378.

Zavar, S., 2017. A novel three component synthesis of 2-amino-4H-chromenes derivatives using nano ZnO catalyst. Arab. J. Chem. 10, S67–S70.

Zhang, H., Banfield, J.F., 2005. Size dependence of the kinetic rate constant for phase transformation in TiO_2 nanoparticles. Chem. Mater. 17 (13), 3421–3425.

Zhang, Y., Chen, Y., Westerhoff, P., Hristovski, K., Crittenden, J.C., 2008a. Stability of commercial metal oxide nanoparticles in water. Water Res. 42 (8–9), 2204–2212.

Zhang, Y., Yang, M., Portney, N.G., Cui, D., Budak, G., Ozbay, E., Ozkan, M., Ozkan, C.S., 2008b. Zeta potential: a surface electrical characteristic to probe the interaction of nanoparticles with normal and cancer human breast epithelial cells. Biomed. Microdevices 10 (2), 321–328.

Zhang, X., Qu, Z., Li, X., Zhao, Q., Wang, Y., Quan, X., 2011. Low temperature CO oxidation over Ag/SBA-15 nanocomposites prepared via in-situ "pH-adjusting" method. Catal. Commun. 16 (1), 11–14.

Zhang, H., Ji, Z., Xia, T., Meng, H., Low-Kam, C., Liu, R., Pokhrel, S., Lin, S., Wang, X., Liao, Y.P., Wang, M., 2012a. Use of metal oxide nanoparticle band gap to develop a predictive paradigm for oxidative stress and acute pulmonary inflammation. ACS Nano 6 (5), 4349–4368.

Zhang, J., Zhang, Y., Chen, Y., Du, L., Zhang, B., Zhang, H., Liu, J., Wang, K., 2012b. Preparation and characterization of novel polyethersulfone hybrid ultrafiltration membranes bending with modified halloysite nanotubes loaded with silver nanoparticles. Ind. Eng. Chem. Res. 51 (7), 3081–3090.

Zhang, C., Hu, Z., Deng, B., 2016. Silver nanoparticles in aquatic environments: physiochemical behavior and antimicrobial mechanisms. Water Res. 88, 403–427.

Zhang, J., Guo, W., Li, Q., Wang, Z., Liu, S., 2018. The effects and the potential mechanism of environmental transformation of metal nanoparticles on their toxicity in organisms. Environ. Sci. Nano 5 (11), 2482–2499.

Zhao, L., Peralta-Videa, J.R., Ren, M., Varela-Ramirez, A., Li, C., Hernandez-Viezcas, J.A., Aguilera, R.J., Gardea-Torresdey, J.L., 2012. Transport of Zn in a sandy loam soil treated with ZnO NPs and uptake by corn plants: electron microprobe and confocal microscopy studies. Chem. Eng. J. 184, 1–8.

Zheng, Y.R., Gao, M.R., Gao, Q., Li, H.H., Xu, J., Wu, Z.Y., Yu, S.H., 2015. An efficient $CeO_2/CoSe_2$ nanobelt composite for electrochemical water oxidation. Small 11 (2), 182–188.

Zhou, D., Jin, S., Li, L., Wang, Y., Weng, N., 2011. Quantifying the adsorption and uptake of CuO nanoparticles by wheat root based on chemical extractions. J. Environ. Sci. 23 (11), 1852–1857.

Zhu, M.T., Feng, W.Y., Wang, B., Wang, T.C., Gu, Y.Q., Wang, M., Wang, Y., Ouyang, H., Zhao, Y.L., Chai, Z.F., 2008a. Comparative study of pulmonary responses to nano-and submicron-sized ferric oxide in rats. Toxicology 247 (2–3), 102–111.

Zhu, X., Zhu, L., Duan, Z., Qi, R., Li, Y., Lang, Y., 2008b. Comparative toxicity of several metal oxide nanoparticle aqueous suspensions to Zebrafish (*Danio rerio*) early developmental stage. J. Environ. Sci. Health A 43 (3), 278–284.

Zhu, X., Tian, S., Cai, Z., 2012. Toxicity assessment of iron oxide nanoparticles in zebrafish (*Danio rerio*) early life stages. PLoS ONE. 7(9).

Zhu, S., Xue, M.Y., Luo, F., Chen, W.C., Zhu, B., Wang, G.X., 2017. Developmental toxicity of Fe_3O_4 nanoparticles on cysts and three larval stages of *Artemia salina*. Environ. Pollut. 230, 683–691.

Zuverza-Mena, N., Medina-Velo, I.A., Barrios, A.C., Tan, W., Peralta-Videa, J.R., Gardea-Torresdey, J.L., 2015. Copper nanoparticles/compounds impact agronomic and physiological parameters in cilantro (*Coriandrum sativum*). Environ Sci Process Impacts 17 (10), 1783–1793.

Further reading

Adams, J., Wright, M., Wagner, H., Valiente, J., Britt, D., Anderson, A., 2017. Cu from dissolution of CuO nanoparticles signals changes in root morphology. Plant Physiol. Biochem. 110, 108–117.

Amde, M., Liu, J.F., Tan, Z.Q., Bekana, D., 2017. Transformation and bioavailability of metal oxide nanoparticles in aquatic and terrestrial environments. A review. Environ. Pollut. 230, 250–267.

Ansari, M.A., Khan, H.M., Khan, A.A., Cameotra, S.S., Saquib, Q., Musarrat, J., 2014. Interaction of Al_2O_3 nanoparticles with *Escherichia coli* and their cell envelope biomolecules. J. Appl. Microbiol. 116 (4), 772–783.

Bădin, L., Daraio, C., Simar, L., 2012. How to measure the impact of environmental factors in a nonparametric production model. Eur. J. Oper. Res. 223 (3), 818–833.

Baek, Y.W., An, Y.J., 2011. Microbial toxicity of metal oxide nanoparticles (CuO, NiO, ZnO, and Sb_2O_3) to *Escherichia coli*, *Bacillus subtilis*, and *Streptococcus aureus*. Sci. Total Environ. 409 (8), 1603–1608.

Bian, S.W., Mudunkotuwa, I.A., Rupasinghe, T., Grassian, V.H., 2011. Aggregation and dissolution of 4 nm ZnO nanoparticles in aqueous environments: influence of pH, ionic strength, size, and adsorption of humic acid. Langmuir 27 (10), 6059–6068.

Borrego, B., Lorenzo, G., Mota-Morales, J.D., Almanza-Reyes, H., Mateos, F., López-Gil, E., de la Losa, N., Burmistrov, V.A., Pestryakov, A.N., Brun, A., Bogdanchikova, N., 2016. Potential application of silver nanoparticles to control the infectivity of Rift Valley fever virus in vitro and in vivo. Nanomedicine 12 (5), 1185–1192.

Buchman, J.T., Hudson-Smith, N.V., Landy, K.M., Haynes, C.L., 2019. Understanding nanoparticle toxicity mechanisms to inform redesign strategies to reduce environmental impact. Acc. Chem. Res. 52 (6), 1632–1642.

Cai, X., Lee, A., Ji, Z., Huang, C., Chang, C.H., Wang, X., Liao, Y.P., Xia, T., Li, R., 2017. Reduction of pulmonary toxicity of metal oxide nanoparticles by phosphonate-based surface passivation. Part. Fibre Toxicol. 14 (1), 13.

Exbrayat, J.M., Moudilou, E.N., Lapied, E., 2015. Harmful effects of nanoparticles on animals. J. Nanotechnology. 2015, 861092.

Fernández-Cruz, M.L., Lammel, T., Connolly, M., Conde, E., Barrado, A.I., Derick, S., Perez, Y., Fernandez, M., Furger, C., Navas, J.M., 2013. Comparative cytotoxicity induced by bulk and nanoparticulated ZnO in the fish and human hepatoma cell lines PLHC-1 and Hep G2. Nanotoxicology 7 (5), 935–952.

Garner, K.L., Suh, S., Keller, A.A., 2017. Assessing the risk of engineered nanomaterials in the environment: development and application of the nanoFate model. Environ. Sci. Technol. 51 (10), 5541–5551.

Ge, Y., Schimel, J.P., Holden, P.A., 2012. Identification of soil bacteria susceptible to TiO_2 and ZnO nanoparticles. Appl. Environ. Microbiol. 78 (18), 6749–6758.

Graves, J.L., Thomas, M., Ewunkem, J.A., 2017. Antimicrobial nanomaterials: why evolution matters. Nanomaterials 7 (10), 283.

Haris, Z., Ahmad, I., 2017. Impact of metal oxide nanoparticles on beneficial soil microorganisms and their secondary metabolites. Int. J. Life Sci. Sci. Res. 3, 1020–1030.

Holden, P.A., Nisbet, R.M., Lenihan, H.S., Miller, R.J., Cherr, G.N., Schimel, J.P., Gardea-Torresdey, J.L., 2013. Ecological nanotoxicology: integrating nanomaterial hazard considerations across the subcellular, population, community, and ecosystems levels. Acc. Chem. Res. 46 (3), 813–822.

Hu, G., Cao, J., 2019. Metal-containing nanoparticles derived from concealed metal deposits: an important source of toxic nanoparticles in aquatic environments. Chemosphere 224, 726–733.

Kalhapure, R.S., Sonawane, S.J., Sikwal, D.R., Jadhav, M., Rambharose, S., Mocktar, C., Govender, T., 2015. Solid lipid nanoparticles of clotrimazole silver complex: an efficient nano antibacterial against Staphylococcus aureus and MRSA. Colloids Surf. B: Biointerfaces 136, 651–658.

Karimi, S., Troeung, M., Wang, R., Draper, R., Pantano, P., 2018. Acute and chronic toxicity of metal oxide nanoparticles in chemical mechanical planarization slurries with *Daphnia magna*. Environ. Sci. Nano 5 (7), 1670–1684.

Karunakaran, G., Suriyaprabha, R., Manivasakan, P., Rajendran, V., Kannan, N., 2014. Influence of nano and bulk SiO_2 and Al_2O_3 particles on PGPR and soil nutrient contents. Curr. Nanosci. 10 (4), 604–612.

Krishnaraj, C., Ramachandran, R., Mohan, K., Kalaichelvan, P.T., 2012. Optimization for rapid synthesis of silver nanoparticles and its effect on phytopathogenic fungi. Spectrochim. Acta A Mol. Biomol. Spectrosc. 93, 95–99.

Krstić, P.S., Wells, J.C. (Eds.), 2010. Nanotechnology for Electronics, Photonics, and Renewable Energy. In: vol. 78. Springer, New York.

Kumar, V., Sharma, N., Maitra, S.S., 2017. In vitro and in vivo toxicity assessment of nanoparticles. Int. Nano Lett. 7 (4), 243–256.

Leareng, S.K., Ubomba-Jaswa, E., Musee, N., 2020. Toxicity of zinc oxide and iron oxide engineered nanoparticles to *Bacillus subtilis* in river water systems. Environ. Sci. Nano 7 (1), 172–185.

Liu, R., Rallo, R., Weissleder, R., Tassa, C., Shaw, S., Cohen, Y., 2013. Nano-SAR development for bioactivity of nanoparticles with considerations of decision boundaries. Small 9 (9–10), 1842–1852.

López-Luna, J., Camacho-Martínez, M.M., Solís-Domínguez, F.A., González-Chávez, M.C., Carrillo-González, R., Martinez-Vargas, S., Mijangos-Ricardez, O.F., Cuevas-Díaz, M.C., 2018. Toxicity assessment of cobalt ferrite nanoparticles on wheat plants. J. Toxic. Environ. Health A 81 (14), 604–619.

Lowry, G.V., Hotze, E.M., Bernhardt, E.S., Dionysiou, D.D., Pedersen, J.A., Wiesner, M.R., Xing, B., 2010. Environmental occurrences, behavior, fate, and ecological effects of nanomaterials: an introduction to the special series. J. Environ. Qual. 39 (6), 1867–1874.

Madhura, L., Singh, S., Kanchi, S., Sabela, M., Bisetty, K., 2019. Nanotechnology-based water quality management for wastewater treatment. Environ. Chem. Lett. 17 (1), 65–121.

Nekrasova, G.F., Ushakova, O.S., Ermakov, A.E., Uimin, M.A., Byzov, I.V., 2011. Effects of copper (II) ions and copper oxide nanoparticles on *Elodea densa Planch*. Russ. J. Ecol. 42 (6), 458.

Noventa, S., Hacker, C., Rowe, D., Elgy, C., Galloway, T., 2018. Dissolution and bandgap paradigms for predicting the toxicity of metal oxide nanoparticles in the marine environment: an in vivo study with oyster embryos. Nanotoxicology 12 (1), 63–78.

Oberdörster, G., 2010. Safety assessment for nanotechnology and nanomedicine: concepts of nanotoxicology. J. Intern. Med. 267 (1), 89–105.

Pradhan, A., Seena, S., Pascoal, C., Cássio, F., 2011. Can metal nanoparticles be a threat to microbial decomposers of plant litter in streams? Microb. Ecol. 62 (1), 58–68.

Prosser, J.I., Bohannan, B.J., Curtis, T.P., Ellis, R.J., Firestone, M.K., Freckleton, R.P., Green, J.L., Green, L.E., Killham, K., Lennon, J.J., Osborn, A.M., 2007. The role of ecological theory in microbial ecology. Nat. Rev. Microbiol. 5 (5), 384–392.

Rashidi, L., Khosravi-Darani, K., 2011. The applications of nanotechnology in food industry. Crit. Rev. Food Sci. Nutr. 51 (8), 723–730.

Reidy, B., Haase, A., Luch, A., Dawson, K.A., Lynch, I., 2013. Mechanisms of silver nanoparticle release, transformation and toxicity: a critical review of current knowledge and recommendations for future studies and applications. Materials 6 (6), 2295–2350.

Rotini, A., Tornambè, A., Cossi, R., Iamunno, F., Benvenuto, G., Berducci, M.T., Maggi, C., Thaller, M.C., Cicero, A.M., Manfra, L., Migliore, L., 2017. Salinity-based toxicity of CuO nanoparticles, CuO-bulk and Cu ion to *Vibrio anguillarum*. Front. Microbiol. 8, 2076.

Rubilar, O., Rai, M., Tortella, G., Diez, M.C., Seabra, A.B., Durán, N., 2013. Biogenic nanoparticles: copper, copper oxides, copper sulphides, complex copper nanostructures and their applications. Biotechnol. Lett. 35 (9), 1365–1375.

Rui, M., Ma, C., Hao, Y., Guo, J., Rui, Y., Tang, X., Zhao, Q., Fan, X., Zhang, Z., Hou, T., Zhu, S., 2016. Iron oxide nanoparticles as a potential iron fertilizer for peanut (*Arachis hypogaea*). Front. Plant Sci. 7, 815.

Sawhney, A.P.S., Condon, B., Singh, K.V., Pang, S.S., Li, G., Hui, D., 2008. Modern applications of nanotechnology in textiles. Text. Res. J. 78 (8), 731–739.

Seabra, A.B., Haddad, P., Duran, N., 2013. Biogenic synthesis of nanostructured iron compounds: applications and perspectives. IET Nanobiotechnol. 7 (3), 90–99.

Sendra, M., Moreno-Garrido, I., Yeste, M.P., Gatica, J.M., Blasco, J., 2017. Toxicity of TiO_2, in nanoparticle or bulk form to freshwater and marine microalgae under visible light and UV-A radiation. Environ. Pollut. 227, 39–48.

Soltani, F., Yavari, K., Sadeghi, M., Samani, A.B., Simindokht, S.A., 2018. Toxicity of nano and bulk forms of cerium oxide in different cell lines. Iran. J. Pharmacol. Ther. 16 (1), 1–6.

Tsai, S.J., Huang, R.F., Ellenbecker, M.J., 2010. Airborne nanoparticle exposures while using constant-flow, constant-velocity, and air-curtain-isolated fume hoods. Ann. Occup. Hyg. 54 (1), 78–87.

Tyner, K.M., Wokovich, A.M., Doub, W.H., Buhse, L.F., Sung, L.-P., Watson, S.S., Sadrieh, N., 2009. Comparing methods for detecting and characterizing metal oxide nanoparticles in unmodified commercial sunscreens. Nanomedicine 4 (2), 145–159.

Worth, A.P., 2019. Types of toxicity and applications of toxicity testing. In: The History of Alternative Test Methods in Toxicology. Academic Press, pp. 7–10.

Zhao, X., Ng, S., Heng, B.C., Guo, J., Ma, L., Tan, T.T.Y., Ng, K.W., Loo, S.C.J., 2013. Cytotoxicity of hydroxyapatite nanoparticles is shape and cell dependent. Arch. Toxicol. 87 (6), 1037–1052.

Kumar, A. and Sharma, P., 2013. Impact of climate variation on agricultural productivity and food security in rural India (No. 2013-43). Economics Discussion Papers.

CHAPTER 8

Health and safety hazards of nanomaterials

Umair Riaz[a], Shazia Iqbal[b], Laila Shahzad[c], Tayyaba Samreen[b], and Waleed Mumtaz Abbasi[d]
[a]Soil and Water Testing Laboratory for Research, Bahawalpur, Pakistan
[b]Institute of Soil & Environmental Sciences, University of Agriculture, Faisalabad, Pakistan
[c]Sustainable Development Study Center, Government College University, Lahore, Pakistan
[d]Department of Soil Science, Faculty of Agriculture and Environmental Sciences, The Islamia University of Bahawalpur, Bahawalpur, Pakistan

1. Introduction and general background

Nanomaterials are now a highly promising topic, and because of their small size and composition, they possess diverse applications in different fields of life, such as chemistry, physics, biology, electronics, biomedical, engineering cosmetics, and agriculture (Varghese et al., 2019; Hadef, 2018) (Table 1). Nano-materials refer to materials that have a range of 1–100 nm in one or more dimensions. These can be sourced from a variety of engineered, targeted materials and in naturally occurring forms (Maurya et al., 2020). The term "nano" originates from the Greek word "nanos" (or Latin "nanus"), meaning "dwarf," but scientifically "nano" means one in a billion (Varghese et al., 2019). Nanomaterials have properties considerably different to and improved over their coarser counterparts. With the advancement in technology in the 21st century, the upgradation of devices into nanometer sizes is required for better improvement in performance. This raises certain issues regarding new materials for achieving specific functionality and selectivity. "Nanotechnology" is a new scientific domain, defined as the design, fabrication, and application of nanomaterials and the fundamental understanding of the relationships between material dimensions and physical properties or phenomena. It creates metastable phase nanostructures with nonconventional properties such as magnetism and superconductivity. Another very significant feature of nanotechnology is the miniaturization of current and new instruments that will significantly influence today's world. Computers are examples of possible miniaturization or nanotechnology that, with infinitely high power, can compute algorithms, and can mimic human brains. Nanoscaled electronics regularly monitor our local environment; nanorobots can remove chemical toxins and repair internal damage in human bodies. Biosensors are another nanomaterial tool that warns us about the onset of disease at an early stage and, if possible, at a molecular level.

Nanomaterials: Synthesis, Characterization, Hazards and Safety
https://doi.org/10.1016/B978-0-12-823823-3.00012-4

Table 1 Nanotechnology application in different areas.

Nanomaterial	Areas	Application
ZnO, SiO_2 and nano-material-coated fertilizer	Agriculture	Nanofertilizers
CeO_2	Automotive	Catalysts, lightweight construction, painting, tires, sensors, windshield and body coatings
TiO_2	Chemical	Composite materials, adhesives, magnetic fluids, fillers for paints, impregnation of papers
TiO_2, Ag	Construction	Flame retardants, insulation, mortar, surface coatings
TiO_2, ZnO, Fullerene C_{60}	Cosmetics	Sunscreen, lipsticks, skin creams, toothpaste
Quantum dots	Electronics	Antistatic coatings, conductive coatings, fiber optics, filters, data memory, displays, laser diodes, optical switches, transistors
TiO_2, SiO_2, Ag, Quantum dots	Engineering	Lubricant-free bearings, machines, protective coatings for tools
Carbon nanotubes	Energy	Batteries, capacitors, fuel cells, lighting, solar cells
Metals and metal oxides	Environment	Environmental monitoring, fuel changing catalysts, green chemistry, soil and groundwater remediation, toxic exposure sensors
TiO_2, Ag, nano clay	Food and drink	Additives, juice clarifiers, packaging, storage life sensors
TiO_2, copper-coated silica nanoparticles	Household	Ceramic coatings for irons, cleaners for glass, ceramics, and metals, odor removers
Fullerenes (Carbon-60, 70, 80 derivatives, multiwall nanotubes)	Medicine	Drug delivery systems, contrast medium, rapid testing systems, prostheses and implants, antimicrobial agents, in-body diagnostic systems
SiO_2, TiO_2, iron nanoparticles	Military	Neutralization materials for chemical weapons, bulletproof protection
Carbon nanotubes, nano clays, fullerenes, silica nanoparticles	Sports	Antifouling coatings for boats, antifogging coatings for glasses, goggles, golf clubs, ski wax, tennis rackets, tennis balls
TiO_2, Al_2O_3, SiO_2, ZnO, carbon nanotubes, polybutylacrylate	Textiles	Surface coatings, smart clothes (antiwrinkle, stain-resistant, temperature controlled)

Modified from Naidu, K.S.B., 2020.Engineered nanoparticles: hazards and risk assessment upon exposure-a review. Curr. Trends Biotechnol. Pharm. 14(1), 111–122. https://doi.org/10.5530/ctbp.2020.1.11.

It targets the specific drugs that automatically attack the diseased cells on-site (Shodhganga, 2020).

An exciting challenge exists concerning natural nanomaterials versus "synthetic" nanomaterials. The existence of "naturally" created types of nanoparticle has been documented from the very beginning. Natural nanomaterials include particulate components of combustion processes associated with commercial products. These were not identified by the public previously as being in the nanoscale range, e.g., carbon black particles (natural nanoscale carbon particulates), nanoscale titanium dioxide particulates, or zinc oxide particles for health-related sunscreen applications (Warheit, 2018).

Although nanotechnology is a research area in the modern era, nanomaterials are known to have been used for centuries. In the medieval era, nanomaterials were used in the colored glass of cathedral windows. The Romans used metal nanoparticles in glass artifacts, which gave beautiful colors. The Chinese used gold nanoparticles as an inorganic dye in their ceramic porcelains to introduce red color more than 1000 years ago. The study of nanometer-scale materials can be traced back for centuries. The current nanotechnology fever is at least partially driven by the constant shrinking of devices in the semiconductor industry. The continuous diminution in device dimensions has followed the well-known Moore's law, predicted in 1965 (Pokropivny et al., 2007).

Because of the commercializing of nanomaterials, apprehensions have been raised about health and safety issues as well as their impact on the environment. It is a challenge for regulators and companies to certify the development of safe nanomaterial products for consumers. Therefore, potential hazards assessment of nanomaterials has become a new area for health risk assessments. During the past 10 years, numerous research challenges and strategies have been projected to direct and simplify the "verification," i.e., the safe exposures and handling of nanomaterials and nanotechnology. A significant problem is that all the different nanoparticle types cannot be efficiently evaluated for safety effects promptly because of their numerous variations within specific nanoparticle types, vast numbers of different nanoparticle types, expensive testing, and short time for testing for each nanomaterial. This process requires broad engagement with potential customers and stakeholders regarding safety and health issues. For this purpose, several authors, researchers, many workshops, and millions of dollars/euros of grant funds have been focused on providing "guidance" on developing research on health and safety strategies and challenges. This has also provided support for research validation efforts between laboratories and given funding for studying the hazards of different nanoparticle exposures on human health and the environment. It is expected that an understanding of both environmental hazards and health effects (pulmonary, oral, or dermal) by nanoparticle exposure could lead to improved environmental and health risk evaluations (Warheit, 2018). This chapter provides a brief introduction to nanomaterials, their effects on human health, and safety guidelines in the use of nanomaterials in different countries (Table 2).

Table 2 Nanomaterials hazards to human health.

Hazard type	Route of exposure	Nanomaterial	Damage	Reference
Dermal	Skin penetration through wounds	Carbon nanotubes	Oxidative stress, pro-inflammatory cytokines, decreased viability	USNIOSH (2009, 2013) (https://www.nano.gov/NIOSH) NCRPM (2017) (https://ncrponline.org)
		Raw SWCNT	Dermal irritation	USNIOSH (2009, 2013) (https://www.nano.gov/NIOSH)
Gastrointestinal	Unintentional hand-to-mouth transfer of materials, inhalation exposure		Ingestion	USNIOSH (2009) (https://www.nano.gov/NIOSH)
Respiratory	Inhalation exposure	Carbon nanotubes and carbon nanofibers	Pulmonary effects including inflammation, granulomas, and pulmonary fibrosis	USNIOSH (2013) (https://www.nano.gov/NIOSH)
		Titanium dioxide (TiO_2) dust and ultrafine particles	Lung tumor	NCRPM (2017) (https://ncrponline.org)
Radioactive	Any kind of contact	Engineered radioactive nanoparticles in medical diagnostics, medical imaging, toxic kinetics, and environmental health	Affects operational health physics and internal dosimetry	NCRPM (2017) (https://ncrponline.org)
Fire and explosion	Dust explosion	Engineered carbon nanoparticles Aluminum nanoparticles and titanium nanoparticles	Explosion hazards	Turkevich et al. (2016) NCRPM (2017) (https://ncrponline.org)

2. Safety guidelines for nanomaterials usage

2.1 World Health Organization guidelines

The World Health Organization (WHO) established guidelines for protecting against the potential risk of industrial nanomaterials on workers at the end of 2017 (WHO, 2018). The primary guideline used is a precautionary approach; the precaution is reduction in exposure, because recent studies showed the ability of nanomaterials to cross cell barriers and interrelate with cellular structures (Verma and Stellacci, 2010). Another important guiding principle was the hierarchy of controls. This addressed the idea that when there is a choice between control measures, it is preferable to choose those measures that are closer to the problem root than measures that exert a high load on workers, such as the use of personal protective equipment (PPE). Recommendations by the WHO were rated as "conditional" or "strong" depending on the quality of the scientific values, preferences, evidence, and costs. The WHO guidelines contain the following recommendations for safe handling of manufactured nanomaterials (MNMs):

(1) Assess health hazards of manufactured nanomaterials:
- The WHO recommends up-to-date safety data sheets with manufactured nanomaterials' specific hazard information or indicating which toxicological endpoints did not have adequate testing available (strong recommendation, moderate-quality evidence).
- Assigning hazard classes to all manufactured nanomaterials according to the Globally Harmonized System of Classification and Labelling of Chemicals for use in safety data sheets (strong recommendation, moderate-quality evidence).
- For the granular bio-persistent particles groups and respirable fibers, the Guideline Development Group suggests the use of the available classification of manufactured nanomaterials for the provisional classification of nanomaterials of the same group (conditional recommendation, low-quality evidence).

(2) Assess exposure to manufactured nanomaterials:
- The WHO suggests assessing workers' exposure in workplaces with methods similar to those used for the proposed specific occupational exposure limit value of the manufactured nanomaterials (conditional recommendation, low-quality evidence).
- Because there are no specific regulatory occupational exposure limit values for manufactured nanomaterials in workplaces, the WHO suggests assessing whether workplace exposure exceeds a proposed occupational exposure limit value for the manufactured nanomaterials. A list of proposed occupational exposure limit values is provided in an annex to the guidelines. The chosen occupational exposure limit should be at least as protective as a legally mandated occupational exposure limit for the bulk form of the material (conditional recommendation, low-quality evidence).

- The WHO suggests a step-wise approach for inhalation exposure in workplaces with availability of specific occupational exposure limits for manufactured nanomaterials:
  (1) Assess the potential exposure.
  (2) Conduct basic exposure assessment.
  (3) Conduct a comprehensive exposure assessment such as those proposed by the Comité Européen de Normalisation (Cen, or the European Committee for Standardization) or Organisation for Economic Cooperation and Development (OECD) (conditional recommendation, moderate-quality evidence).
- The WHO found a small indication to recommend a method of dermal exposure assessment.

(3) Control exposure to manufactured nanomaterials:
- The WHO recommends control of exposure for preventing inhalation exposure to lowering it as much as possible (strong recommendation, moderate-quality evidence).
- The WHO recommends a reduction of exposure to a range of manufactured nanomaterials. In the absence of toxicological information, the WHO recommends applying the highest level of controls to avoid exposure of workers (strong recommendation, moderate-quality evidence).
- The WHO recommends taking control measures based on the principle of the hierarchy of controls. It means the first control measure should be to eradicate the source of exposure, then implement control measures that are dependent on worker involvement, with personal protective equipment used as a last option only (strong recommendation, moderate-quality evidence).
- The WHO recommends avoiding dermal exposure by occupational hygiene measures like use of appropriate gloves and surface cleaning (conditional recommendation, low-quality evidence).
- The WHO proposes the use of nanomaterials control banding to select exposure control measures when the assessment and measures by workplace safety experts are not accessible, but the WHO cannot recommend one method of control banding over another due to lack of studies (conditional recommendation, very low-quality evidence).

2.2 United States

- The Environmental Protection Agency (EPA) regulates nanomaterials under the same provisions as other hazardous chemical substances (Vance et al., 2015). The EPA manages nanomaterials under the Toxic Substances Control Act, and has permitted limited manufacture of new chemical nanomaterials through the use of consent orders or Significant New Use Rules (SNURs). Other statutes falling into the EPA's jurisdiction may apply, such as the Federal Insecticide, Fungicide, and Rodenticide Act (if bacterial claims are being made), the Clean Air Act, or the Clean Water Act (USNIOSH, 2016, https://www.nano.gov/NIOSH). In 2011, the EPA issued a SNUR on multiwalled carbon nanotubes, codified as 40 C.F.R. 721.10155.

- The Food and Drug Administration regulates nanomaterials under the Federal Food, Drug, and Cosmetic Act when used as food additives, drugs, or cosmetics (Vance et al., 2015).
- The Consumer Product Safety Commission requires testing and certification of many consumer products for compliance with consumer product safety requirements, and cautionary labeling of hazardous substances under the Federal Hazardous Substances Act (USNIOSH, 2016 https://www.nano.gov/NIOSH).
- The Occupational Safety and Health Administration also has recording and reporting requirements for occupational injuries and illness under 29 C.F.R. 1904 for businesses with more than 10 employees, and protection and communication regulations under 29 C.F.R. 1910.
- The General Duty Clause of the Occupational Safety and Health Act requires all employers to keep their workplace free of serious recognized hazards.
- Companies producing new products containing nanomaterials must use the Hazard Communication Standard to create safety data sheets containing 16 sections for downstream users such as customers, workers, disposal services, and others. It may require toxicological or other testing, and all data or information provided must be vetted by properly controlled testing. The ISO/TR 13329 standard (IOS, 2012) provides guidance specifically on the preparation of safety data sheets for nanomaterials.
- The National Institute for Occupational Safety and Health does not issue regulations, but conducts research and makes recommendations to prevent worker injury and illness. State and local governments may have additional regulations (USNIOSH, 2016, https://www.nano.gov/NIOSH).

2.3 United Kingdom

- In the United Kingdom, powders of nanomaterials may fall under the Chemicals (Hazard Information and Packaging for Supply) Regulations 2002, as well as the Dangerous Substances and Explosive Atmosphere Regulations 2002 if they are capable of fueling a dust explosion (UKHSE, 2010, https://www.gov.uk/government/organisations/health-and-safety-executive).

2.4 European Union

The European Commission classified nanomaterials as hazardous chemical substances and they are regulated under the European Chemical Agency's Classification, Labeling, and

Packaging (CLP) regulations as well as the Registration, Evaluation, Authorization, and Restriction of Chemicals (REACH) regulations (Vance et al., 2015).

- Under the REACH regulations, companies are responsible for collecting information on the use and properties of substances that they manufacture or import at or above 1 ton in quantity per year, including nanomaterials (USNIOSH, 2016, https://www.nano.gov/NIOSH).
- For biocidal materials, under the Biocidal Products Regulation (BPR), there are special requirements for nanomaterials-containing cosmetics, when almost 50% of their prime particles are nanoparticles (Vance et al., 2015).

3. Physico-chemical determinants of toxicity

As explained earlier, toxicity depends upon various factors to highlight its effects; however, three significant parameters are dose, dimension, and durability of nanomaterials (Prigodich et al., 2012). Some recent studies have emphasized a greater correlation between physico-chemical properties of nanomaterials, e.g., size, mass, bulk, number, charge, aggregation, and hydrophobicity, individually or together, to determine the health impacts.

3.1 Size and surface area dependent toxicity of nanomaterials

Particle size plays a significant role when interacting with biological systems. A decrease in the size of nanoparticles may increase the surface area, which enhances its reactivity to the binding materials (Verma and Stellacci, 2010). Similarly, toxicological studies have shown that small size nanoparticles cause more adverse health effects than larger size particles (Chithrani et al., 2009; Cherukuri et al., 2010; Cui et al., 2005); due to prolonged retention in organs like the lungs, greater inflammation, increased translocation, and more significant epithelial effects occur (Chithrani et al., 2009). In general, surface area of any nanoparticles is linked with the entrance of macromolecules in all biological systems (Donaldson and Tran, 2004; Donaldson and Stone, 2003). It is seen in studies that, as the size of nanoparticles decreases, this increases the surface area, which further increases the dose-dependent oxidation and the ability of these particles to damage the DNA (Petersen and Nelson, 2010).

3.2 Dose-dependent toxicity of nanoparticles

Here, dose is defined as any amount of a substance that reaches or enters a biological structure. It has a direct link with the exposure of a substance to the duration of interaction in any relevant environmental medium (air, water, or food). The dose of a nanomaterial is generally not correlated with adverse health effects; it was reported that at higher doses, some large size particles were nontoxic and vice versa. A study based on nanoparticles of TiO_2 showed that exposure to 20 nm particle size at a low dose caused severe lung tumor than the exposure to 300 nm particle size at high dose (Champion and Mitragotri, 2006). This indicates that the real measure of effects is neither the surface area of a particle or its mass dose (Chithrani et al., 2009).

3.3 Shape and aspect ratio of nanoparticles

The shape and size of nanoparticles has been a subject of interest for nanotechnologist due to the development of efficient and targeted nanomaterial systems. Nanomaterials have varied shapes including rings, planes, fibers, tubes, etc. (Hoet et al., 2004; Lovric et al., 2005) (Fig. 1). The shape-based toxicity of nanoparticles has been reported for gold, carbon, nickel, silica, titanium, etc. This shape-dependent toxicity of many nanoparticles stimulates the membrane wrapping processes during phagocytosis (Oberdorster, 2001). It was stated that phagocytosis of spherical shape nanoparticles is faster and easier than of rod shape (Lippmann, 1990). Similarly, it was noted that shape-based toxicity in silica nanomaterials had distinct characteristics, as crystalline silica is a so-called carcinogen while amorphous silica is in use, such as in food additives (Maynard et al., 2004).

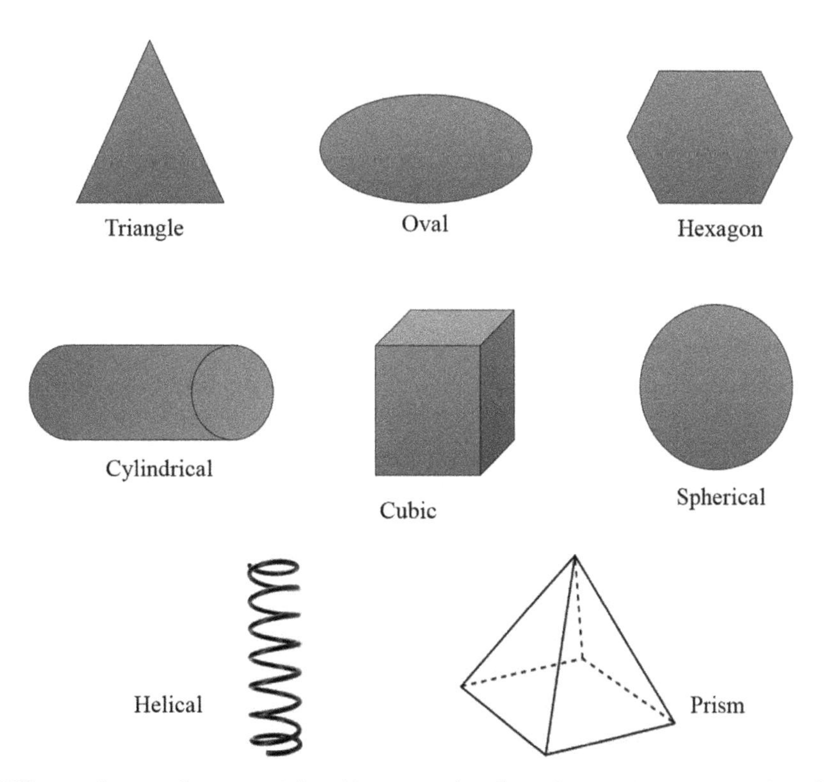

Fig. 1 Different shapes of nanoparticles. *(Concept taken from Gatoo, M.A., Naseem, S., Arfat, M.Y., Mahmood Dar, A., Qasim, K., Zubair, S., 2014. Physicochemical properties of nanomaterials: implication in associated toxic manifestations. BioMed Res. Int. 2014, 498420.)*

Moreover, the aspect ratio of nanoparticles is directly proportional to their toxic nature; the higher the aspect ratio, the higher will be toxicity. It was described in asbestos-based toxicity that longer fibers were more toxic and had lethal effects. Fibers of asbestos longer in size (>10 μm) were responsible for lung cancers, while smaller size particles (>2 μm) instigated asbestosis (Oberdorster et al., 2005). A study conducted on the fibers of TiO_2 in mice showed higher toxicity with a length of 15 mm than 5 mm (Oberdorster, 2002). New bioengineered and long aspect ratio nanoparticles of carbon, popularly called carbon nanotubes (CNTs), have attracted scientists due to their possible adverse health properties (Oberdorster et al., 1994; Aggarwal et al., 2009; Hamilton et al., 2009; Chen et al., 2006).

3.4 Concentration and aggregation dependent toxicity

In terms of concentration and aggregation of nanoparticles, there are many contradictions in studies due to the level of toxicity at different concentrations and in different mediums. Some studies reporting toxic effects contradict others on the same nanoparticles that show them causing less toxicity (Yang et al., 2008; Wick et al., 2007). This is because the level of aggregation in nanoparticles also influences their level of toxicity; three things make a difference: size, surface charge, and composition of NPs. Agglomerated CNTs are reported to have more adverse effects on lung scarring than dispersed nanotubes (Wick et al., 2007). Therefore, it should be considered while describing the concentration of NPs that higher concentration promotes aggregation in particles, which reduces the toxicity (Cherukuri et al., 2004; Takenaka et al., 2001).

3.5 Crystalline structure and chemistry-based toxicity of NPs

Although the size of nanoparticles is of significance, we cannot merely ignore the studies that have reported the importance of chemical composition-based toxicity. The chemistry of nanoparticles is pertinent, considering molecular cell interactions and oxidative stress. Based on the chemistry, nanoparticles differ in their cellular uptake and catalyzing assembly of reactive oxygen species (ROS) (Xia et al., 2006). Particles showing the same composition can have a different crystalline structure (Cherukuri et al., 2004). The crystalline structure of NPs also influences their toxicity. Taking, for example, the crystalline structures of titanium dioxide (rutile and anatase), rutile nanoparticles were responsible for damaging DNA in an oxidative state and absence of light, while anatase NPs of the same size (200 nm) were benign (Cherukuri et al., 2004). Besides crystalline structures, the interfering medium (air or water) also matters. The nanoparticles of zinc sulfide (ZnS) in the presence of water reorganize their crystalline structure and form a solid ZnS (Zhang et al., 2003).

3.6 Surface coating-based toxicity in nanoparticles

Surfactants play a critical role in determining the toxicity of nanomaterials due to their close contact with biological units. It is because of the presence of ozone, transition metals, and oxygen on the surfaces of nanoparticles, which leads to the formation of ROS and resulting inflammation (Petersen and Nelson, 2010). The surface coating on nanoparticles can reduce the toxicity of noxious particles and enhance the toxic effects of less damaging particles. For instance, nanoparticles of nickel ferrite ($NiFe_2O_4$) with and without the application of surface oleic acid display different levels of cytotoxicity (Yin et al., 2005).

4. Human health and effects of nanoparticles

There are abundant nanoparticles in nature produced by natural processes, e.g., volcanic eruption, forest fires, erosion, photochemical reactions, and many more, although a large number of nanoparticles are bioengineered by humans for multiple purposes and uses. Both natural and human-made nanoparticles can be a toxin in nature with varying levels and cause numerous health impacts. Many studies have reported the potential toxicity upon exposure in human beings (Cherukuri et al., 2004; Chen et al., 2006; Takenaka et al., 2001; Shah, 1998; Nohynek et al., 2007; Husgafvel-Pursiainen, 2004). By the inhalation of nanoparticles, diseases like bronchitis, lung cancer, asthma, emphysema, and neurodegenerative conditions such as Alzheimer's and Parkinson's are reported. In addition, nanoparticles that reach the circulatory system may cause heart disease, cardiac arrest, blood clots, and many more. Exposure to nanomaterials is also linked with autoimmune diseases like scleroderma and rheumatoid arthritis. The overall effects of nanoparticles on the human body are shown in Fig. 2.

5. Environment and nanoparticles

As described earlier, nanotechnology is an emerging and applied science with a vast array of applications. Although the environment is of grave concern when problems and solutions are considered by humans, less has been done to protect and conserve the natural environment, which is critically effected by demanding anthropogenic activities. Human beings are using and producing many materials initially for beneficial purposes, but once these materials become part of environmental pathways, it affects them in different ways. For instance, some metals are needed by the human body in trace amounts, but the problem begins when bioaccumulation occurs and changes the level of toxicity. Some materials are highly toxic in the environment, even in low concentrations, while others are not harmful even in higher concentrations (Shah, 1998).

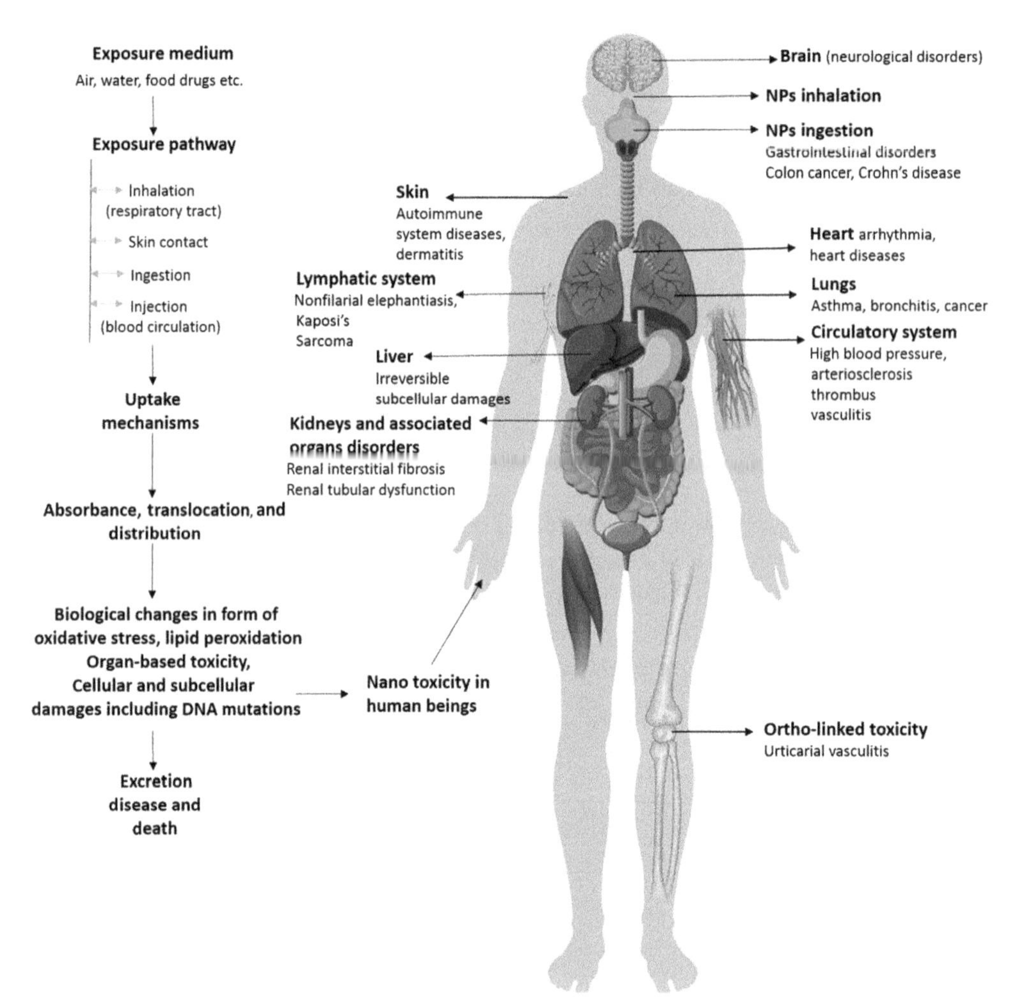

Fig. 2 Schematic illustrations of exposure pathways of nanoparticles and resulting toxicity in human beings.

In the environmental field, one of the critical applications of nanotechnology is in the water sector. Water is a resource that is essential for all life activities and has been badly misused over time. The growing population places ever-increasing demands on freshwater, and as a result, the reservoirs of fresh water are declining fast. The rest of the saltwater in the world cannot be used without treatment. The use of advanced technology based on nanomaterials like CNT membranes could minimize the cost of desalination. Similarly, nanosensors can be used for detecting water pollutants, and nanofillers can be effectively used for cleaning contaminated water (Mansoori et al., 2008; Darnault et al., 2005).

For the problem of air pollution, nanotechnology can provide promising solutions to long-term complications. Filtration techniques using nanofilters can be used in buildings, automobile exhaust pipes, and factory smokestacks to purify the air. Similarly, nanosensors can be used for the detection of gas leakage at slight concentrations (Mansoori et al., 2008; Ahmadpour et al., 2003).

Hence, nanotechnology can be used in providing efficient, cost-effective, and clean treatments for many enormous environmental problems (Mansoori et al., 2008; Shahsavand and Ahmadpour, 2004).

6. Future potential of nanoparticles

Based on the broad applications of nanotechnology, there is and will be a massive market for nanoparticles in all fields of life. On the one side, it will improve the quality of life while, on the other hand, with limited research on the nature of nanomaterials, it will have a significant impact on the life of human beings. Human wellbeing will be improved by the use of nanomaterials in biomedical, medicine, cosmetics, fabrics, electronics, animal production, and many more areas. This section will cover how the use of nanomaterials will enhance its prospects.

6.1 Biomedical science

A wide range of nanomaterials are beginning to change the medical sciences through prevention, disease treatment, diagnosis, and innovative drug delivery. The applications of nanoparticles in medical sciences are recognized to have pronounced biocompatibility; this is mainly due to their suitable diameter (range), which enables them to enter the bloodstream (Manolova et al., 2008). Furthermore, recent studies have shown that therapeutics based on these nanomaterials have tremendous potential in the treatment of inflammation, infections, diabetes, tumors, neurodegenerative diseases, and many more (Ibanez et al., 2015; Liu et al., 2015; Lv et al., 2014). Especially in the treatment of different types of cancers and tumors, it is considered that the use of nanomaterials with biochemical, spectroscopic, and optical methods will be revolutionized, while avoiding the side effects caused by previous techniques (Huang et al., 2017). Similarly, the use of nanofiber scaffolds for the treatment of regeneration of central nervous system cells has been reported (Ellis-Behnke et al., 2006). Nanopowders can kill 90% of bacteria and viruses within a few minutes, as these have antimicrobial properties (Li et al., 2006).

6.2 Electronic equipment and coating materials

The last few decades have seen a revolution in electronic devices due to using electric circuits of the smallest sizes and higher efficiency. The primary goal of microelectronics would be a fabrication of nanoscopic electric circuits. In this development, CNTs would be exciting options for overdoped semiconductor crystals due to specific properties (Thompson and Parthasarathy, 2006; Buzea and Robbie, 2005). Likewise, nanocrystalline materials can significantly enhance the display of screens by enhancing the resolutions and reducing the cost (Carey, 2003). Nanomaterials have become popular in their use as thin coating materials in optoelectronic devices, microelectronics, textile industry, architectural glass, and many more, due largely to their resistance properties.

6.3 Environmental remediation

The use of nanoparticles for environmental problems is still a new area of research interest, yet it is growing fast. The application of nanoparticles in treating air, water, and land pollution and converting these pollutants into less harmful substances is underway. Likewise, the use of nanoparticles in enhancing food productivity by limiting saline-sodic soils is also an area of great research interest (Hogan, 2004).

7. Hazard potential of nanomaterials to plants

Risk analyses based on comparison of nanomaterials (no observed adverse effect level) NOAEL values with their concentrations in wastewater treatment plant (WWTP) biosludge and biosludge-amended soil shows that, with a few exceptions, which include dissolving nanomaterials, there is a low risk of plant phytotoxicity. Mean concentrations of Ag- and Ce-based nanomaterials in biosludge-amended soil (assuming the unlikely scenario of 100% ENM persistence) were approximately 400–20,000 times lower and 200–11,000 times lower, respectively, than their lowest LOAEL values for root exposure in soil (Giese et al., 2018). As TiO_2 nanomaterials had no adverse plant effects up to a concentration of 750 mg/kg in soil, it is unlikely that a biosludge concentration of 170 mg/kg poses a risk to plants (Sun et al., 2014). Among C-based nanomaterials, CNTs and CB (which exhibited an inverse dose-response relationship of higher toxicity at lower concentrations) (Wang et al., 2017) pose a low hazard potential, as they were measured at concentrations approximately 1.5 times and 5000–50,000 times above the LOAEL value of CNTs and CB in biosludge, respectively (Sun et al., 2014; Lazareva and Keller, 2014). Even where adverse effects increased with increasing concentration in soil, the LOAEL concentration for CNTs is roughly 300 times below the concentration of CNTs measured in biosludge (Sun et al., 2014) and therefore would not have a high hazard potential. With Cu-based nanomaterials, their concentration measured in biosolids was ~200–5000 times lower than the lowest LOAEL value for root exposure in soil, indicating a nearly nonexistent risk of phytotoxicity (Lazareva and Keller, 2014). However, plants were adversely affected by short-term foliar exposures to CuO NMs (Xiong et al., 2017) and a $Cu(OH)_2$ pesticide spray, therefore this route of exposure could present a hazard to plants (Zhao et al., 2017). Only Zn-based nanomaterials were measured in biosludge at higher concentrations than the lowest level that induced toxicity from root exposure in soil (eight times higher). However, soil characteristics were shown to play a more dominant role than that of ZnO NM exposure in plant responses, and the reported adverse effects were changes to antioxidant enzyme activity levels rather than to physiological growth parameters (García-Gómez et al., 2017).

There remains a lack of understanding about plant responses to long-term, low-dose exposures to nanomaterials, including exposures that occur over successive plant generations. Therefore, it is not possible to model accurately the future severity and types of effects that might occur as a result of current agricultural plant exposures to nanomaterials. Instead, we propose that environmental and biological monitoring (biomonitoring) would present an acceptable solution for recording plant exposures to nanomaterials before other regulatory requirements are elaborated and put into practice. Wastewater treatment plant biosludge and biosludge-amended soils are ideal matrices for environmental monitoring because they form the main point of agricultural plant exposure to nanomaterials originating from consumer and industrial sources (Paterson et al., 2011). Biomonitoring goes further by providing information on the bioavailability of a given substance through the measurement of specific biomarkers in target species. Several ENM-specific plant biomarkers have been identified in a recent meta-analysis of the literature on omics-level plant responses to nanomaterials by indicating promising advances in this area (Ruotolo et al., 2018).

8. Conclusion

Nanoparticles no doubt possess different physico-chemical properties, and their diverse nature makes their exposure a challenging domain to human health. Although the exposure to these nanoparticles is not new: since the era of industrialization, human beings have been exposed to anthropogenic and natural nanoparticles. The real concern should be to minimize the adverse effects and enhance the positive outputs of these nanomaterials to human health. Therefore, to understand the nature and potential toxic effects of these materials, a multidisciplinary approach is needed. In-depth research is needed with a mutual dialogue of scientists, epidemiologists, medical specialists, and policymakers to highlight the toxic effects caused by nanoparticles, and their safe use and disposal. Therefore, it is concluded that the higher potentials of nanoparticles cannot be ignored; instead, there should be a policy to guide on the level of toxicity and their threshold for safe use.

References

Aggarwal, P., Hall, J.B., McLeland, C.B., Dobrovolskaia, M.A., McNeil, S.E., 2009. Nanoparticle interaction with plasma proteins as it relates to particle biodistribution, biocompatibility and therapeutic efficacy. Adv. Drug Deliv. Rev. 61 (6), 428–437.

Ahmadpour, A., Shahsavand, A., Shahverdi, M.R., 2003. Current application of nanotechnology in environment. In: Proceedings of the 4th Biennial Conference of Environmental Specialists Association, Tehran.

Buzea, C., Robbie, K., 2005. Assembling the puzzle of superconducting elements. Supercond. Sci. Technol. 18, R1–R8.

Carey, J.D., 2003. Engineering the next generation of large-area displays: prospects and pitfalls. Proc. R. Soc. Lond. A Math. Phys. Sci. Eng. 361, 2891–2907.

Champion, J.A., Mitragotri, S., 2006. Role of target geometry in phagocytosis. Proc. Natl. Acad. Sci. U. S. A. 103, 4930–4934.

Chen, Z., Meng, H.A., Xing, G.M., 2006. Acute toxicological effects of copper nanoparticles in vivo. Toxicol. Lett. 163 (2), 109–120.

Cherukuri, P., Bachilo, S.M., Litovsky, S.H., Weisman, R.B., 2004. Near-infrared fluorescence microscopy of single-walled carbon nanotubes in phagocytic cells. J. Am. Chem. Soc. 126, 15638–15639.

Cherukuri, P., Glazer, E.S., Curley, S.A., 2010. Targeted hyperthermia using metal nanoparticles. Adv. Drug Deliv. Rev. 62 (3), 339–345.

Chithrani, B.D., Ghazani, A.A., Chan, W.C.W., 2009. Determining the size and shape dependence of gold nanoparticle uptake into mammalian cells. Nano Lett. 6 (4), 662–668.

Cui, D., Tian, F., Ozkan, C.S., Wang, M., Gao, H., 2005. Effect of single wall carbon nanotubes on human HEK293 cells. Toxicol. Lett. 155, 73–85.

Darnault, C., Rockne, K., Stevens, A., Mansoori, G.A., Sturchio, N., 2005. Fate of environmental pollutants. Water Environ. Res. 177, 2576–2658 In this issue.

Donaldson, K., Stone, V., 2003. Current hypotheses on the mechanisms of toxicity of ultrafine particles. Ann. Ist. Super. Sanita 39, 405–410.

Donaldson, K., Tran, C.L., 2004. An introduction to the short-term toxicology of respirable industrial fibers. Mutat. Res. 553, 5–9.

Ellis-Behnke, R.G., Liang, Y.X., You, S., Tay, D.K., Zhang, S., So, K.F., Schneider, G.E., 2006. Nano neuro knitting: peptide nanofibers scaffold for brain repair and axon regeneration with functional return of vision. Proc. Natl. Acad. Sci. U. S. A. 103, 5054–5059.

García-Gómez, C., Obrador, A., González, D., Babín, M., Fernández, M.D., 2017. Comparative e_ect of ZnO NPs, ZnO bulk and $ZnSO_4$ in the antioxidant defences of two plant species growing in two agricultural soils under greenhouse conditions. Sci. Total Environ. 589, 11–24.

Giese, B., Klaessig, F., Park, B., Kaegi, R., Steinfeldt, M., Wigger, H., Gleich, A., Gottschalk, F.R., 2018. Release and concentrations of engineered nanomaterial in the environment. Sci. Rep. 8, 1565.

Hadef, F., 2018. An introduction to nano-materials. In: Dasgupta, N., Ranjan, S., Lichtfouse, E. (Eds.), Environmental Nanotechnology. Environmental Chemistry for a Sustainable World. Springer, Cham, p. 14.

Hamilton, R.F., Porter, D., Buford, M., Wolfarth, M., Holian, A., 2009. Particle length-dependent titanium dioxide nano-materials toxicity and bioactivity. Part. Fibre Toxicol. 6, 35.

Hoet, M., Bruske-Hohlfeld, I., Salata, O.V., 2004. Nanoparticles—known and unknown health risks. J. Nanobiotechnology 2, 12–27.

Hogan, J., 2004. Smog-busting paint soaks up noxious gases. New Scientist. Retrieved from: http://www.newscientist.com/.

Huang, Y., Chao-Qiang, F., Dong, H., Wang, S., Yang, X., Yang, S., 2017. Current applications and future prospects of nano-materials in tumor therapy. Int. J. Nanomedicine 12, 1815–1825.

Husgafvel-Pursiainen, K., 2004. Genotoxicity of environmental tobacco smoke: a review. Mutat. Res. 567, 427–445.

Ibanez, I.L., Notcovich, C., Catalano, P.N., Bellino, M.G., Duran, H., 2015. The redox-active nanomaterial toolbox for cancer therapy. Cancer Lett. 359 (1), 9–19.

IOS, 2012. ISO/TR 13329:2012: Nano-Materials—Preparation of Material Safety Data Sheet (MSDS). International Organization for Standardization Accessed 25-03-2020.

Lazareva, A., Keller, A.A., 2014. Estimating potential life cycle releases of engineered nanomaterials from wastewater treatment plants. ACS Sustain. Chem. Eng. 2, 1656–1665.

Li, Y., Leung, P., Yao, L., Song, Q.W., Newton, E., 2006. Antimicrobial effect of surgical masks coated with nanoparticles. J. Hosp. Infect. 62, 58–63.

Lippmann, M., 1990. Effects of fiber characteristics on lung deposition, retention, and disease. Environ. Health Perspect. 88, 311–317.

Liu, L., Lv, H.Y., Teng, Z.Y., Wang, C.Y., Wang, G.X., 2015. Glucose sensors based on core-shell magnetic nano-materials and their application in diabetes management: a review. Curr. Pharm. Des. 21 (37), 5359–5368.

Lovric, J., Bazzi, H.S., Cuie, Y., Fortin, G.R.A., Winnik, F.M., Maysinger, D., 2005. Differences in subcellular distribution and toxicity of green and red emitting CdTe quantum dots. J. Mol. Med. 83 (5), 377–385.

Lv, X.N., Wang, P., Bai, R., 2014. Inhibitory effect of silver nano-materials on transmissible virus-induced host cell infections. Biomaterials 35 (13), 4195–4203.

Manolova, V., Flace, A., Bauer, M., Schwarz, K., Saudan, P., Bachmann, M.F., 2008. Nanoparticles target distinct dendritic cell populations according to their size. Eur. J. Immunol. 38 (5), 1404–1413.

Mansoori, G.A., Bastami, T.R., Ahmadpour, A., Eshaghi, Z., 2008. Environmental application of nanotechnology. In: Annual Review of Nano Research. second ed. World Scientific Publisher & Co.

Maurya, P., Singh, S., Naik, R.R., Shakya, A.K., 2020. Biohazards of nano-materials. In: Integrative Nanomedicine for New Therapies.https://doi.org/10.1007/978-3-030-36260-7_3.

Maynard, A.D., Baron, P.A., Foley, M., Shvedova, A., Kisin, E.R., Castranova, V., 2004. Exposure to carbon nanotube material: aerosol release during the handling of unrefined single-walled carbon nanotube material. J. Toxicol. Environ. Health A 67, 87–107.

NCRPM, 2017. Radiation Safety Aspects of Nanotechnology. National Council on Radiation Protection and Measurements. 2017-03-02. pp. 2–6, 88–90, 119–130. https://ncrponline.org. Accessed 27 March 2020.

Nohynek, G.J., Lademann, J., Ribaud, C., Roberts, M.S., 2007. Grey goo on the skin? Nanotechnology cosmetic and sunscreen safety. Crit. Rev. Toxicol. 37, 251–277.

Oberdorster, G., 2001. Pulmonary effects of inhaled ultrafine particles. Int. Arch. Occup. Environ. Health 74, 1–8.

Oberdorster, G., 2002. Toxicokinetics and effects of fibrous and nonfibrous particles. Inhal. Toxicol. 14, 29–56.

Oberdorster, G., Ferin, J., Lehnert, B.E., 1994. Correlation between particle size, in vivoparticle persistence and lung injury. Environ. Health Perspect. 102 (Suppl 5), 173–179.

Oberdorster, G., Oberdörster, E., Oberdörster, J., 2005. Nanotoxicology: an emerging discipline evolving from studies of ultrafine particles. Environ. Health Perspect. 113, 823–839.

Paterson, G., Macken, A., Thomas, K.V., 2011. The need for standardized methods and environmental monitoring programs for anthropogenic nanoparticles. Anal. Methods 3, 1461–1467.

Petersen, E.J., Nelson, B.C., 2010. Mechanisms and measurements of nonmaterial-induced oxidative damage to DNA. Anal. Bioanal. Chem. 398 (2), 613–650.

Pokropivny, V., Lohmus, R., Hussainova, I., Pokropivny, A., Vlassov, S., 2007. Introduction in Nano-Materials and Nanotechnology. (A Dissertation) University of Tartu 225 p.

Prigodich, A.E., Randeria, P.S., Briley, W.E., 2012. Multiplexed nanoflares: MRNA detection in live cells. Anal. Chem. 84 (4), 2062–2066.

Ruotolo, R., Maestri, E., Pagano, L., Marmiroli, M., White, J.C., Marmiroli, N., 2018. Plant response to metal-containing engineered nanomaterials: an omics-based perspectiveEnviron Sci. Technol. 52 (5), 2451–2467

Shah, C.P., 1998. Public Health and Preventive Medicine in Canada. University of Toronto Press, Toronto, Canada.

Shahsavand, A., Ahmadpour, A., 2004. The role of nanotechnology in environmental culture development. In: Proceedings of the First International Seminar on the Methods for Environmental Culture Development, Tehran.

Shodhganga, 2020. Nano-Materials—General Introduction. https://shodhganga.inflibnet.ac.in/bitstream/10603/22774/9/09_chapter1.pdf. Accessed 9 April 2020 (Chapter 1).

Sun, T.Y., Gottschalk, F., Hungerbühler, K., Nowack, B., 2014. Comprehensive probabilistic modelling of environmental emissions of engineered nanomaterials. Environ. Pollut. 1987 (185), 69–76.

Takenaka, S., Karg, E., Roth, C., Schulz, H., Ziesenis, A., Heinzmann, U., Schramel, P., Heyder, J., 2001. Pulmonary and systemic distribution of inhaled ultrafine silver particles in rats. Environ. Health Perspect. 109 (Suppl. 4), 547–551.

Thompson, S.E., Parthasarathy, S., 2006. Moore's law: the future of Si microelectronics. Mater. Today 20–25.

Turkevich, L.A., Joseph, F., Dastidar, A.G., Paul, O., 2016. Potential explosion hazard of carbonaceous nanoparticles: screening of allotropes. Combust. Flame 167, 218–227.

UKHSE, 2010. Fire and Explosion Properties of Nanopowders. U.K. Health and Safety Executive. pp. 2, 13–15, 61–62, https://www.gov.uk/government/organisations/health-and-safety-executive. Accessed 1 April 2020.

USNIOSH, 2009. Building a Safety Program to Protect the Nanotechnology Workforce: A Guide for Small to Medium-Sized Enterprises. U.S. National Institute for Occupational Safety and Health. March 2016https://www.nano.gov/NIOSH. Accessed 25 March 2020.

USNIOSH, 2013. Current Intelligence Bulletin 65: Occupational Exposure to Carbon Nanotubes and Nanofibers. U.S. National Institute for Occupational Safety and Health. April 2013, https://www.nano.gov/NIOSH. Accessed 25 March 2020.

USNIOSH, 2016. Building a Safety Program to Protect the Nanotechnology Workforce: A Guide for Small to Medium-Sized Enterprises. U.S. National Institute for Occupational Safety and Health. March 2016. https://www.nano.gov/NIOSH. Accessed 25 March 2020.

Vance, M.E., Todd, K., Vejerano, E.P., McGinnis, S.P., Hochella, J.M.F., David, R., Hull, M.S., 2015. Nanotechnology in the real world: redeveloping the nano-material consumer products inventory. Beilstein J. Nanotechnol. 6 (1), 1769–1780.

Varghese, R.J., Sakho, E.H.M., Parani, S., Thomas, S., Oluwafemi, O.S., Wu, J., 2019. Introduction to nano-materials: synthesis and applications. In: Thomas, S., Sakho, E.H.M., Sakho, J.W. (Eds.), Nano-Materials for Solar Cell Applications. pp. 75–95. https://doi.org/10.1016/B978-0-12-813337-8.00003-5.

Verma, A., Stellacci, F., 2010. Effect of surface properties on nanoparticle-cell interactions. Small 6 (1), 12–21.

Wang, Y., Chang, C.H., Ji, Z., Bouchard, D.C., Nisbet, R.M., Schimel, J.P., Gardea-Torresdey, J.L., Holden, P.A., 2017. Agglomeration determines effects of carbonaceous nanomaterials on soybean nodulation, dinitrogen fixation potential, and growth in soil. ACS Nano 11, 5753–5765.

Warheit, D.B., 2018. Hazard and risk assessment strategies for nanoparticle exposures: how far have we come in thepast 10 years? [version 1; referees: 2 approved] F1000Res. 7 (F1000 Faculty Rev), 376. https://doi.org/10.12688/f1000research.12691.1.

WHO, 2018. WHO Guidelines on Protecting Workers From Potential Risks of Manufactured Nano-Materials. WHO. https://www.who.int/en/. Accessed 1 April 2020.

Wick, P., Manser, P., Limbach, L.K., 2007. The degree and kind of agglomeration affect carbon nanotube cytotoxicity. Toxicol. Lett. 168 (2), 121–131.

Xia, T., Kovochich, M., Brant, J., Hotze, M., Sempf, J., Oberley, T., Sioutas, C., Yeh, J.I., Wiesner, M.R., Nel, A.E., 2006. Comparison of the abilities of ambient and manufactured nano particles to induce cellular toxicity according to an oxidative stress paradigm. Nano Lett. 6, 1794–1807.

Xiong, T., Dumat, C., Dappe, V., Vezin, H., Schreck, E., Shahid, M., Pierart, A., Sobanska, S., 2017. Copper oxide nanoparticle foliar uptake, phytotoxicity, and consequences for sustainable urban agriculture. Environ. Sci. Technol. 51, 5242–5251.

Yang, S., Wang, X., Jia, G., 2008. Long-term accumulation and low toxicity of single-walled carbon nanotubes in intravenously exposed mice. Toxicol. Lett. 181 (3), 182–189.

Yin, H., Too, H.P., Chow, G.M., 2005. The effect of particle size and surface coating on the cytotoxicity of nickel ferrite. Biomaterials 26, 5818–5826.

Zhang, H., Gilbert, B., Huang, F., Banfield, J.F., 2003. Water-driven structure transformation in nanoparticles at room temperature. Nature 424 (6952), 1025–1029.

Zhao, L., Hu, Q., Huang, Y., Keller, A.A., 2017. Response at genetic, metabolic, and physiological levels of maize (*Zea mays*) exposed to a $Cu(OH)_2$ nanopesticide. ACS Sustain. Chem. Eng. 5, 8294–8301.

CHAPTER 9

Protection and hazard controls for exposure reduction measurements

Ayesha Baig[a], Muhammad Zubair[a], Shafaqat Ali[b], Mujahid Farid[c], and Muhammad Bilal Tahir[d]
[a]Department of Chemistry, University of Gujrat, Gujrat, Pakistan
[b]Department of Environmental Sciences and Engineering, Government College University, Faisalabad, Pakistan
[c]Department of Environmental Sciences, University of Gujrat, Gujrat, Pakistan
[d]Department of Physics, Khwaja Fareed University of Engineering and Information Technology, RYK, Pakistan

1. Introduction

Several nanomaterials (NMs) have gained popularity in the scientific innovations due to adjustable biological, chemical, and physical properties and improved functionality compared to their bulk counterparts. Nanomaterials have wide diversity, and are therefore classified on the basis of their structure, shape, origin, and size. Due to increased progress in development and economic applications of nanomaterials, concerns associated with their toxicity are unavoidable. The higher consumption of constructed nanoparticles (NPs) increases the hazards associated with their release into the environment, e.g., the absorbance and translocation of nanomaterials within plants (Abd-Elsalam et al., 2020). As advancement in nanoscience continues to drive transformations in drug delivery, energy efficiency, consumer products, medical imaging, and several other important applications, responsible implementation of such technologies has remained a critical societal issue. Particularly, the potentially enhanced human exposure to ENMs through inhalation, dermal penetration, or ingestion due to increased prevalence of these nanomaterials in commercial applications is a very important concern (Zhang et al., 2020). Indeed, the lack of appropriate characterization and detection techniques and lack of validated and reproducible methods for toxicological studies have been recognized as major bottlenecks for sustainable and safe use of nanomaterials (Johnston et al., 2020). Indeed, due to the occurrence of numerous accidents, research topics in the field of safety in the academic institutes have recently been prioritized.

The small size of nanomaterials results in nanotoxicity, which usually enables passage through physiological barriers and fast distribution in the human body (Dolez, 2015). Exclusive physicochemical properties of nanoparticles make the evaluation of their specific toxic effects complicated and complex. Thus it is essential to gain specific data on the distinct mechanisms involved in nanomaterial toxicity and action. One important mechanism of nanotoxicity is oxidative stress and generation of ROS (reactive oxygen

Nanomaterials: Synthesis, Characterization, Hazards and Safety
https://doi.org/10.1016/B978-0-12-823823-3.00009-4

species) (Jain et al., 2018). The small size of nanoparticles facilitates the active chemical species to translocate from organismal barriers like lung, skin, organs, and body tissues. Thus, organelle damage, irreversible oxidative stress, cancer, and asthma can be caused by nanoparticles, depending on their composition. Uptake by the reticuloendothelial system, generation of neo-antigens, nucleus, and neuronal tissue that cause dysfunction and possible organ enlargement are commonly occurring chronic toxic effects of NPs (Jeevanandam et al., 2018). In order to avoid or minimize the toxicity of nano-materials, proper safety precautions or preventive measures should be adopted. Laboratories and rooms where NPs are handled must be properly labeled. In particular, when NPs are handled openly (as dry powder), adequate protective measures (gloves, respiratory mask, lab coats) must be implemented. Staff who work even for a short period of time have to be properly instructed, according to their tasks they have to perform and their place of work (Dhawan et al., 2011).

In order to avoid or minimize potential hazards of engineered nanomaterials (ENMs) in consumer or commercial products, laws and regulations are implemented in many countries of the world. In regulatory toxicology, the environmental hazard assessment data is obtained more often by using single species toxicity tests and these tests are also utilized for hazard identification of ENMs (Boyle et al., 2020). Recent trends in nanotechnology have led toward the development of viral nanoparticles (VNPs) for targeted therapy and imaging. VNPs are basically virus-like particles (both animal and plant viruses)—bacteriophages—and can cause infectious or noninfectious effects in the human body. VNPs are self-assembling, dynamic systems that form monodisperse, polyvalent, and highly symmetrical/rod-shaped structures. A wide range of conjugation chemistries should be implemented for endowing these viral NPs with various functions and making them less likely to cause pathogenic and undesirable side effects in humans (Steinmetz, 2010). Additionally, these emerging VNPs and nanozymes should be subjected to the rigorous toxicity tests for establishing benign mechanisms of dosage levels and application. Extensive research is essential in nanotoxicology and by government agencies, strict laws are necessary to identify and avoid using toxic NPs (Jeevanandam et al., 2018).

2. Exposure of chemicals

The nanoparticles can be ingested, inhaled, deposited in lungs or on the skin surface, potentially transfer within the body and may have nanostructure-associated toxicity. Exposure studies are absolutely essential for selecting and developing desired toxicity screening tests (Haynes and Asmatulu, 2013). At present, there are comparatively few environments where there is occurrence of chemical exposure. However, if NMs-containing products are commercialized as anticipated, exposure potential will increase dramatically over the forthcoming decades. Thus, potential human exposure should be estimated and future use should be made after consideration of developing toxicity screening (Oberdörster et al., 2005).

2.1 Pulmonary exposure

Due to aerosolization and small size, exposure of respirable aggregated or singlet nanoparticles by inhalation route is of concern.

2.1.1 Inhalation

Inhalation is a preferred method of respiratory tract exposure to collect dose response data and for hazard identification. The specific biokinetics of the phagocytic, transcytotic, and endocytotic processes of nanoscale particles depend on in vivo surface modifications and chemistry. With continued exposure, inhalation of the ^{13}C based nanoparticles (i.e., 35 nm in size) results in the translocation of a significant increment to the olfactory bulb (Papp et al., 2008). There should be well-controlled particle size and generated aerosol concentration, and special attempts should be made to reproduce the conditions of human exposure for the particular types of airborne nanomaterials (Oberdörster et al., 2005).

2.1.2 Intratracheal instillation

In an appropriate vehicle, intratracheal instillation of the suspended nanomaterials is considered as an acceptable strategy for pulmonary exposure, to analyze or investigate the toxic test material. Special efforts should be applied to disaggregation of these nanomaterials suspended in the vehicle. Due to the high cost of well-characterized, homogenous material and difficulties associated with the generation of aerosols of C-nanotubes to ease an evaluation of inhaled nanomaterial, many in vivo pulmonary studies consider intratracheal instillation as an exposure methodology. SWCNTs (single-walled carbon nanotubes) have been evaluated in rodents by using this technique.

2.1.3 Laryngeal and pharyngeal aspiration

Laryngeal and pharyngeal aspiration is an effective exposure method for relatively even distribution of particles throughout the lungs. However, during this pulmonary exposure, a risk associated with this process is the unintentional release of small food particles from the oral cavity of the human body. Therefore, the intake of food must be stopped at least one night before aspiration exposure and it will definitely be helpful in prevention of inflammation caused by pulmonary exposure to NiO nanoparticles (Hadrup et al., 2020).

2.1.4 Factors affecting pulmonary toxicity

The factors influencing pulmonary toxicity of nanoparticles are:

(a) size distribution and particle number;

(b) the degree of agglomeration or aggregation of ENMs, which strongly influences the properties of deposition of nanoscale particles in the lungs;
(c) particular dose of nanoparticles to the target tissues in the human body;
(d) nanoparticle synthesis method; liquid (precipitated/colloidal) or gas (fumed) phase synthesis;
(e) postsynthetic changes in NPs which influence their aggregation behaviors;
(f) particle electrostatic attraction potential and shape;
(g) surface charges on particles; and
(h) surface treatment on particles, particularly for ENMs.

2.2 Oral exposure

There is a possibility of inadvertent ingestion or appearance of a nanomaterial in the water supply during its production, application, or disposal, etc. Therefore, the consequences of oral exposure to nanomaterials should be investigated. In particular, the liver, GALT, and mesenteric lymph nodes should be analyzed for the presence of nanoparticles (Shatkin and Kim, 2015). If absorption of nanomaterial from the gastrointestinal tract is nearly zero, then there is no need for evaluation of systemic effects of oral exposure. However, if there is a significant amount of nanomaterial absorbed, systemic toxicity evaluation is recommended by using multiple functional and histology assays.

2.3 Injection

Some nanomaterials are being considered as drug delivery systems. In such cases, there should be a potential toxicity evaluation of these nanomaterials after their injection. After exposure, there should be injection of a tagged nanoparticle and, for a week, its distribution to different body organs (spleen, liver, heart, lung, kidney, bone marrow) (Fig. 1) and elimination in feces should also be monitored.

2.4 Dermal exposure

The skin, or integument, is unique, represents the largest organ of the body, and is a potential pathway for exposure to toxicants such as novel nanoparticles during their preparation process. Within the avascular epidermis, the skin provides an environment where these nanoparticles can be potentially lodged and are not susceptible to being removed by phagocytosis. Due to particle size, nanoparticles may travel across the skin (stratum corneum layers) at varying rates or they may undergo a sequestration within the skin epidermis, increasing their exposure durations to the viable keratinocytes of epidermis. Larger particles of Ti (titanium) or ZnO (zinc oxide) used in skin care consumer products have the potential to penetrate layers (stratum corneum) of rabbit skin with enhanced absorption due to water and oily vehicles (Oberdörster et al., 2005).

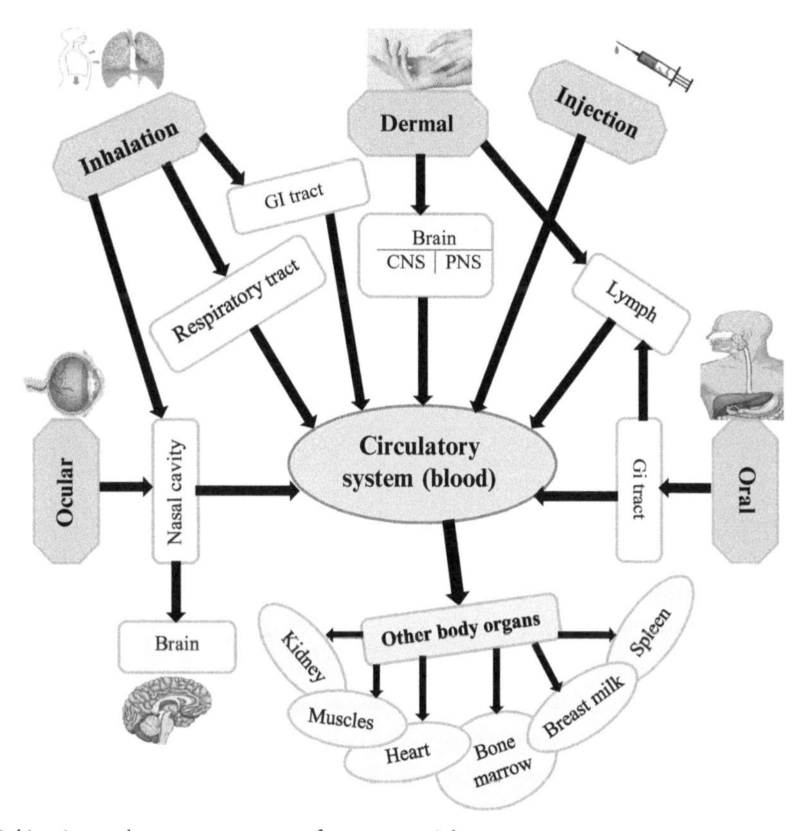

Fig. 1 Biokinetics and exposure routes of nanomaterials.

Intentional exposure of dermal layers to nanomaterials may include the application of creams/lotions containing ZnO or TiO_2 nanoscale particles as a sunscreen component or as fibrous materials with such nanomaterials used for their stain/water repellent characteristics. These metal-oxide based nanomaterials also cause inflammation by the secretion of a pro-inflammatory cytokines, i.e., IL-1β (Leso et al., 2018).

Concerns regarding dermal penetration are:

(a) toxicity of photo-activated nanoparticles;

(b) metabolism to smaller sized particles with potentially enhanced toxicity;

(c) skin/any other organ toxicity; and

(d) accumulation/aggregation in the skin or any other body organs, which results in hazardous and highly toxic effects after an exposure for a longer duration.

Skin biopsy should be taken for TEM (transmission electron microscopy), which will localize the penetrated nanoparticles within the skin and identify the cellular changes as well. Clinical chemistry, local lymph nodes evaluation, immune-toxicology tests, and hematology must be done for evaluation of potential toxicity of nanomaterials.

Table 1 Possible modes of action and group-specific categories for NPs related to hazard or risk assessment (Kuempel et al., 2012).

Type of particles	Solubility	Toxicity	Effects	Mode of action
Copper oxide (I) Zinc oxide	Highly soluble	Poor	Acute lungs effect systemic toxicity	Toxic ion reaches the systemic tissues
Carbon black Titanium dioxide	Poor soluble	Low	Fibrosis and lung inflammation; lung cancer (rats)	Toxicity with respect to the total retained/deposited particle doses in the target respiratory tract
Nickel oxide (III) Crystalline silica Chromium oxide (III)	Poor soluble	High	Fibrosis and lung inflammation; lung cancer	Same as poorly soluble low toxicity reactive surfaces (e.g., reactive oxygen species)
Carbon nanofibers Carbon nanotubes	Not soluble	Highest	Possible cancer, lung fibrosis and mesothelioma	Biopersistence or migration of NPs from lung tissues to pleural and into alveolar walls

2.5 Ocular exposure

Extensive studies have been made on the toxicity and hazardous effects of nanomaterials to the human body organs but a highly vulnerable organ, i.e., the eye, is almost always neglected. Although the eye is a small part of the human body, the ocular system is responsible for obtaining about 90% of the outside information from the surrounding environment of a human. Ocular toxicity by exposure to nanomaterials can result in severe eye damage and diseases like dry eye syndrome, conjunctiva chemosis, conjunctiva hyperemia, argyrosis, etc. (Zhu et al., 2019). The eye is a very sensitive and accessible organ for both intentional and unintentional exposure to nanomaterials. Thus, the nanoparticles' interactions are used for determination of biocompatibility, hazardous effects, biological performance, and stability of the toxic or nontoxic nanomaterials and this strategy is also helpful in developing benign and safe NMs (Mehra et al., 2016). Different types of nanoparticles, their solubility, toxicity, effects and modes of action are given in Table 1.

3. Chemical hazards

Nanomaterials are generally considered harmful because they become transparent to the cell-dermis. The toxicity of nanomaterials also appears predominant due to their enhanced surface activity and high surface area. Nanomaterials have been indicated to be carcinogenic and also cause irritation (Thomas et al., 2013). If inhaled, the low mass of these nanoparticles allows them to accumulate in the lungs, and there is no way to eliminate them from the body. Their interaction with the blood/liver could also result in harmful effects (Alagarasi, 2011). For many insoluble NMs, it cannot be excluded that an intake by the inhalation of particles of nano-size might pose hazards in the workplace, irrespective of the classification these substances on the basis of their chemical composition (Heinemann and Schäfer, 2009). The longer the exposure time of a researcher to ENMs, the higher are the chances for increased amounts of nonmaterial intake and risk of exposure. But in the case of exposure to highly toxic ENMs, one dose is enough to cause very serious health hazards. When laboratories have been fully classified, and complete hazard portfolio has been developed, a set of strategical and technical measures are established to ensemble the nano classified labs. This hazard portfolio gives a strategy for hazard management and a systematic overview of hazards in a company. It aids with prioritization of mitigation of various hazards and also identifies cases that require a risk assessment (Novello et al., 2020). Generally, it is assumed that when ENMs are present in powder form, the potential for occurrence of major accidents is relatively increased, because they are easily dispersible than the suspended ENMs. This is because in the regulation of major accidents, only the first 30 min after the accident are of interest, from which acute damage to the environment and acute loss of lives is calculated (Nowack et al., 2014). The potential is also enhanced by storing explosive or flammable compounds in the vicinity of ENMs.

3.1 Protection/safety measures

The fundamental safety concerns are:

(a) the nature and extent of hazardous health effects that can be caused by the possible exposures to nanomaterials and the particular types of toxic nanoparticles associated with such harmful effects; and

(b) the dosing patterns and doses of nanomaterials required to avoid toxicity or for the safe usage of these dangerous nano-sized substances (Warheit et al., 2007).

3.2 Occupational human health and safety measures from toxicity of hazardous nanomaterials

The growing usage of ENMs in consumer products creates a very important requirement in predicting and managing as well as understanding health hazards associated with human exposure, mainly in the workplace, where nanomaterials are manufactured and then incorporated into consumer products. For workers who handle or work with these NMs, inhalation of the NPs is a pathway of occupational human exposures of the highest concern, which is then followed by exposure through skin and ingestion (Stebounova et al., 2012). The important issue regarding human exposure to ENMs is due to their size or other novel physicochemical properties, which may cause unpredictable biologically harmful effects. NMs also pose relatively higher hazards or risks for catalytic reactions and risks of explosion/fire than the larger counterparts of such nanomaterials.

3.3 Occupational human exposure and maintenance of hygiene in industries

Owing to concerns about occupational exposure to ENMs, the National Institute for Occupational Safety and Health (NIOSH) and many other private and public institutes are encouraging a number of collaborative efforts to develop the safe handling recommendations for workers potentially exposed to ENMs. One such recommendation is developing occupational health surveillance programs that are inclusive of ENMs (Hallock et al., 2009). Occupational health surveillance is the systematic collection of nanomaterial exposure and collected health data for a particular group of workers with the goal of early detection of disease and ultimately, prevention of disease (Stebounova et al., 2012). Additionally, occupational health surveillance programs can be used in order to analyze whether the hierarchy of controls for prevention of injury and illness are effective (Rodríguez-Ibarra et al., 2020). The development of such occupational health surveillance programs requires answers to the critical questions about nanoparticle properties relevant to toxicity, how and where exposure is occurring in the industrial processes, personal protective equipment, engineering controls, and the health effects of exposure. Benchmark occupational exposure controls or limits for nanomaterials were suggested by the British Standards Organization (BSO) in January 2008. Conducting exposure assessments for ENMs utilizing a framework of lifecycle analysis contributes to scientific understanding of the exposure to nanomaterials and it is a necessary component for developing risk management and risk assessment strategies for nanomaterials (Stebounova et al., 2012). While there are still several hurdles in the understanding of exposure to nanomaterials—including their uncertain hazard potential, undetermined metrics of exposure, i.e., complicated by lack of standardized methodologies of measurement for nanomaterials, and uncertainty of what health outcomes and biological intermediates (i.e., biomarkers) are associated with the exposure to nanoscale materials—exposure assessment is an important first step in designing adequate and appropriate risk management programs. However, best control methods for safe handling of various toxic nanomaterials are given in Table 2.

3.4 Classification of the nanomaterials

Nanomaterials are mainly classified into three major categories:

(a) nanoclays
(b) nanoemulsions
(c) nanoparticles.

Table 2 Summary of best practices used for safe handling of nanomaterials.

Control methods	Tasks, process, and equipment
Substitution	Change design to minimize or eliminate the hazardous materials.
Elimination	Replace highly toxic material with material having comparatively low toxicity.
Technical measures	Create physical barriers between a person and nanomaterial. For this purpose, a valid control strategy is the usage of respiratory protective equipment, down-flow booths, and/or air filters of ultra-low penetration.
Organizational measures	In order to minimize the number of people exposed to toxic nanomaterials, segregate the work areas of the workers dealing with nanomaterials (Haynes and Asmatulu, 2013). Air concentration levels should be monitored. Regularly clean the work areas of workers handling nanomaterials. Health surveillance programs should also include the workers potentially exposed to nanomaterials with their detailed exposure situations.
Respiratory protection	Synthesis should be carried out in reactor/furnace: exhaust the gases from reactor, always purge it before opening and provide exhaust ventilation points for emission. Use biosafety cabinet, fume hood, or other exhausted enclosure. Avoid use of nanomaterials on the open lab-bench.
Dermal protection	Use sturdy gloves on hands for the dry or powder-form particulates of nanomaterials (Groso et al., 2010). If NPs are in suspension form, use solvent-resistant gloves. Use appropriate eye protection. Use lab coats, preferably disposable lab coats.
Prevention of fire and/or explosion	Handling should be carried out in an inert atmosphere. Conditions facilitating electrostatic charging, ignition sources, or low spark equipment should be removed from workplace. Materials should be made soluble by wetting the workplace. Antistatic bags should be used for storing flammable or explosive materials (Dhawan et al., 2011).
Engineering	Use enclosure/isolation, filtration, ventilation, and collection.
Personal protective equipment	Use goggles, earplugs, clothing, gloves, and respirators.
Administration	Adhere to policies, shift design, procedures, and all rules and regulations (Haynes and Asmatulu, 2013).

3.5 Nanoclays

Manufacturing of both nanoclays and organoclays is done by utilizing the hydrophilic (charged) nature of clay molecules like aryl/alkyl-ammonium, imidazolium, or phosphonium in the solid/aqueous state. The ion exchange reactions have two important consequences:

(a) The gap between the single sheets is broadened, thus enabling the movement of the chain of organic cations between these single sheets.

(b) The outer surface characteristics of each single sheet are modified from hydrophilic to organophilic/hydrophobic.

3.6 Nanoemulsions

Dispersion of the polymer, solid material, and droplets in the form of viscous liquid leads to a very soft and interesting material. The internal or dispersed phase is known as the discontinuous phase, the dispersion medium is the outer phase, which is also known as external or continuous phase. The emulsifying agent in such nanoemulsions is known as an interphase/intermediate phase. Nanoemulsions can be easily synthesized by using two processes or methods: low (solvent displacement, phase inversion composition, and by the phase inversion temperature) and higher energy emulsification (micro fluidizer, higher pressure homogenization, and ultrasonification) (Mageswari et al., 2016). Important characteristics and usefulness of NPs have proved to be in different aspects, such as stability, polydispersity, size, and dimensions of the biomolecules (poly nucleic acids, proteins) and NPs.

3.7 Nanoparticles

Nanoparticles can exist in the form of composites/nanostructures. All the nanomaterials can be manufactured from blocks of one, two, as well as of three dimensions (Fig. 2). 1D nanomaterials are materials with only one dimension in nm scale and electrons can move freely in only the *X*-direction, e.g., thin films, circuitry of computer chips, surface coatings, hard, thin coatings on eyeglasses, antireflection, etc. 2D nanomaterials are materials with two dimensions in nm scale and the free electrons move in both *X* and *Y*-directions, e.g., 2D nanostructured films, asbestos fibers, etc. (Buzea et al., 2007). 3D nanomaterials have all three dimensions in nm scale and their electrons move freely in the *X*, *Y*, and *Z*-directions, e.g., a number of thin films that are deposited under special conditions that ultimately generate free NPs, colloids, and atomic size porosity. On the basis of nature of material fabrication, NPs can be classified into inorganic and organic NPs. Inorganic NPs are shaped by the precipitation process of inorganic salts, which then interconnect with chemical molecules by metallic and covalent bonding.

Organic NPs assemble themselves in three-dimensional geometries. Thus, for the fabrication of organic NPs, self-assembly and reactivity of the zwitterionic particles

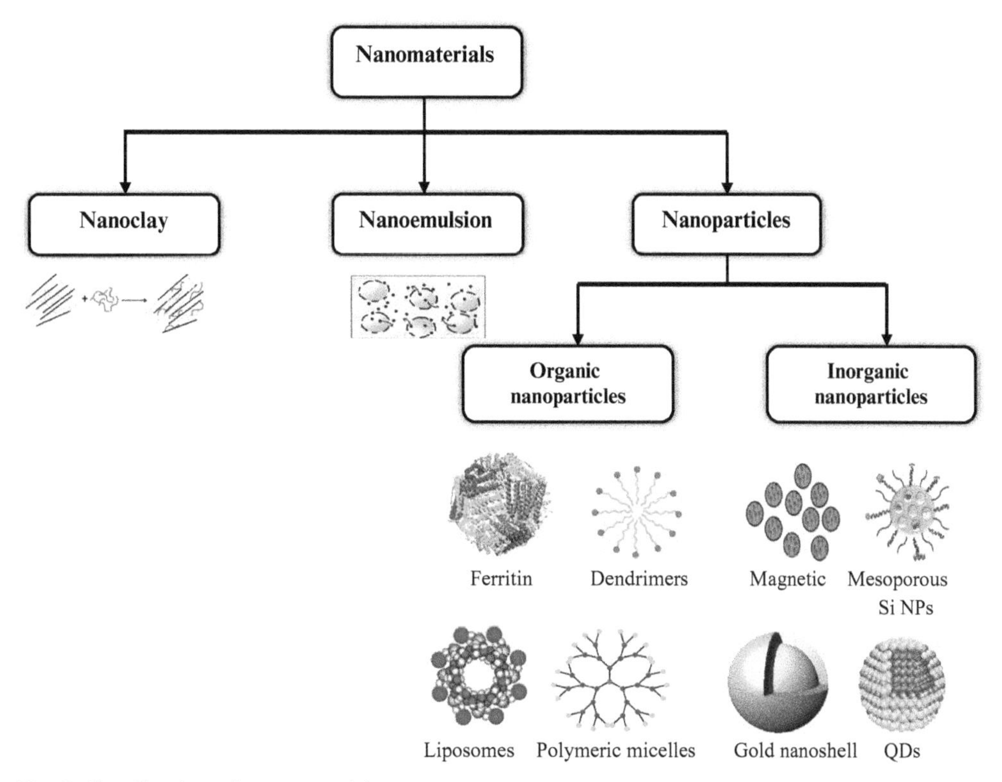

Fig. 2 Classification of nanomaterials.

toward nonpolar as well as polar areas are the key elements (Mageswari et al., 2016). Organic nanoparticles can be easily synthesized by using both synthetic and natural organic molecules like complex structure (e.g., viruses), lipid bodies, milk emulsion, and protein aggregates. Nanocomposites are found as a microcrystalline matrix, i.e., micro-nano type, in which NPs or inclusions (plate-like, fiber-like, spherical) of a second phase are dispersed in both intra/inter granular spaces or in intergranular regions of matrix grains (Kumar et al., 2019). On the basis of matrix material, nanocomposites can be categorized as: ceramic matrix nanocomposite, polymer matrix nanocomposite, and metal matrix nanocomposite (Khan et al., 2016), as shown in Fig. 3.

4. Reactivity of toxic nanomaterials

In recent years, global research on the biosafety of nanomaterials has demonstrated that nanomaterials have a negative impact on living organisms—importantly, on humans. Nanoparticles can destroy the health and ecology of organisms by entering the interior of cells and even by penetrating the blood-brain barrier. Therefore, while developing

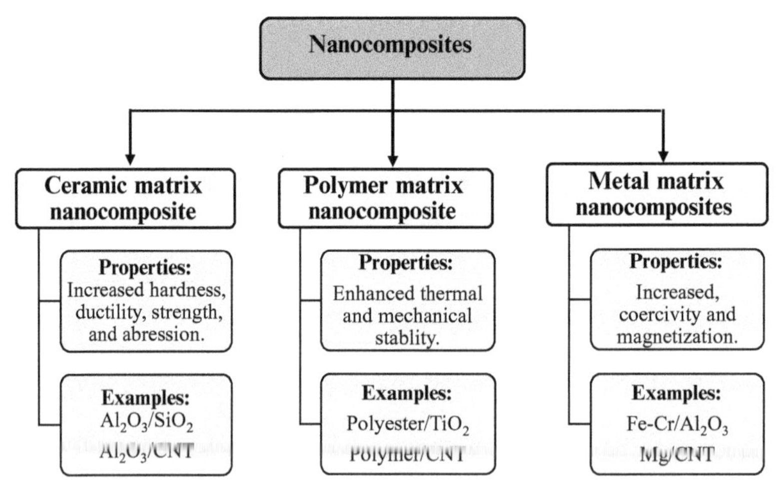

Fig. 3 Classification of nanocomposites.

nanotechnology, human beings are seeing potential hazards of nanomaterials to biological health and the ecological environment. In order to understand potential risks to the environment and humans, it is necessary to understand NMs' release, transform, and transport into the matrix. Nanoparticles-containing products enter the atmosphere through intentional or unintentional routes. The intentional routes include waste water discharge from industry and households, wastewater treatment, land fill, and environmental cleanup. Consumption of NPs-containing products, deposition due to air pollution, and accidental spills can cause add to the unintentional routes of NMs exposure (Baysal and Saygin, 2020). By using such pathways or exposure routes, NPs may be released into various environmental areas or compartments such as soil, air, and aqueous systems (sediment, fresh water, sea water, etc.).

Four types of metal nanomaterials are widely used for different biomedical applications, namely platinum (Pt) and gold (Au) NPs as well as silver (Ag) and copper (Cu) nanoparticles.

4.1 Toxicity of platinum (Pt) nanoparticles

After human exposure, small-sized platinum (Pt) nanoparticles, released from lysosomes and then transferred to nucleus, could destroy DNA. The potential influence of Pt nanoparticles is also determined in cellular redox-systems by using DCF assay.

4.2 Toxicity of gold (Au) nanoparticles

It is believed that gold nanoparticles can induce liver necrosis and gene changes by their accumulation in the spleen and liver (Sharma et al., 2018). Bovine serum albumin-protected nanoclusters of gold (Au) exhibit lower efficiency in cases of renal clearance by almost 1% of gold (Au) discharge. Compared to protection with BSA, GSH-protected gold (Au) NPs can be cleared by about 36% after 24 h. Both of these forms of Au NPs are almost equally harmful but the toxicity of gold nanoparticles can be efficiently reduced by their modification with GSH.

4.3 Toxicity of silver (Si) nanoparticles

In vitro experiments widely use silver (Ag) NPs; these silver (Ag) NMs also cause different levels of hazard/toxicity to the environment as well as organisms, such as apoptosis, DNA damage, and oxidative stress. Toxic effects of Ag NPs having variable size of 4, 20, and 70 nm on the macrophage line U937 in the human body system have been studied and it is concluded that Ag nanoparticles of about 4 nm particle size promoted the secretion of inflammatory factor IL-8 and oxidative stress (Yang et al., 2019).

4.4 Toxicity of copper (Cu) nanoparticles

It has been demonstrated that >95% copper (Cu) released into the surrounding environment would enter into aquatic sediments, soil, and water, where it could reach potentially hazardous levels, i.e., >50–500 μg/L. While about 97%–99% of Cu NPs were sequestered in the plant leaves, a very small fraction of about 1%–3% were transported and distributed to root tissues of the plants through the plant phloem loading. Similarly, toxic effects at much lower concentrations of Cu NPs are also observed in aquatic organisms like marine amphipods and in freshwater daphnids (Yang et al., 2019).

5. Reduction/control

5.1 Toxicity controls for quantum dot (QD) materials

The many exceptional qualities and superior optical properties of NMs themselves, and potential for semiconductor fluorescent quantum dots (QDs) as efficient materials in the biomedical field, in vivo imaging, and biology labeling, make these materials widely applicable. But tiny QDs invade the body and affect cellular function, ecological environment and also cause negative impacts on the individual. **Quantum Dots** emit light when they are near radiations, and this is a way to show/detect nearby radiological materials. Therefore, quantum dots could be a radiation detector for screening applications and security purposes (Kangas and Pitkänen, 2019). Thus, research on the safety of QDs has received great attention and it is an important issue to be solved for the usage of QDs. The biocompatibility of Cd QDs with the highest luminous efficiency has restricted their clinical applications. Yang et al. demonstrated that the enhanced use of QDs in the field of biomedicine is due to their greater potential and unique nature for such applications. However, due to the toxic effects of complex mechanisms of QDs, current research is confined to animals and cells. The synthesis of QDs is a diversified process due to a lot of factors affecting their toxicity. For controlling such water soluble and toxicity problems of QDs, reaction from modifying the raw materials should be properly controlled, and many other experiments are required on reducing the toxicity of QDs. Thus, for the application of QDs on a large scale in various fields, the following aspects need to be focused on: improvement in synthesis process, standardized evaluation mechanisms, standards for quantification of toxic size of QDs, enhancement of the persuasion of toxicological data of QDs, and use of nontoxic or QDs with relatively low toxicity instead of toxic or highly toxic QDs (Yang et al., 2019).

5.2 Safer design approaches for controlling toxicity of ENMs

5.2.1 Doping process

Numerous research studies have demonstrated that doping is a very useful method for reducing cytotoxicity of industrially important engineered nanomaterials like SiO_2, CuO, and ZnO nanoparticles. The doping process is an effective yet facile method for modifying a material's crystal lattice or structure by adding some impurities, to obtain improved or desired electro-optical, catalytic, magnetic, physical, or chemical characteristics. Dopants like aluminum (Al), titanium (Ti), and iron (Fe) are uniformly incorporated into the host material's crystal lattice for changing the binding energy of the metal ions and oxygen atoms, or for the reduction of reactive chemical groups' density, which are present on the particle surface. The working mechanism of the doping process in the decrease or reduction of engineered nanomaterial cytotoxicity is based on the physicochemical characteristics of the changing nanoparticle, which involves either release of toxic ions or decrease in nanoparticle dissolution (Oksel et al., 2016), modifications of the reactive surfaces in order to reduce ROS (reactive oxygen species) generation, or perturbation of the cellular membranes, which ultimately leads to cell death or inflammation (Hwang et al., 2018). FSP (flame spray pyrolysis) is a well-established method for the doping process of the NMs, which then employs a rapid combustion or explosion during preparation or synthesis procedure. Through a liquid precursor, a self-sustaining flame with large temperature gradient and high local temperature allows for formation of the homogenous crystalline nanomaterials from gas or droplet to particle. This is a suitable process for industrial applications because of facile one-step synthesis and the potential to produce doped nanomaterials (Hwang et al., 2018).

5.2.2 Surface coating of the materials

Coating is one of the major surface modification strategies used to design safer engineered nanomaterials. Unlike doping, which may have disadvantages like irreversible chemical modifications that can alter intrinsic properties of ENMs, surface coating can be a reversible process through noncovalent modification. In addition, the dispersion state of the nanomaterials plays a key role in the determination of their bioreactivity, bioavailability, pro-inflammatory, and potential toxicological responses. Thus, toxicity of the nanomaterials can be influenced with alteration in the dispersion state by different noncovalent coatings (Hwang et al., 2018).

5.2.3 Adjustment of the materials' surface chemistry and characteristics

Adjustments of hydrophobicity and charge density are the mechanisms by which the modifications in surface chemistry can ameliorate toxicity of nanomaterials and improve functionality in biomedical applications, e.g., for targeted drug delivery. Such properties of surface chemistry can be adjusted by the covalent bonding of functional groups present on the ENM's surface. Functional groups include cationic, anionic, and nonionic groups, which can impact the hydrophobicity and surface charge density. Covalent functionalization is an effective and facile method to create safer and functional CNTs for commercial applications. Moreover, the hydration process opens epoxy rings in pristine GO, which increases density of carbon radicals that leads to the extensive peroxidation of lipid, followed by the failure of membrane integrity, and ultimately results in cell death. Evidently, for environmental and human exposure, a safer design option is reduced GO (Hwang et al., 2018).

5.2.4 Adjustment or modifications in the aspect ratio

Optimization of aspect ratio of nanomaterials is another effective method in order to reduce their cytotoxicity. Synthesis and preparation of nanomaterials in rod or wire shapes could diversify their uses and develop new applications (Ramsden, 2018). Although with both low and high aspect ratios, nanomaterials can be taken up in the cells, studies have concluded that fiber-like particles, having high aspect ratios, can cause damage to intracellular organelles, may not be fully engulfed by the macrophages, and ultimately result in cytotoxicity and inflammation. Therefore, the process of designing safer nanomaterials can be made possible with the help of size manipulation of nanowires and nanorods to produce materials with the optimal aspect ratios, which will not trigger inflammatory responses (Hwang et al., 2018).

5.3 Nanomaterial regulations

Nanomaterials possess properties such as greater bioavailability, high reactivity and bioactivity, and tissue as well as cellular and other organ penetration ability. These unique characteristics of nanomaterials make them superior in biomedical applications. However, these merits are also avenues for potential toxicity. Thus, regulations via rules, legislation, and laws have been implemented by many government organizations to avoid or minimize the risks associated with nanomaterials (Hull and Bowman, 2018). However, there are no specific internationally agreed upon legal definitions or protocols for production, labeling, or handling, international regulation, evaluating environmental impact, and testing toxicity of NPs. Medical standards related to medical governance, ethics, and environmental safety have been modified to cover the introduction of nanomaterials into the biomedical field. Currently, the European Union and United States have guideline legislation and strong regulatory bodies to control potential risks of NMs. The European Commission has developed many pieces of technical guidance and legislation for the European Union with specific references to nanomaterials (Groso et al., 2016). Such legislation has been employed in EU countries to ensure conformity across the legislative areas and to ensure that a nanomaterial in one sector will be treated the same when it is utilized in another sector (Jeevanandam et al., 2018). According to the European Commission, the term "nanomaterial" means: incidental, manufactured, and natural particles, where for 50% or more than 50% of particles in number size distribution, in unbound state or as agglomerate or as an aggregate, one or more than one external dimensions is in the size range of 1–100 nm. As the specifications of products and materials meet the substance definitions of CLP (European classification and labeling of chemicals) and REACH (European chemical agency), the provisions in such regulations apply. It can be recognized that cosmetics face moderations and regulations from SCCS (Scientific Committee on Consumer Safety), REACH (Bowman et al., 2018) and ICCR (International Cooperation on Cosmetic Regulation).

6. Conclusion

Materials engineered to nano-scale have unique optical, magnetic, electrical, and other properties. These emergent properties have the potential for great impacts in electronics, medicine, and other fields. Nanomaterials can be used to develop nanomedicines that can target specific organs or cells in the body, such as cancer cells, and enhance the effectiveness of therapy. Recently, nanomaterials have also been used in cement and clothing, to make them lighter and stronger. Their size makes them extremely useful in environmental remediation of toxic material and water cleanup. Nanomaterials provide great benefits; however, we are not fully aware of the potential effects on human health and the environment. Even well-known materials, such as silver, may pose a hazard when engineered to nano size. Nano-sized particles can enter the human body through inhalation, ingestion, or dermally. Researchers consider nanomaterial a "double-edged sword" because the properties such as size, shape, and high reactivity that make nanomaterial potentially beneficial are the same properties that make them hazardous for biological systems and the environment. This work suggests that minimizing exposure and inhalation of nanoparticles is a sustainable principle until further long-term assessments have been conducted. Nanomaterials are advanced materials that are replacing conventional materials because of their unique properties. These materials have a significant place in the areas of energy, drug delivery, nanomedicines, cosmetics, sunscreen, and other consumer products. Above all, presently there is no or limited assessment of the health and environmental risks of these nanomaterials. To control the risk of nanomaterials, researchers are concerned with and committed to prevention through design of any potential hazards in the production, use, or disposal of nanoscale products and devices by anticipating them in advance.

References

Abd-Elsalam, K.A., Kasem, K., Almoammar, H., 2020. Carbon nanomaterials (CNTs) phytotoxicity: Quo vadis? In: Abd-Elsalam, K.A. (Ed.), Carbon Nanomaterials for Agri-Food and Environmental Applications. Elsevier (Chapter 24).

Alagarasi, A., 2011. Introduction to nanomaterials. In: National Center for Environmental Research. pp. 141–198.

Baysal, A., Saygin, H., 2020. Smart nanosensors and methods for detection of nanoparticles and their potential toxicity in air. In: Nanomaterials for Air Remediation. Elsevier.

Bowman, D.M., May, N.D., Maynard, A.D., 2018. Nanomaterials in cosmetics: regulatory aspects. In: Analysis of Cosmetic Products. Elsevier.

Boyle, D., Clark, N.J., Handy, R.D., 2020. Toxicities of copper oxide nanomaterial and copper sulphate in early life stage zebrafish: effects of pH and intermittent pulse exposure. Ecotoxicol. Environ. Saf. 190, 109985.

Buzea, C., Pacheco, I.I., Robbie, K., 2007. Nanomaterials and nanoparticles: sources and toxicity. Biointerphases 2, MR17–MR71.

Dhawan, A., Shanker, R., Das, M., Gupta, K.C., 2011. Guidance for safe handling of nanomaterials. J. Biomed. Nanotechnol. 7, 218–224.

Dolez, P.I., 2015. Progress in personal protective equipment for nanomaterials. In: Nanoengineering. Elsevier.

Groso, A., Petri-Fink, A., Magrez, A., Riediker, M., Meyer, T., 2010. Management of nanomaterials safety in research environment. Part. Fibre Toxicol. 7, 40.

Groso, A., Petri-Fink, A., Rothen-Rutishauser, B., Hofmann, H., Meyer, T., 2016. Engineered nanomaterials: toward effective safety management in research laboratories. J. Nanobiotechnol. 14, 21.

Hadrup, N., Saber, A.T., Kyjovska, Z.O., Jacobsen, N.R., Vippola, M., Sarlin, E., Ding, Y., Schmid, O., Wallin, H., Jensen, K.A., 2020. Pulmonary toxicity of Fe_2O_3, $ZnFe_2O_4$, $NiFe_2O_4$ and $NiZnFe_4O_8$ nanomaterials: inflammation and DNA strand breaks. Environ. Toxicol. Pharmacol. 74, 103303.

Hallock, M.F., Greenley, P., Diberardinis, L., Kallin, D., 2009. Potential risks of nanomaterials and how to safely handle materials of uncertain toxicity. J. Chem. Health Saf. 16, 16–23.

Haynes, H., Asmatulu, R., 2013. Nanotechnology safety in the aerospace industry. In: Asmatulu, R. (Ed.), Nanotechnology Safety. Elsevier, Amsterdam (Chapter 7).

Heinemann, M., Schäfer, H., 2009. Guidance for handling and use of nanomaterials at the workplace. Hum. Exp. Toxicol. 28, 407–411.

Hull, M.S., Bowman, D.M., 2018. Nanotechnology environmental health and safety—learning from the past, preparing for the future. In: Hull, M.S., Bowman, D.M. (Eds.), Nanotechnology Environmental Health and Safety. third ed. William Andrew Publishing, Boston (Chapter 1).

Hwang, R., Mirshafiee, V., Zhu, Y., Xia, T., 2018. Current approaches for safer design of engineered nanomaterials. Ecotoxicol. Environ. Saf. 166, 294–300.

Jain, A., Ranjan, S., Dasgupta, N., Ramalingam, C., 2018. Nanomaterials in food and agriculture: an overview on their safety concerns and regulatory issues. Crit. Rev. Food Sci. Nutr. 58, 297–317.

Jeevanandam, J., Barhoum, A., Chan, Y.S., Dufresne, A., Danquah, M.K., 2018. Review on nanoparticles and nanostructured materials: history, sources, toxicity and regulations. Beilstein J. Nanotechnol. 9, 1050–1074.

Johnston, L.J., Gonzalez-Rojano, N., Wilkinson, K.J., Xing, B., 2020. Key challenges for evaluation of the safety of engineered nanomaterials. NanoImpact 18, 100219.

Kangas, H., Pitkänen, M., 2019. Nanomaterials in industry—how to assess the safety? In: Thomas, S., Grohens, Y., Pottathara, Y.B. (Eds.), Industrial Applications of Nanomaterials. Elsevier (Chapter 1).

Khan, W.S., Hamadneh, N.N., Khan, W.A., 2016. Polymer nanocomposites–synthesis techniques, classification and properties. In: Science and Applications of Tailored Nanostructures.p. 50.

Kuempel, E., Castranova, V., Geraci, C., Schulte, P., 2012. Development of risk-based nanomaterial groups for occupational exposure control. J. Nanopart. Res. 14, 1029.

Kumar, P.S., Pavithra, K.G., Naushad, M., 2019. Characterization techniques for nanomaterials. In: Nanomaterials for Solar Cell Applications. Elsevier.

Leso, V., Fontana, L., Iavicoli, I., 2018. Nanomaterial exposure and sterile inflammatory reactions. Toxicol. Appl. Pharmacol. 355, 80–92.

Mageswari, A., Srinivasan, R., Subramanian, P., Ramesh, N., Gothandam, K.M., 2016. Nanomaterials: classification, biological synthesis and characterization. In: Ranjan, S., Dasgupta, N., Lichtfouse, E. (Eds.), Nanoscience in Food and Agriculture 3. Springer International Publishing, Cham.

Mehra, N.K., Cai, D., Kuo, L., Hein, T., Palakurthi, S., 2016. Safety and toxicity of nanomaterials for ocular drug delivery applications. Nanotoxicology 10, 836–860.

Novello, A.M., Buitrago, E., Groso, A., Meyer, T., 2020. Efficient management of nanomaterial hazards in a large number of research laboratories in an academic environment. Saf. Sci. 121, 158–164.

Nowack, B., Mueller, N.C., Krug, H.F., Wick, P., 2014. How to consider engineered nanomaterials in major accident regulations? Environ. Sci. Eur. 26, 2.

Oberdörster, G., Maynard, A., Donaldson, K., Castranova, V., Fitzpatrick, J., Ausman, K., Carter, J., Karn, B., Kreyling, W., Lai, D., 2005. Principles for characterizing the potential human health effects from exposure to nanomaterials: elements of a screening strategy. Part. Fibre Toxicol. 2, 8.

Oksel, C., Subramanian, V., Semenzin, E., Ma, C.Y., Hristozov, D., Wang, X.Z., Hunt, N., Costa, A., Fransman, W., Marcomini, A., 2016. Evaluation of existing control measures in reducing health and safety risks of engineered nanomaterials. Environ. Sci. Nano 3, 869–882.

Papp, T., Schiffmann, D., Weiss, D., Castranova, V., Vallyathan, V., Rahman, Q., 2008. Human health implications of nanomaterial exposure. Nanotoxicology 2, 9–27.

Ramsden, J.J., 2018. The safety of nanofacture and nanomaterials. In: Ramsden, J.J. (Ed.), Applied Nanotechnology. third ed. William Andrew Publishing (Chapter 14).

Rodríguez-Ibarra, C., Déciga-Alcaraz, A., Ispanixtlahuatl-Meráz, O., Medina-Reyes, E.I., Delgado-Buenrostro, N.L., Chirino, Y.I., 2020. International landscape of limits and recommendations for occupational exposure to engineered nanomaterialsToxicol Lett. 322, 111–119

Sharma, S., Mehta, S.K., Parmar, A., Sachar, S., 2018. Understanding toxicity of nanomaterials in the environment: crucial tread for controlling the production, processing, and assessing the risk. In: Hussain, C.M. (Ed.), Nanomaterials in Chromatography. Elsevier (Chapter 18).

Shatkin, J.A., Kim, B., 2015. Cellulose nanomaterials: life cycle risk assessment, and environmental health and safety roadmap. Environ. Sci. Nano 2, 477–499.

Stebounova, L.V., Morgan, H., Grassian, V.H., Brenner, S., 2012. Health and safety implications of occupational exposure to engineered nanomaterials. Wiley Interdiscip. Rev. Nanomed. Nanobiotechnol. 4, 310–321.

Steinmetz, N.F., 2010. Viral nanoparticles as platforms for next-generation therapeutics and imaging devices. Nanomedicine 6, 634–641.

Thomas, S.P., Al-Mutairi, E.M., De, S.K., 2013. Impact of nanomaterials on health and environment. Arab. J. Sci. Eng. 38, 457–477.

Warheit, D.B., Borm, P.J., Hennes, C., Lademann, J., 2007. Testing strategies to establish the safety of nanomaterials: conclusions of an Ecetoc workshop. Inhal. Toxicol. 19, 631–643.

Yang, L., Luo, X.-B., Luo, S.-L., 2019. Assessment on toxicity of nanomaterials. In: Luo, X., Deng, F. (Eds.), Nanomaterials for the Removal of Pollutants and Resource Reutilization. Elsevier (Chapter 9).

Zhang, T., Gaffrey, M.J., Thomas, D.G., Weber, T.J., Hess, B.M., Weitz, K.K., Piehowski, P.D., Petyuk, V.A., Moore, R.J., Qian, W.-J., Thrall, B.D., 2020. A proteome-wide assessment of the oxidative stress paradigm for metal and metal-oxide nanomaterials in human macrophages. NanoImpact 17, 100194.

Zhu, S., Gong, L., Li, Y., Xu, H., Gu, Z., Zhao, Y., 2019. Safety assessment of nanomaterials to eyes: an important but neglected issue. Adv. Sci. 6, 1802289.

CHAPTER 10

Nanomaterial safety regulations

Maria Batool[a], Muhammad Faizan Nazar[a], Muhammad Bilal Tahir[b], Muhammad Sagir[c], and Saira Batool[d]
[a]Department of Chemistry, Faculty of Science, University of Gujrat, Gujrat, Pakistan
[b]Department of Physics, Khwaja Fareed University of Engineering and Information Technology, RYK, Pakistan
[c]Department of Chemical Engineering, Khwaja Fareed University of Engineering and Information Technology, RYK Pakistan
[d]Department of Physics, Faculty of Sciences, University of Punjab, Lahore, Pakistan

1. Nanomaterials

Nanoparticles can occur in nature, can be produced by accident as a side product, or can be manufactured purposefully. Many researchers have shed light on the issue of toxicity related to different nanoparticles (Gupta and Gupta, 2005; Kreyling et al., 2004; Rothen-Rutishauser et al., 2006) due to its high market value and excessive use (Taylor, 2002). A global review focused on health and safety research risks related to nanomaterials and nanotechnology resulted in a report (EMERGNANO) in an effort to evaluate worldwide developments in relation to nanotechnology risk management (Aitken et al., 2009). Nanomaterials are categorized further in relation to this into the following different types, along with their potential risk levels.

1.1 Nanofibers

Nanofibers come into the category of nanomaterials with one dimension in a larger size range (not in the nanoscale range) and two dimensions in the nanoscale range. According to the World Health Organization (WHO), a nanofiber has length greater than 5 μm and width less than 3 μm. The highest concern of researchers is about the toxicity or carcinogenicity of manufactured nanofibers, as their morphology very closely resembles asbestos. Asbestos fibers inhalation causes progressive fibrotic disease of the lung (asbestosis) and can cause lung cancer. In nanofibers, carbon nanotubes are responsible for most of the health hazards, which is why it is the subject of intensive research. Morphology and toxicity relationships of carbon nanotubes are observed to be the same (Magrez et al., 2006; Poland et al., 2008) as for asbestos fibers. This increases the need to design less harmful nanofibers (Magrez et al., 2009).

1.2 Nanopowder

Nanopowders, due to their smaller size and high surface area (Garcia-Garcia et al., 2005), exhibited good transmission into (Rothen-Rutishauser et al., 2007) and through epithelial cells (Yacobi et al., 2008) and to other body parts. Transmission of stable agglomerates

Nanomaterials: Synthesis, Characterization, Hazards and Safety
https://doi.org/10.1016/B978-0-12-823823-3.00006-9

and aggregates does not happen through specific transfer routes, thus health is affected by them in the same way as by ambient air pollution particles (Brook et al., 2004). Occupational safety is heavily compromised due to less control over manufacturing and other processes when nanopowder is supplied by other laboratories or external sources. Moreover, consumers use nanopowder more often in congested places.

1.3 Nanoparticles in colloidal suspension

Nanoparticles in colloidal suspensions are used in many applications and investigations. Compared to bare nanoparticles, there are significantly fewer nanoparticles in a suspension as colloidal stability needs to be maintained in this by surface modification or derivatization. This ultimately increases the difficulty for assessing colloidal toxicology. The nature of nanoparticles, along with dispersant in suspension, affects their risk level.

Airborne droplets can be a route for nanoparticles into the lungs, so aerosols should only be released into closed containers. The risk of exposure is decided on the base of the nanoparticles' state.

There is significantly less risk of exposure when the nanomaterial is incorporated into a solid matrix than when it is associated with aerosol formation during mixing or sonication when dealing with nanoparticles in a solution. Dry nanoparticles in the form of a powder are potentially the biggest contributor to nanomaterial exposure, as nanoparticles can travel great distances due to their high mobility. There are three primary exposure routes of nanomaterial: into the dermal system, pulmonary system, and gastrointestinal system through absorption, inhalation, and ingestion, respectively.

The many potential health effects (Fig. 1) due to nanomaterial exposure include bronchitis, asthma, emphysema, Parkinson's disease, Alzheimer's disease, heart attack, and cancer of the lungs, liver, and colon.

2. Exposure routes

2.1 Inhalation

Air-suspended nanoparticles provide the major cause of exposure due to the majority of particles being in the breathable size range. The lungs provide a route for bioaccumulation of nanoparticles into the body as, once they enter into the lungs, they can translocate into other areas. Exposure effects can be divided into three types depending upon the size range:

1. Nanoparticles with size above 2500 nm will tend to settle in the upper airway.
2. Nanoparticles with size less than 2500 nm will penetrate deep into the alveolar region.
3. Nanoparticles with size less than 100 nm will diffuse into the lungs, as in this range, particles behave more like gas than solid particles.

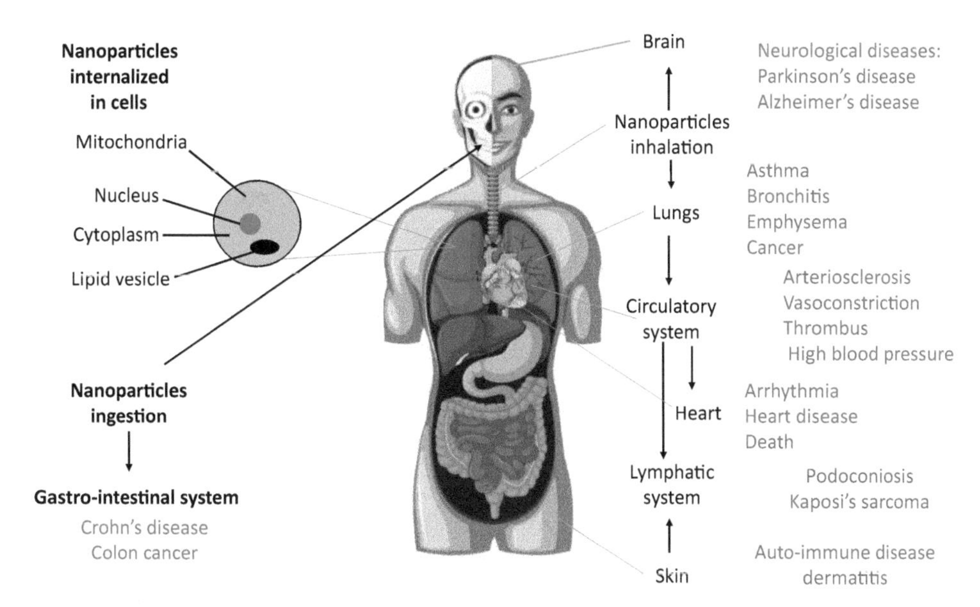

Fig. 1 Diseases linked to nanomaterial exposure.

2.2 Dermal exposure

Nanoparticle exposure to the skin causes skin irritation; for example, dermatitis (skin disease) is known to be caused by nickel. Their small size allows the passage of nanoparticles through the skin pores or hair follicles. Penetration into deeper skin layers mostly happens through injured skin. Increased oxidative stress and cytokine inflammation due to nanoparticles exposure can lead to cell death.

2.3 Ingestion

Nanoparticles can pass into the gastrointestinal tract by unintentional oral consumption. This unintentional consumption can be due to unwashed hands or contaminated drinks or food. The adverse effects of ingested nanoparticles are still not completely understood. After ingestion, nanoparticles can gather in the spleen, liver, and kidney, which can lead to permanent damage.

3. Protective measures

The exposure routes considered most important are inhalation and skin contact. Measures are structured in consequence.

Protective measures are classified as technical, organizational, and personal for different laboratories.

3.1 Technical measures

These measures involve elimination or substitution of toxic nanomaterials. However, the feasibility of this method is questionable as material to be substituted may also be toxic. This measure involves alteration of some aspects of the procedure, such as use of a liquid suspension or less toxic solvent in place of loose powder. Technical measures at laboratory level involve eradication of hazardous material on the spot, or use of F7 filters.

3.2 Organizational measures

Organizational protective measures are largely the same in all laboratories:

1. Each laboratory must have an accountable person who can supervise all the protective measures in the laboratory.
2. Entry of pregnant women should be prohibited in nanochemistry laboratories, except when work approval is issued by an organization physician.
3. Laboratory audits in regard of safety measures should be performed regularly by specialists.
4. Laboratory staff working on a permanent basis should be subject to medical surveillance of potential target organs.

3.3 Personal measures

Personal protective measures involve particular equipment depending on the level of different hazards—for example, use of mask with air respirator for workers working for longer periods of time and use of P3 and FFP3 filtering masks for workers working for shorter periods of time.

3.4 Cleaning management

Regular cleaning staff should be allowed to clean only in less hazardous nanomaterials laboratories. More hazardous nanomaterials laboratories must be cleaned by trained workers wearing standard protective equipment under the supervision of nano-experts.

4. Workplace exposure

Workplace exposure is related to areas of nanomaterials production, handling, processing, or usage by professionals. It includes hospitals, construction sites, manufacturing factories, chemical laboratories, and pharmaceutical laboratories. Workers should ask their employer to determine usage of nanomaterials in their workplace. Chemicals and material used at the workplace should be assessed in relation to nanomaterials with potential health or safety hazard in particular.

Workforces should be informed and trained if exposed to nanomaterials, at least on the following points:

- Nanomaterials and their related processes that come into use of employees should be identified.
- Workers should be aware of all the results of nanomaterials exposure assessments.
- Employees should have proper knowledge in engineering and administrative controls. Personal protective equipment (PPE) should be provided to reduce exposure to nanomaterials.
- In cases of accidental spill or release, emergency measures should be taken.

OSHA proposed a recommended exposure limit (REL), which suggested the exposure limit of nanomaterials to workers. The exposure limit should not exceed 1 $\mu g/m^3$, 0.3 mg/cm^3, and 2.4 mg/m^3 for carbon nanotubes/nanofibers, titanium-based nanoparticles, and fine-sized titanium particles, respectively.

Owners should evaluate workers' exposure to nanomaterial and their control measures to check existing effectiveness of control measures and required changes to improve. This can be accomplished by determining physical state of nanomaterial, exposure routes, workers' exposure frequency, and ongoing processes, by identifying required changes to improve, and by sampling methods to assess the duration and quantities of airborne nanoparticles.

5. Consumer exposure

Consumer goods are commonly sold products for personal use instead of commercial use. Nowadays, most of these goods contain nanomaterials; thus, the extent to which the community is exposed to these goods and their resulting effects are discussed here. Apart from inhalation, other exposure routes are also considered important in consumer exposure. Therefore, consumer exposure routes are described here briefly.

5.1 Consumer exposure directions

5.1.1 Dermal exposure track

Consumer product exposure through direct body contact such as through textiles, shoes, belts, and jewelry can cause exposure to nanomaterials. It can also occur by deposition of particles from airborne substances. Cosmetics are considered the major cause of dermal exposure to nanomaterials. Specific guidance on performance of nanomaterials risk assessment of cosmetics was published by the European Scientific Committee on Consumer Safety (SCCS) (2012). One hundred percent absorption is considered as default absorption when no data are provided regarding the amount of nanoparticles absorbed. An evaluation was done on dermal absorption of titanium nanoparticles, which proved that titanium nanoparticles do not enter the epidermis or dermis layer of skin, as no appropriate amount of it was detected in the epidermis or dermis layer of skin. It has been detected only in the outer stratum corneum layers.

In addition, the effect of UV sunburn was studied in compromised skin. In contrast, another study showed penetration of nano-scale SiO_2 into the skin up to the dermis layer. It was also detected in dendritic cells. A detailed investigation regarding nano-SiO_2 penetration, extent of penetration, and possible effects was carried out (Hoet, 2016), whereas the release of nano-silver and nanoTiO_2 from industries (Benn and Westerhoff, 2008; Windler et al., 2012) is less well-studied. Studies also show simulated release of nano-silver particles during consumer use. The released amount of silver was significant, while only a minor amount of TiO_2 was detected. However, overall absorption of nanoparticles through skin was recorded as low compared to other routes (Poland et al., 2007).

5.1.2 Oral exposure and uptake

In oral exposure, uptake of nanoparticles happens through cosmetics and dishwashing products, and may occur through chewing or licking of toys in the case of children. Clearance of nanoparticles from lungs with the mucus, which then ends up being swallowed, is another route of oral exposure. Food is the most significant source for oral exposure to nanomaterial. A risk assessment for oral exposure through food has been reported in the scientific opinion of EFSA (2016).

Assessment of nano-silver particles was required as its release from plastic to food was reported, but it was not evaluated by EFSA. Its use in plastic was therefore prohibited due to fear of unknown risks (Mackevica et al., 2016). Nanomaterial assessment in oral exposure was also performed by using artificial fluid, in order to track changes of nanomaterial during the course of its transfer through the gastrointestinal tract (Peters et al., 2012).

5.1.3 Inhalation exposure

Exposure through inhalation is considered one of the main concerns in consumer exposure to nanomaterials; this happens through inhalation of spray products, makeup powder, and wear residues of articles or paints. Usually, exposure by inhalation remains within the normal range of concentration, but exposure in higher concentration range in shorter duration should be considered seriously. A German-funded project regarding this issue addressed the exposure of nanomaterials through spray products (Riebeling et al., 2016).

5.1.4 Other exposure routes

Other exposures routes in consumer exposure may include exposure via use of mascara, as it can contain nano-carbon black, and tattooing, which can cause incursion of nanoparticles through dermal routes.

6. Legislation and regulation

In the context of the rapid growth and diversity in applications of nanomaterials in products, increased release of nanomaterials into the environment occurs through various exposure directives. European Parliament Resolution (2009) recognized the need for formulation of law in relation to nanomaterials release limit value and quality standards of air and water. However, there is no solid proof regarding the exact concentration of nanomaterials in the environment or their impacts on the environment. There are some regulations, which are discussed below (Ganzleben et al., 2011).

6.1 Regulation no. 2008/98/EC on waste framework

Waste is the main source of nanomaterial so release of it in the environment needs to be controlled by regulations that assesses release, treatment, and discarding of waste containing nanomaterials. Altogether, there is a need for determination of the kind, quality, and volume of nanowaste in a specific waste stream.

6.2 Regulation no. 2008/98/EC on waste

Waste framework directives do not set out specific criteria regarding waste materials containing nanomaterial; it is therefore treated as general waste to assess possible risks associated with it. Thus regulation (2008/98/EC) was formulated, which defines waste containing nanomaterial, and contains details that tackle disposal of this specific waste material and formulate key points for its management.

6.3 Regulation no. 1272/2008 on classification

This regulation is for legislation of chemical, it classifies chemical waste of labelling and packaging as nonhazardous or hazardous. This regulation was for categorizing waste as hazardous according to their bulk form or nonhazardous if not enough knowledge regarding this material is present. Therefore, steps will be taken only on the basis of the fact that either it is hazardous or not, without any specific requirements.

6.4 Regulation (EC) no 1907/2006 on assessment

This regulation is related to REACH, which is an abbreviation of Registration, Evaluation, Authorization, and restriction of chemicals. It is a detailed report of a chemical's possible effects on the environment during its life cycle. This is especially designed for materials that are utilized and commercialized in large amounts (10 tons) per year. Their exposure limits and possible impacts should therefore be evaluated critically.

6.5 Directive 2000/53/EC

This directive covers manufacture of vehicles involving use of nanomaterial in body parts, coatings, exhaust systems, lubrication, and electrical equipment. Directive 2000/53/EC focuses on reuse of vehicle waste to reduce exposure in the environment.

6.6 Landfill directive 1999/31/EC

Products are dumped as waste at the end phase of their life along with toxic nanomaterial. Directive 1999/31/EC8 sets rules for waste dumping into the environment with the aim of reducing toxicity in surface water, groundwater, soil, and air.

6.7 WEEE directive 2002/96/EC

New-generation computers and other electronics are active source of nanomaterials. Thats why Waste Electronic and Electrical Equipment (WEEE) Directive was formulated to check the contribution of nanomaterials through electronics in the environment. This is accompanied by reusing, recycling, and other methods.

6.8 Packaging and packaging waste directive 1994/62/EC

Nanomaterials entry into the environment through packaging materials has been increasing, as 400–500 packaging nanomaterials are in commercial use. Moreover, there is uncertainty in relation to possible emission during recycling of packaging waste materials; thus Directive 1994/62/EC was designed to take account of prevention of packaging waste and its release into the environment.

6.9 Sewage sludge directive 86/27/EEC

The purpose of this directive is to decrease release of nanomaterials in the agricultural field through household sewage containing nanoparticles from detergent, fabrics, lubricants, and paints.

6.10 Directive 91/271/EEC

This directive deals with urban wastewater treatment, which includes its collection, treatment, and disposal. Urban wastewater is a source of nanomaterial through household wastewater containing detergents, cleaning products, fuels, and paint material.

6.11 Directive 98/83/EC

Industrial development has increased the release of nanomaterial into drinking water, thus increasing the risk to human health. This release can be intentional, through wastewater plants, or can occur through diffusion. Directive 98/83/EC was formulated to set quality standards on the supply of drinking water.

7. Handling of nanoparticles

The need to address nanotoxicology came under the spotlight (Groso et al., 2010) when seven women working in a paint factory in China were confirmed to have lung cancer (Song et al., 2009). However, the cause of their illness was ambiguous—it may have been due to nanoparticles or other chemicals (A. Maynard). It was proposed (Maynard, 2009a, b) that this could be avoided by addressing nanotoxicology critically. It was stressed that a proper risk assessment should be carried out when there is possibility of exposure of workers for longer periods (Groso et al., 2010).

Nanoparticle exposure limit should be restricted, preferably by substituting hazardous substances with a nonhazardous material or through a process that eradicates the possible risks to health. If this is not possible, then exposure can be avoided by adopting appropriate protection measures. There are control measures that are used commonly to tackle nanomaterial exposure (Amoabediny et al., 2009).

7.1 Engineering control

Substitution of particularly hazardous substances and processes for less harmful use of chemicals can decrease or complete remove hazardous material from the workplace, in order to eradicate or reduce potential exposure. Substitution should be controlled and appropriate to achieve the desired results, which involves research and experimentation. Substitution encourages reuse of existing material, thus reducing waste material, and eventually their disposal and management costs will decrease. Exposure to airborne nanoaerosols can be limited through a variety of techniques similar to those used to limit exposure to normal aerosols. These techniques include isolation of the generation source of nanoaerosol from workers. This is accomplished through isolation booths, closed containers such as sealed reactor vessels, closed storage containers, pump enclosures, valve isolation, and glove boxes. The organization should install fume hoods, biological safety cabinets, and laminar flow, and should encourage use of glove boxes and glove bags among workers.

7.1.1 Fume hood

The fume hood is the most extensively used engineering control and is preferred to other engineering controls. The use of fume hoods has been reported with different materials of different phases but its use for solutions was highest, which indicates the role of the fume hood as a filter or barrier against vapors of nanomaterials. It is also observed to increase the risk of inhalation due to air turbulence generated through a fume hood exhaust system.

7.1.2 Glove box

Use of gloves boxes in North American and European organizations during nanomaterial operations is 64% and 45%, respectively, while in Asian countries, its reported use is 36%.

Work with nanowires, nanocrystals, and carbon nanotubes is reported in the presence of glove boxes.

7.1.3 Ventilation

Use of ventilation is recommended on the basis of the physical properties of the nanomaterial—that is, whether it behaves like gas or not—as it is strongly suggested that airborne, nonagglomerated nanoparticles, which behave like gases, diffuse rapidly over long distances. This is why ventilation must be targeted according to the gaseous and particulate properties of the nanoparticles.

7.1.4 Safety cabinets

Biological safety cabinets are intended to protect the operator from harmful materials by keeping laboratory environment clean, preventing exposure of working materials to infectious aerosols and protecting against exposure to infectious material through splashing.

Biological safety cabinets of class II type B2 prevent the release of nanomaterials into the environment as they have HEPA filters. Although commercial use of these cabinets is limited, they are used by many organizations working particularly with nanopowders.

7.2 Administrative controls

Administrative control works as an additional control method when other methods are not seen to provide the desired results. These include short work period duration, modification of work environment, and maintenance of hygiene measures. Administration controls also include practice of good personal hygiene, such as hand washing after each handling, change of clothes, and showering at the end of the working day. These practices will limit the exposure of nanoparticles by their transfer to clothing and skin. Consumption of food and smoking should be prohibited in areas where nanomaterials are handled. Written operating procedures should be suggested by specific trained personnel for timely inspection of ongoing processes, and manufacturing and exposure control equipment. The number of potentially exposed workers should be small.

7.3 Personal protective equipment

Nanoparticle exposure can be limited by wearing adequate PPE. Protective gadets used in wet chemistry laboratories are recommended, which may include long shoes with low permeability material, full trousers, long-sleeved shirts, chemical splash goggles, and laboratory coats. Nonwoven fabric is recommended as protective clothing as it has proved very efficient against nanoparticle penetration, while cotton fabrics are prohibited.

7.4 Nanomaterial waste disposal

Current legislation regarding nanoparticles is formulated on the basis of existing laws and regulations for chemicals. Most organization use conventional safety methods throughout their nanomaterial's life cycle. Lack of modification in existing regulations made organization suggest all nanomaterial waste as potentially hazardous wastes. Waste management also dealing with these nanomaterial in the context that all the nanomaterials are hazardous thus causing discardation of all nanomaterials which is not correct approach so there is high need for modification in old regulation which will be able for better and distinct classification of nanomaterilas in relation of their hazardous nature.

8. WHO guidelines

The need for these guidelines was felt due to health hazards associated with inhalation, ingestion, or skin absorption of nanomaterials in the workplace. Thus, a project team was assigned to construct guidelines for exposure and risk handling.

8.1 Scope of the guidelines

The group assigned by the WHO has formulated recommendations on the basis of the following key issues and questions, thus leading to the improved health of workers:

1. Which nanomaterials are risky?
2. Which specific hazard classes should be assigned to nanomaterials?
3. What are the forms and routes of exposure for nanomaterials?
4. What are typical exposure situations?
5. How is exposure measurement carried out and what are the exposure assessment techniques?
6. Which occupational exposure limit (OEL) values should be used for nanomaterials?
7. Is control banding significant enough for safe treatment of nanomaterials?
8. What are the specific risk-mitigation techniques for nanomaterials?
9. What training is required for workers to limit risks from potential exposure?
10. What are the most appropriate health surveillance approaches to detect and limit the risk of potential exposure?
11. How should workers and their representatives behave in risk assessment?

8.2 WHO guidelines relating to this topic

Air quality guidelines already consider nanoscale airborne particles hazardous, although no specific exposure has been suggested regarding this. Air quality is checked by classification of small particles according to their size; it includes particles in ranges smaller than 10 μm (PM10), smaller than 2.5 (PM2.5), and smaller than 100 nm. However, these guidelines only address intentionally produced nanoparticles.

The WHO, through the Coorporation of Healthy Environment Units in the Department of Public Health, Environmental and Social Determinants of Health, has attained approval to develop guidelines for possible risk assessment. As a result, the WHO Guideline Steering Group and a Guideline Development Group (GDG) have been established. The GDG contains experts and end-users who are considered responsible for designing evidence-based recommendations.

8.3 Guiding principles

The guiding principles are based on the following approaches.

8.3.1 Precautionary approach

If the GDG decides there is a health concern, but there are not enough scientific data to allow an evaluation of possible risk factors, then a precautionary approach should be adopted to reduce or prevent exposure as far as possible. For example, workers should not be exposed in the absence of toxicological information. There can be a more tailored control strategy only if toxicological information is available.

8.3.2 Hierarchy of controls

A hierarchy of controls includes reduction or elimination of hazardous material and its exposure. At first, hazardous material should be eliminated. If this is not possible, then it should be substituted with a less hazardous material. In the second step, engineering control (isolation, local exhaust ventilation, or dust suppression) should be applied. If this is not possible, then the last recommended step is application of administrative controls, which includes worker training or scheduling. PPE can be used.

8.4 WHO GDG recommendations

1. The GDG recommends the involvement of workers in health and safety issues, thus leading to optimal control of health and safety risks.
2. The GDG recommends education and training of employees on the risks of nanomaterials to which they are potentially exposed so that they can protect themselves.
3. The GDG recommends classification of all nanomaterials according to the Globally Harmonized System (GHS) of Classification and Labeling of Chemicals, which can also be used in safety data sheets.
4. The GDG recommends renewal of safety data sheets with nanomaterials' risk level information.
5. The GDG suggests assessment of workers with methods similar to those proposed by specific OELs for nanomaterials exposure in the workplace.
6. In the absence of OELs for nanomaterials in workplaces, the GDG suggests a stepwise approach based on assessment of the potential for nanomaterial exposure, followed by basic exposure assessment and comprehensive exposure assessment, as proposed by OECD or CEN.

7. On the basis of a precautionary approach, the GDG recommends controlling the exposure limit of nanomaterials.
8. The GDG recommends lessening of exposure to nanomaterials that have been detected consistently in workplaces, such as during cleaning and maintenance.
9. The GDG recommends appliance of control measures on the basis of the principle of hierarchy of controls. According to this, first there should be elimination of the source of exposure followed by implementation of control measures. These control measures include involvement of workers to reduce exposure to nanomaterials. Engineering control is recommended when exposure to nanoparticles is high but available toxicological information is low.
10. The GDG suggests the use of control banding for selection of exposure control measures in the workplace when no appropriate control measures has been provided by workplace safety experts.
11. The GDG is restricted in some actions such as recommendation of one method of control banding over another. Similarly, the GDG cannot suggest health scrutiny programs for specific nanomaterial over health scrutiny programs already in use, owing to lack of evidence.

9. Conclusion

In spite of the high associated risk of nanomaterials, it is certain that nanotechnology is the most emerging and revolutionary technology of this era. Therefore, adaptation of precautionary measures in the light of WHO guidelines is the best strategy to take, as a complete ban of nanomaterial use is highly absurd and impractical to implement due to its high demand in the present era.

References

Aitken, R., Hankin, S., Ross, B., Tran, C., Stone, V., Fernandes, T., … Wilkins, T., 2009. EMERG-NANO: a review of completed and near completed environment, health and safety research on nanomaterials and nanotechnology Defra Project CB0409. Institute of Occupational Medicine Report TM/09/01.

Amoabediny, G., Naderi, A., Malakootikhah, J., Koohi, M., Mortazavi, S., Naderi, M., Rashedi, H., 2009. Guidelines for safe handling, use and disposal of nanoparticles. In: Paper Presented at the Journal of Physics: Conference Series.

Benn, T.M., Westerhoff, P., 2008. Nanoparticle silver released into water from commercially available sock fabrics. Environ. Sci. Technol. 42 (11), 4133–4139.

Brook, R.D., Franklin, B., Cascio, W., Hong, Y., Howard, G., Lipsett, M., … Smith Jr., S.C., 2004. Air pollution and cardiovascular disease: a statement for healthcare professionals from the expert panel on population and prevention science of the American Heart Association. Circulation 109 (21), 2655–2671.

EFSA P. (EFSA Panel on Plant Protection Products and Their Residues), 2016. Guidance on the establishment of the residue definition for dietary risk assessment. EFSA J 14 (12), 4549 129 pp.

Ganzleben, C., Pelsy, F., Hansen, S.F., Corden, C., Grebot, B., Sobey, M., 2011. Review of Environmental Legislation for the Regulatory Control of Nanomaterials. For DG Environment of the European Commission under Contract No 070307/2010/580540/SER/D.

Garcia-Garcia, E., Gil, S., Andrieux, K., Desmaele, D., Nicolas, V., Taran, F., … Couvreur, P., 2005. A relevant in vitro rat model for the evaluation of blood-brain barrier translocation of nanoparticles. Cell. Mol. Life Sci. 62 (12), 1400–1408.

Groso, A., Petri-Fink, A., Magrez, A., Riediker, M., Meyer, T., 2010. Management of nanomaterials safety in research environment. Part. Fibre Toxicol. 7 (1), 40.

Gupta, A.K., Gupta, M., 2005. Synthesis and surface engineering of iron oxide nanoparticles for biomedical applications. Biomaterials 26 (18), 3995–4021.

Hoet, P., 2016. Opinion of the Scientific Committee on Consumer Safety (SCCS)—revision of the opinion on the safety of the use of silica, hydrated silica, and silica surface modified with alkyl silylates (nano form) in cosmetic products. Regul. Toxicol. Pharmacol. 74, 79–80.

Kreyling, W.G., Semmler, M., Möller, W., 2004. Dosimetry and toxicology of ultrafine particles. J. Aerosol Med. 17 (2), 140–152.

Mackevica, A., Olsson, M.E., Hansen, S.F., 2016. Silver nanoparticle release from commercially available plastic food containers into food simulants. J. Nanopart. Res. 18 (1), 5.

Magrez, A., Kasas, S., Salicio, V., Pasquier, N., Seo, J.W., Celio, M., … Forró, L., 2006. Cellular toxicity of carbon-based nanomaterials. Nano Lett. 6 (6), 1121–1125.

Magrez, A., Horváth, L., Smajda, R., Salicio, V., Pasquier, N., Forró, L., Schwaller, B., 2009. Cellular toxicity of TiO2-based nanofilaments. ACS Nano 3 (8), 2274–2280.

Maynard, A., 2009a. New Study Seeks to Link Seven Cases of Occupational Lung Disease With Nanoparticles and Nanotechnology. Institute of Occupational Medicine, Edinburgh, Scotland.

Maynard, A., 2009b. Nanoparticle Exposure and Occupational Lung Disease—Six Expert Perspectives on a New Clinical Study. Safenano Community, p. 18.

Peters, R., Kramer, E., Oomen, A.G., Herrera Rivera, Z.E., Oegema, G., Tromp, P.C., … Weigel, S., 2012. Presence of nano-sized silica during in vitro digestion of foods containing silica as a food additive. ACS Nano 6 (3), 2441–2451.

Poland, C., Read, S., Varet, J., Carse, G., Christensen, F., Hankin, S., 2007. Dermal absorption of nanomaterials. In Vitro *268* (146), 185.

Poland, C.A., Duffin, R., Kinloch, I., Maynard, A., Wallace, W.A., Seaton, A., … Donaldson, K., 2008. Carbon nanotubes introduced into the abdominal cavity of mice show asbestos-like pathogenicity in a pilot study. Nat. Nanotechnol. 3 (7), 423.

Riebeling, C., Luch, A., Götz, M.E., 2016. Comparative modeling of exposure to airborne nanoparticles released by consumer spray products. Nanotoxicology 10 (3), 343–351.

Rothen-Rutishauser, B.M., Schürch, S., Haenni, B., Kapp, N., Gehr, P., 2006. Interaction of fine particles and nanoparticles with red blood cells visualized with advanced microscopic techniques. Environ. Sci. Technol. 40 (14), 4353–4359.

Rothen-Rutishauser, B., Mühlfeld, C., Blank, F., Musso, C., Gehr, P., 2007. Translocation of particles and inflammatory responses after exposure to fine particles and nanoparticles in an epithelial airway model. Part. Fibre Toxicol. 4 (1), 9.

SCCS, 2012. Scientific Committee on Consumer Safety (SCCS) Opinion on Fragrance Allergens in Cosmetic Products. 26 June 2012.

Song, Y., Li, X., Du, X., 2009. Exposure to nanoparticles is related to pleural effusion, pulmonary fibrosis and granuloma. Eur. Respir. J. 34 (3), 559–567.

Taylor, J.M., 2002. New Dimensions for Manufacturing: A UK Strategy for Nanotechnology. Office of Science and Technology, London (United Kingdom).

Windler, L., Lorenz, C., von Goetz, N., Hungerbuhler, K., Amberg, M., Heuberger, M., Nowack, B., 2012. Release of titanium dioxide from textiles during washing. Environ. Sci. Technol. 46 (15), 8181–8188.

Yacobi, N.R., DeMaio, L., Xie, J., Hamm-Alvarez, S.F., Borok, Z., Kim, K.-J., Crandall, E.D., 2008. Polystyrene nanoparticle trafficking across alveolar epithelium. Nanomedicine 4 (2), 139–145.

CHAPTER 11

Conclusion

Nanomaterials are metals possessing grain size in the order of a billionth of a meter. Classification of nanoparticles is done in relation to their dimensionality, morphology, composition, uniformity, and agglomeration. Their classification in relation to their dimensions is advantageous as dimensionality plays an important role in determining their nanotoxicology. One-dimensional nanoparticles are particles with one dimension at nanoscale. Thin film is one of the most common examples of one-dimensional nanoparticles. Two-dimensional nanoparticles are particles with two dimensions at nanoscale. Nanostructured films and nanopore filters are common examples of two-dimensional nanomaterials. Three-dimensional nanomaterials are materials that have all three dimensions at nanoscale. Thin films, colloids, and nanoparticles with various morphologies come into the three-dimensional nanomaterial category. Nanoparticles are also classified on the basis of morphological characteristics such as flat morphology, spherical morphology, and facet ratio. A classification according to facet ratio has nanotubes and nanowires of zigzag and belt-like morphology in the high facet ratio class, while nanoparticles of spherical, oval, cubic, prism, and pillar-like morphology come into the class of low facet ratio. Nanoparticles are also classified on the basis of the number of constituents present in the product: there can be a single constituent material in the final product or it can be composed of several materials.

Photochemical reactions, volcanic eruptions, forest fires, simple erosion, plants, and animals are major sources of nanoparticles in nature. Nanoparticles also contribute to air pollution through the release of waste material from industries. Different hybridized nanostructures have been created through incorporation of specific species into two-dimensional nanosheets, thus resulting in enhanced properties and synergistic functionalities in order to achieve excellent performance. Novel approaches have been adopted to create novel products such as by addition of two-dimensional nanosheets with other zero or one-dimensional nanostructures or by mixing the same or different two-dimensional nanosheets, thus resulting in hybridized two-dimensional nanomaterials. This will cause band-structure engineering to increase charge transfer between adjacent layers instead of within the layers. Band structure can also be altered by in-plane elemental doping for tuning electronic, optical, and magnetic features.

Physicochemical characteristics of nanomaterial determine their toxic expressions. This requires the research community to take their role into account in evaluation of related toxicity issues. Synthesis and characterization of nanomaterials plays an important

Nanomaterials: Synthesis, Characterization, Hazards and Safety
https://doi.org/10.1016/B978-0-12-823823-3.00011-2

part in material science and engineering. Nanomaterials' properties can be tuned on the basis of their morphology; therefore, a critical study on the morphology of nanomaterials is required. Thus, various characterization techniques for evaluating various morphologies such as atomic force microscope, scanning electron microscope, transmission electron microscope, etc. are employed.

Metal oxides are now part of research in almost every field of science, especially materials science. Metal oxide nanoparticles are involved in creating toxicity through various methods. Metal oxides are affected by their size, dissolution, and exposure routes. Several nanotechnology-based products have been advertised related to electronic, paint, sports, and stain-resistant fabric industries. Nanomaterials have been used in sun creams and medical products. The potential exposure of workers and the general public to synthesized nanoparticles has been greatly increased and is predicted to rise further in the future. Researchers have described the toxicity issues of different nanoparticles. The most commonly used term in risk assessment is "exposure dose," which is related to exposure level and its duration. In outdated risk assessment, nanomaterial exposure doses are measured in relation to threshold limit values (TLVs). The British Standards Institute projected a regulation that if a material comes into the category of being hazardous in its highest amount, then its nano form will have a threshold limit value 10 times smaller.

The hazardous effects of nanoparticles include air pollution, water pollution, and land pollution, and in humans they include lung inflammation, heart failure, kidney failure, and skin infections. The use of filter masks, gloves, protective clothes, fume hoods, and biological safety cabinets must be mandatory to reduce exposure to nanomaterials. Due to the high potential of graphene's properties, several companies (2D Carbon Tech, ACS Material, Advanced Graphene Products, among others) have focused on its production and are investing in this nanomaterial. For instance, the company Saint Jean Carbon announced the manufacturing of lithium-graphene-based batteries in 2017. However, there have been warnings regarding the high expectations some companies have about their products, which is why information is key before making any investments. The electronic and sensor market is one that promises a particularly high compound annual growth rate, so a lot of productive work can be done on this in the future. In short, there is great potential for development in the nanomaterial sector and their revolutionary applications, but the toxic effects of the manufactured product should be kept in mind.

Index

Note: Page numbers followed by *f* indicate figures and *t* indicate tables.

F

G

H

I

J

L

O

P

Q

R

S

X

Z

CPSIA information can be obtained
at www.ICGtesting.com
Printed in the USA
LVHW060037171121
703502LV00007B/268